Ecology of Threatened Semi-Arid Wetlands

Wetlands: Ecology, Conservation and Management

Volume 2

Series Editor:

Max Finlayson

Institute for Land, Water and Society
Charles Sturt University
Albury, NSW, Australia
mfinlayson@csu.edu.au

Aims & Scope:

The recognition that wetlands provide many values for people and are important foci for conservation worldwide has led to an increasing amount of research and management activity. This has resulted in an increased demand for high quality publications that outline both the value of wetlands and the many management steps necessary to ensure that they are maintained and even restored. Recent research and management activities in support of conservation and sustainable development provide a strong basis for the book series. The series presents current analyses of the many problems afflicting wetlands as well as assessments of their conservation status. Current research is described by leading academics and scientists from the biological and social sciences. Leading practitioners and managers provide analyses based on their vast experience.

The series provides an avenue for describing and explaning the functioning and processes that support the many wonderful and valuable wetland habitats, such as swamps, lagoons and marshes, and their species, such as waterbirds, plants and fish, as well as the most recent research directions. Proposals cover current research, conservation and management issues from around the world and provide the reader with new and relevant perspectives on wetland issues.

For other title published in this series, go to
www.springer.com/series/7215

Salvador Sánchez-Carrillo · David G. Angeler
Editors

Ecology of Threatened Semi-Arid Wetlands

Long-Term Research in Las Tablas de Daimiel

Springer

Editors
Salvador Sánchez-Carrillo
Instituto de Recursos Naturales
CSIC
Madrid
Spain
sanchez.carrillo@ccma.csic.es

David G. Angeler
Department of Aquatic Sciences
and Assessment
Swedish University of Agricultural
Sciences (SLU)
Uppsala
Sweden
david.angeler@vatten.slu.se

ISSN 1875-1261 e-ISSN 1875-127x
ISBN 978-90-481-9180-2 e-ISBN 978-90-481-9181-9
DOI 10.1007/978-90-481-9181-9
Springer Dordrecht Heidelberg London New York

Library of Congress Control Number: 2010930222

Printed on acid-free paper

Springer is part of Springer Science+Business Media (www.springer.com)

Preface

Wetlands are productive and diverse ecosystems which play a critical role in influencing climate change and mitigating its impacts. However, wetlands are one of the world's most threatened ecosystems as over-exploitation of natural resources and conversion to agricultural lands have already resulted in large-scale wetlands loss and degradation. Sound management and conservation schemes require a long-term understanding of the ecology of wetlands. Yet until now, long-term and interdisciplinary wetland research is limited to a few examples from tropical or temperate climates (such as the Florida Everglades, Middle Paraná River, and Czech Biosphere Reserve). Monographic information on wetlands in semi-arid and arid environments is scant. This new book contributes to fill this gap; it provides a unique reference in basic and applied Mediterranean wetland ecology, based on long-term research at the RAMSAR and UNESCO Biosphere site, Las Tablas de Daimiel (Central Spain).

Las Tablas de Daimiel serves as a case study that demonstrates the adverse impacts of human activities on wetlands ecological integrity in Mediterranean Europe where water is fundamentally limiting. This book is based on the collaborative and interdisciplinary research efforts of geologists, hydrologists, ecologists, botanists, planktologists, paleolimologists, and geographers who have intensively studied this wetland during the last 30 years. Much information has been gathered on the structure and function of this unique wetland and how its ecosystem compartments changed in response to cumulative anthropogenic stressors (land use changes, disruption of the natural hydrological cycle, point- and diffuse pollution, exotic species invasions) during the twentieth century. However, in spite of the vast amount of ecological and biogeochemical information shown in this book, further scientific research is required to fill remaining knowledge gaps.

The book scope is clearly scientific although a few of chapters have been written for a broader, non-scientific audience. The complexity of the interacting abiotic and biotic components across different spatial and temporal scales and across various levels of biological hierarchy should be useful for researchers, postgraduate students and wetland resource managers in the Mediterranean and elsewhere. The take-home message of this book is that scientific progress will not be enough for the survival of this unique wetland. An integration of scientific, cultural and historical knowledge in the interaction cycles between ecological, social, political and economic systems should be the ultimate goal. Without this integral approach to understanding ecosystems and their management, sustainable development will not be possible.

Madrid and Uppsala
January 2010

S. Sánchez-Carrillo
D.G. Angeler

Acknowledgments

The editors and authors extend their sincere thanks to many people and Institutions which have supported the research which provides the basis for this book. The support of the staff of Las Tablas de Daimiel National Park headed by María J. Sánchez-Soler, Manolo Carrasco and Carlos Ruiz de la Hermosa was crucial over the years. Without their support this research would have been impossible. Also, the field and laboratory support provided by the researchers and technicians linked to the Aquatic Ecology group of the Institute of Natural Resources and Royal Botanical Garden (CSIC) has been fundamental for this research (Juan Carlos Rodríguez-Murillo, María J. Ortíz, Eva López, José Luis Ayala, Laura Parrilla, Raquel Rey, Palmira Riolobos, Miriam Moreno and Leopoldo Medina). We are grateful to Esperanza Montero for english translating and editing. We are also grateful to State Meteorological Agency (AEMET), Guadiana Water Authority (CHG), Spanish Geological Survey (IGME) and National Statistics Institute (INE) for the invaluable information made available to us. The editors thank Ria Kanters and Catherine Cotton for their editorial assistance in developing the book.

This book has been made possible by the funds provided by the following research grants: TDNP flora and vegetation: current status, evolution and management for sustainability (Ministry of Environment-ICONA 1989), Monitoring of water and biological qualities at TDNP (Ministry of Environment-ICONA 1992–1994), Studying limnology and botany of TDNP (Ministry of Environment-Dirección General de la Naturaleza 1995–1998), Wetlands in semiarid Europe: how to cope with eutrophication (Training and Mobility of Researcher program, European Union 1998–2000), Monitoring and environmental remediation of TDNP (Ministry of Environment-Organismo Autónomo Parques Nacionales 1999–2002), Environmental monitoring of TDNP (Ministry of Environment-Organismo Autónomo Parques Nacionales 2003–2007, 2009), The biodiversity of Mediterranean wetlands: the importance of island biogeographical characteristics and landscape structure HUMACRO (Regional Government of Castilla – La Mancha 2005–2007), Carbon sources and sinks in TDNP (Ministry of Environment-Organismo Autónomo Parques Nacionales 2006–2009), Ecology of large Branchipoda (Anostraca, Notostraca; Crustacea) in temporary wetlands (ECOBRA University of Castilla – La Mancha and Ministry of Education and Science, I3 Programme 2007–2008), Flood pulse effects on carbon biogeochemistry of Mediterranean

Wetlands (CSIC 2007–2008), The influence of large Branchipoda (Anostraca, Notostraca; Crustacea) on ecosystem structuring processes in temporary wetlands (GRANDBEST Ministry of Education and Sciences 2007–2009). Climate change impacts on the trophodynamic coupling of temporary wetland foodwebs (Ministry of Education and Science 2007–2009), Ecological status of the wetlands included in the Upper Guadiana Special Plan (Plan Especial del Alto Guadiana, PEAG; Ministry of Environment and Rural and Marine Affairs-Water Guadiana Authority 2007–2011) and Global change effects on the ecology and biogeochemistry of TDNP (DECAMERON Ministry of Environment and Rural and Marine Affairs-Organismo Autónomo Parques Nacionales 2001/2008 2008–2011).

Finally, the editors thank their families (Raquel and pets, Obelix and Mica [SSC] and María José, Michael and Carmen [DGA]) for their endless patience, support and love.

Madrid and Uppsala
January 2010

S. Sánchez-Carrillo
D.G. Angeler

Contents

Part I Introductory Chapter

1 The Wetland, Its Catchment Settings and Socioeconomic Relevance: An Overview 3
S. Sánchez-Carrillo, D.G. Angeler, S. Cirujano, and M. Álvarez-Cobelas

1.1 Introduction 3
1.2 The Wetland: Las Tablas de Daimiel National Park 6
1.3 The Upper Guadiana Basin and the UNESCO's La Mancha Húmeda Biosphere Reserve 8
1.4 Groundwater Overexploitation and the Wetland Complex 12
1.5 Socioeconomic Aspects in the Upper Guadiana Basin 14
1.6 A Chronological Summary of TDNP Impacts and Its Degradation 16
References 18

Part II Abiotic Environment and Historical Reconstructions

2 Paleoenvironmental Reconstruction of Las Tablas de Daimiel and Its Evolution During the Quaternary Period 23
B. Ruiz-Zapata, M.J. Gil-García, and I. de Bustamante

2.1 Introduction 23
2.2 Material and Methods 25
2.3 Overview of Vegetation Composition in La Mancha Region During Quaternary Period: Palaeo-Sequences Description 27
2.3.1 Landscape Evolution from 325,000 to 21,000 Years BP: LT, TD, TASG-1 and FUENT-1 Profiles 28
2.3.2 Landscape Evolution from 11,000 to 4,000 Years BP: MO and CC-17 Sequences 29
2.3.3 Landscape Evolution from 4,000 Years BP to the Present: CC-17 and Gigüela 4.2 Paleo-records 31

2.4 Summary of Climatic and Wetland Environmental Evolution During the Quaternary Period ... 35
2.4.1 Middle Pleistocene ... 36
2.4.2 Upper Pleistocene ... 37
2.4.3 Holocene ... 37
References ... 38

3 Climate and Hydrologic Trends: Climate Change Versus Hydrologic Overexploitation as Determinants of the Fluctuating Wetland Hydrology ... 45
S. Sánchez-Carrillo and M. Álvarez-Cobelas

3.1 Introduction ... 46
3.2 Material and Methods ... 47
3.3 Long-Term Climate Trends ... 47
3.4 An Historical Analysis of Wetland Hydroperiod and Inundation Patterns: The Importance of Groundwater ... 50
3.5 Long-Term Changes in the Wetland Water Budget: Hydrologic Degradation ... 53
3.6 Causes of Wetland Hydrologic Degradation: Climate Change Versus Hydrologic Overexploitation ... 61
3.7 Evapotranspiration Controls of Wetland Hydrology: A Conceptual Model ... 66
3.8 A Wetland Hydrological Model for Hydroperiod Restoration ... 70
3.9 Future Scenarios of TDNP Hydrology ... 78
References ... 79

4 The Effects of Anthropogenic Stressors on Wetland Loss and Habitat Quality Deterioration in the Upper Guadiana River Basin: A Long-Term Assessment (1970–2000) ... 85
R. Sánchez-Andrés, M.O. Viedma, and S. Sánchez-Carrillo

4.1 Introduction ... 86
4.2 Material and Methods ... 87
4.3 Wetland Cover and Extent of Hydrodynamic Types in the Early 1970s ... 91
4.4 Land Use–Land Cover Changes and Socioeconomic Indicators During 1978–2000 ... 92
4.5 Wetland Losses from 1970 to 2000 ... 95
4.6 Wetland Habitat Quality Evolution During the Period 1978–2000 ... 98
4.7 Patterns on Wetland Loss and Habitat Quality Deterioration: the Influence of the Main Anthropogenic Stressors at Local and Regional Scales ... 99
References ... 105

5 A Story of the Wetland Water Quality Deterioration: Salinization, Pollution, Eutrophication and Siltation 109
M. Álvarez-Cobelas, S. Sánchez-Carrillo, S. Cirujano, and D.G. Angeler

5.1 Introduction 109
5.2 Materials and Methods 111
5.3 Salinization and Major Ionic Composition 111
5.4 Water Pollution 113
5.5 Nutrient Dynamics and Eutrophication 116
5.6 Spatial Heterogeneity of Nutrient Controls 124
5.7 Wetland Sedimentation and Siltation 128
References 131

Part III Ecological Communities

6 Plankton Ecology and Diversity 137
C. Rojo and M.A. Rodrigo

6.1 Introduction 138
6.2 Materials and Methods 139
6.2.1 Field Sampling and Counting Techniques 139
6.2.2 Indexes and Statistical Methods 140
6.3 Diversity of Heterotrophic and Autotrophic Picoplankton 141
6.4 Bacteria and Autotrophic Picoplankton Dynamics 141
6.5 Species Composition and Diversity of Phytoplankton 144
6.6 Phytoplankton Abundance Dynamics 151
6.7 Species Composition and Diversity of Zooplankton 155
6.8 Zooplankton Abundance Dynamics 160
6.9 Bacterioplankton and APP Relationship with Abiotic and Biotic Factors 163
6.10 Environmental Factors Related to Phytoplankton 168
6.11 Environmental Factors Related to Zooplankton and the Food Web 169
References 170

7 Macrophyte Ecology and Its Long-term Dynamics 175
S. Cirujano, M. Álvarez-Cobelas, and R. Sánchez-Andrés

7.1 Introduction 175
7.2 Materials and Methods 176
7.2.1 Surveys and Reconstructions 176
7.2.2 Estimation of Growth and Decay 177
7.2.3 Biomass and Primary Production Estimations 177
7.2.4 Spatial Distribution of Helophytes 178
7.3 Aquatic Flora 179

7.4 Helophyte Cover and Growth 184
7.5 Plant Biomass, Production and Decomposition 189
7.6 Spatial Heterogeneity of Helophytes 191
References 192

8 Fish and Avian Communities: A Testimony of Wetland Degradation 197
M. Álvarez-Cobelas

8.1 Introduction 197
8.2 Fish Communities 198
8.2.1 Material and Methods 198
8.2.2 Results and Discussion 198
8.3 Avian Communities 200
8.3.1 Material and Methods 202
8.3.2 Results and Discussion 202
References 211

Part IV Applied Issues

9 Biomanipulation: A Useful Tool for Wetland Rehabilitation 215
D.G. Angeler

9.1 Introduction 215
9.2 Biomanipulation: Theory and Rationale 217
9.3 Biomanipulation in Wetlands: Applying a Lake Restoration Tool 218
9.4 Biomanipulation in Wetlands: Extending the Paradigm 222
9.5 Biomanipulation: A Model for Las Tablas de Daimiel 224
9.6 Conclusions and Perspectives 226
References 226

10 Analysis of Applied Environmental Management Strategies for Wetland Conservation During the Last 30 Years: A Local History 229
S. Cirujano, M. Álvarez-Cobelas, and C. Ruíz de la Hermosa

10.1 Introduction 229
10.2 Wetland Conservation in Spain Until 1970 230
10.3 Wetland Restoration, Conservation and Management Since 1970: Common Problems Across Spain 230
10.4 Las Tablas de Daimiel: A Case Study 233
References 237

Part V Cultural and Societal Aspects

11 The Man and Las Tablas de Daimiel 241
M. Alvarez-Cobelas, S. Cirujano, and A. Meco

11.1 Early Settlement 241
11.2 From Roman Colonization to Lower Middle Age 243
11.3 Modern Age 244
11.4 The Twentieth Century 246
11.5 Ecosystem Health and the Future of Man in TDNP de Daimiel 250
References 254

Part VI Concluding Chapter

12 Synthesis: The Past, Present and Future of Las Tablas de Daimiel 257
David G. Angeler and Salvador Sánchez-Carrillo

12.1 Background and Current State of Las Tablas de Daimel 258
12.2 Missing Research Links: The Ecological Challenge 262
12.2.1 Research at the Local Scale 263
12.2.2 Research Beyond the Wetland Boundary 267
12.3 The Future: The Socioeconomic Challenge 271
12.4 Conclusion 275
References 275

Index 279

Contributors

Miguel Álvarez-Cobelas
Instituto de Recursos Naturales, CSIC, Serrano 115 dpdo E-28005 Madrid, Spain
malvarez@ccma.csic.es

David G. Angeler
Department of Aquatic Sciences and Assessment, Swedish University of Agricultural Sciences, Ullsväg 31A, 756 51, Uppsala, Sweden
David.Angeler@vatten.slu.se

Irene de Bustamante
Dep. Geología, Universidad de Alcalá de Henares, Carretera Madrid-Barcelona, km. 33,600, E-28871, Alcalá de Henares, Madrid, Spain
irene.bustamante@uah.es

Santos Cirujano
Real Jardín Botánico, CSIC, Pza de Murillo 2, E-28014, Madrid, Spain
santos@rjb.csic.es

María José Gil-García
Dep. Geología, Universidad de Alcalá de Henares, Carretera Madrid-Barcelona, km. 33,600, E-28871, Alcalá de Henares, Madrid, Spain
mjose.gil@uah.es

Carlos Ruíz de la Hermosa
Parque Nacional Las Tablas de Daimiel, Parque del Carmen s/n, E-13250, Daimiel Ciudad Real, Spain
caruiz@oapn.es

Ana Meco
Real Jardín Botánico, CSIC, Pza de Murillo 2, E-28014 Madrid, Spain
anameco78@hotmail.com

María Antonia Rodrigo
Instituto Cavanilles de Biodiversidad y Biología Evolutiva, Universidad de Valencia, Polígono La Coma s/n, E-46980 Paterna Valencia, Spain
maria.a.rodrigo@uv.es

Carmen Rojo
Instituto Cavanilles de Biodiversidad y Biología Evolutiva,
Universidad de Valencia, Polígono La Coma s/n, E-46980 Paterna
Valencia, Spain
carmen.rojo@uv.es

Blanca Ruiz-Zapata
Dep. Geología, Universidad de Alcalá de Henares, Carretera Madrid-Barcelona,
km. 33,600, E-28871, Alcalá de Henares, Madrid, Spain
blanca.ruiz@uah.es

Raquel Sánchez-Andrés
Real Jardín Botánico, CSIC, Pza de Murillo 2, E-28014 Madrid, Spain
rsanchez@rjb.csic.es

Salvador Sánchez-Carrillo
Instituto de Recursos Naturales, CSIC, Serrano 115 dpdo, E-28005 Madrid, Spain
sanchez.carrillo@ccma.csic.es

Olga Viedma
Instituto de Ciencias Ambientales, Universidad de Castilla-La Mancha,
Avda. Carlos III s/n, E-45071 Toledo, Spain
olga.viedma@uclm.es

Part I
Introductory Chapter

Chapter 1
The Wetland, Its Catchment Settings and Socioeconomic Relevance: An Overview

S. Sánchez-Carrillo, D.G. Angeler, S. Cirujano, and M. Álvarez-Cobelas

Abstract Wetland loss is common worldwide. In Spain, around 90% of the surface covered by floodplain wetlands has disappeared during the last century, and only 32 km² remain as of today. Las Tablas de Daimiel National Park (20 km²) is unique in Mediterranean Europe. It is the most representative Spanish floodplain wetland nowadays and depicts the core area of La Mancha Húmeda Biosphere Reserve (8,000 km²), one of the main wetland district occurring in the semi-arid Mediterranean Europe. Despite the ecological richness supported by this wetland, a controversial history of desiccation, groundwater overexploitation and water quality deterioration threatens this wetland since the 1950s. This chapter describes the main features of the wetland as well as of its basin, revising all impacts received and their environmental consequences. Groundwater overexploitation by irrigation farming is discussed as the main cause of wetland degradation in the socioeconomic framework of the Upper Guadiana River Basin.

1.1 Introduction

Wetlands are a common feature of the Spanish landscape (≈1,200 km² or 0.2% of the total country surface; DGCN 1998), and elsewhere, with broad environmental and ecological settings and a wide size range – oligotrophic temporary mountain ponds, permanent karst lagoons, floodplain wetlands, coastal wetlands, ephemeral water bodies, endorheic hypersaline areas, temporary ponds, etc. (Casado and Montes 1995). Currently, out of the thousands of Spain's wetlands, only 63 (281,768 ha) are registered in the Ramsar list (http://195.143.117.139/profile/profiles_spain.htm) and a large proportion of the scientific, cultural and economic value of this natural patrimony still remains to be explored. One of most important

S. Sánchez-Carrillo (✉)
Spanish National Research Council (CSIC), Institute of Natural Resources,
Serrano 115 dpdo, E-28006 Madrid, Spain
e-mail: sanchez.carrillo@ccma.csic.es

S. Sánchez-Carrillo and D.G. Angeler (eds.), *Ecology of Threatened Semi-Arid Wetlands: Long-Term Research in Las Tablas de Daimiel*, DOI 10.1007/978-90-481-9181-9_1,

threats to Spanish wetlands is the lack of water due to the aridity effect in around three fourth of the Iberian Peninsula (Gao and Giorgi 2008). Other environmental pressures such as desiccation, eutrophication and pollution, organic matter accumulation, siltation, salinization and the invasion of exotic species enhanced the degradation processes affecting Spanish wetlands (Cirujano et al. 2009). Between 60% and 65% of the Spanish wetland surfaces have been lost since the nineteenth century (original wetland surface of 2,800 km^2) with most of this disappearance taking place between 1950 and 1990 (DGCN 1998). Whilst mountain ponds and karst lagoons remained quite preserved, floodplain wetlands suffered the largest areal reduction affecting the Spanish inland wetlands. In a very conservative approach DGCN (1998) suggested that 126 km^2 of floodplain wetlands have disappeared, with only 32 km^2 remaining nowadays; unfortunately, the ancient floodplain wetland surface covering Spain could be up to two–threefold of the one considered by that study. Floodplain wetland loss appears to be common worldwide (Finlayson and Spiers 1999). The reason why floodplain wetlands are more consistently reduced than other freshwater wetland types across the world can be found through the wetland landscape location and historical human–development–river relationships (draining for intensive agriculture, river regulation for irrigation and hydropower, etc.; Finlayson and Spiers 1999).

Floodplain wetlands can be found worldwide, where land areas adjacent to rivers or streams are subject to recurring inundation. While riparian wetlands (dominated by forest vegetation) are most frequent in worldwide landscapes, marshes (dominated by graminoids) are less usual throughout floodplains. Inundation of wide floodplain areas combining surface water (river) and groundwater is not very usual (Bradley 1997), and even less in semi-arid climates. Examples of this type of wetland have only been found in Ireland (The Callows at the Shannon river; Heery 1993), Germany (Spreewald at the Spree River; Köhler 1993), Nigeria (Macina or Niger inland delta; John 1986), Botswana (Okavango alluvial fan; Allanson et al. 1990), Irak (the wetland between the Tigris and Eufrates rivers; Richardson et al. 2005) and in South America (the várzea at the mid Amazonas, Junk 1983; the Orinoco floodplains, Hamilton and Lewis 1990; and the Paraná floodplain, Carignan and Neiff 1992). In the semi-arid South Europe it only appears at the Iberian Peninsula, in the area called La Mancha Húmeda Biosphere Reserve (MHBR hereafter; 39°16′N, 3°24′E; Fig. 1.1), a region located in central Spain and covering 8,000 km^2 in one of most arid areas of the Mediterranean basin, with the Las Tablas de Daimiel wetland as the core area (20 km^2). MHBR is an arid landscape featuring vast plains where groundwater surplus discharges on lowland areas, joined seasonally to river discharge, creating extended floodplain wetlands. Nowadays, Las Tablas de Daimiel is the only relict wetland left over from an extended floodplain wetland area covering 150–250 km^2 at the beginning of the twentieth century (Heras et al. 1971; Fig. 1.2). Since the 1950s, the extent of the wetland has been reduced and its ecological integrity strongly threatened. Impact was manifest in many forms, including a controversial history of desiccation in order to both eradicate malaria and to increase farming lands. Later, groundwater overexploitation for irrigation practices and water quality deterioration by wastewater

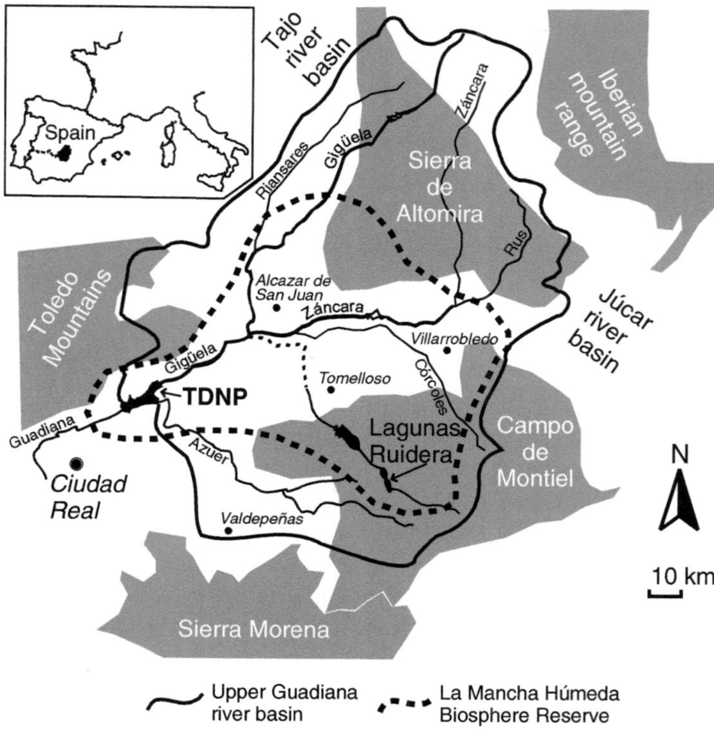

Fig. 1.1 Geographic location of Las Tablas de Daimiel at the Upper Guadiana river basin and La Mancha Húmeda Biosphere Reserve. Highlighted areas (*grey*) represent elevated zones. Only towns over 20,000 population are shown

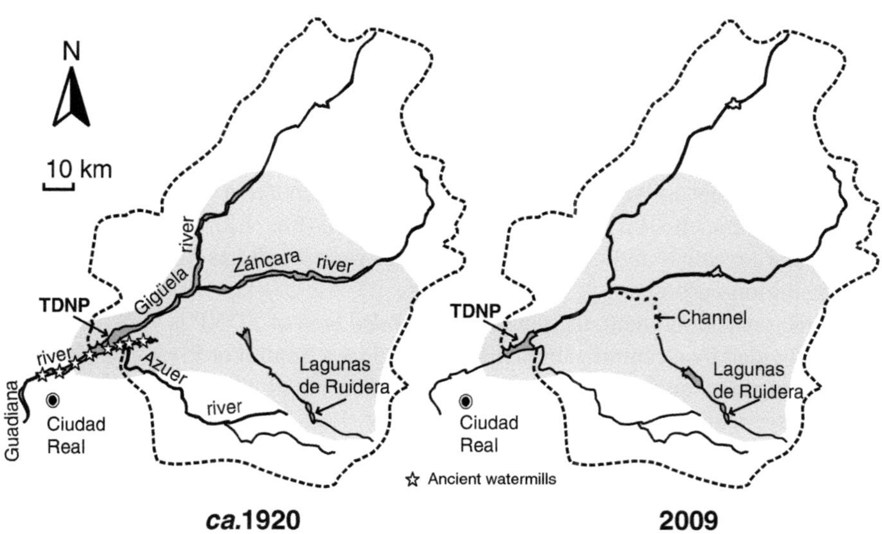

Fig. 1.2 Changes in the extent of floodplain wetlands at La Mancha Húmeda Biosfere Reserve during the twentieth century

discharges through the river, and agriculture runoff pollution aggravated the problem. Assorted hydrological remediation plans were implemented at the wetland scale from the mid-1970s; in spite of this, wetland degradation increased, given that the cumulative impacts on the wetland/watershed scale were never attended. Changes at Las Tablas de Daimiel have been recorded by means of significant long-term research efforts, which can then be used to assess the cumulative effects of global change on floodplain wetland performance and to develop sustainable management strategies, ensuring the survival of this unique wetland ecosystem.

1.2 The Wetland: Las Tablas de Daimiel National Park

Las Tablas de Daimiel National Park (TDNP hereafter) is located in Central Spain in the Castilla-La Mancha Autonomous Community (39°08′N, 3°43′W; Fig. 1.1) within the municipalities of Daimiel and Villarrubia de los Ojos. The wetland extends NE-SW parallel to the foothills of the Toledo Mountains, which represent the sole altitudinal reference in the geographical area (Fig. 1.1). Until the 1970s, wetland occurrence was due to natural flooding in the extended alluvial plain of both the Gigüela and the Guadiana Rivers, the latter flowing from the Ojos del Guadiana, a groundwater source area located around 10 km East of TDNP. The floodplain was also the natural groundwater discharge zone of the Llanura Manchega Occidental aquifer (23 Aquifer or 04.04 Hydrogeologic Unit according to the current official nomenclature of the Geological Survey of Spain–IGME) as the groundwater table was close to the surface. Therefore, TDNP inundation was supported by numerous groundwater surges along the wetland known as "Ojos" or "Ojillos" ('eye's or 'little eyes'). Furthermore, the wetland's flooded area was also maintained by human buildings, such as small water-mill dams which helped to retain water in TDNP. 15 water-mill dams have been identified along the Gigüela and Guadiana Rivers (Álvarez-Cobelas et al. 1996). Thus, the waterscape of TDNP appeared as the result of both natural flooding and human-induced inundation in an area where rainfall is scarce (Álvarez-Cobelas and Cirujano 1996). Until the 1950s, five water-mills artificially supported the inundation at TDNP: Zuacorta, Griñon, Molemocho, La Quebrá and Puente Navarro (Fig. 1.2). Nowadays, none of them are functional and only the Molemocho water-mill has been reconstructed for exhibition purposes.

At the present moment, the potentially flooded area of TDNP is 15.87 km^2. This area is divided by a central gabion dam named Presa Central or Presa del Morenillo which separates the wetland into two inundation areas of 11.82 km^2 and 4.05 km^2, located upstream and downstream, respectively (Fig. 1.3). At the end of TDNP – southwest area – the Puente Navarro dam, a domed concrete dam, controls the water storage in the wetland (Fig. 1.3). This zone is the deepest of the wetland (4.5 m; Álvarez-Cobelas et al. 1996).

The main morphometric features of TDNP are shown in Table 1.1. The number of islands is high (more than 30), the largest being Isla de Algeciras and Isla del

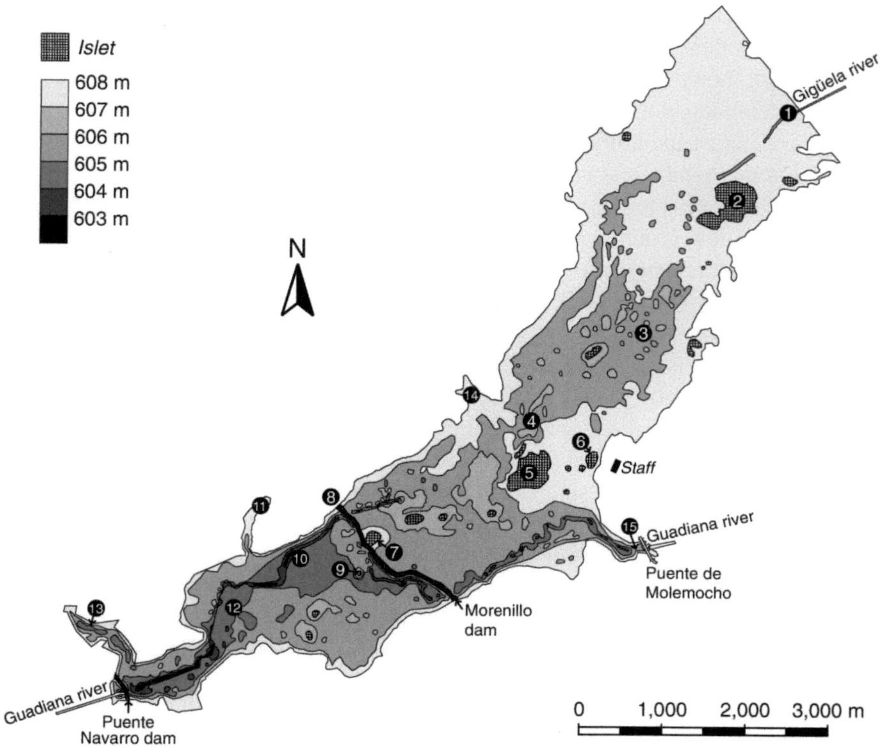

Fig. 1.3 Bathymetry of Las Tablas de Daimiel National Park at its highest inundation. Isolines shown are those of above mean sea level measured at the coast of the city of Alicante (Mediterranean Sea). Some main sites inside Las Tablas de Daimiel are shown by black circles: 1, Pata Gallina; 2, Algeciras island; 3, Tabla Larga – Long Water Table; 4, El Tablazo – Big Water Table; 5, Pan island; 6, Entradilla island; 7, Morenillo island; 8, Quinto de la Torre; 9, Las Cañas island; 10, Vado de los Toros; 11, Cañada del Gato – Cat Glen; 12, Almochinares; 13, Cachón de la Leona; 14, Cañada Lobosa – Lobosa Glen; 15, Molemocho

Table 1.1 Morphometric features of Las Tablas de Daimiel National Park (Álvarez-Cobelas et al. 1996)

Maximum length	10.58 km
Maximum width	2.75 km
Maximum depth	4–5 m
Average depth	0.91 m
Relative depth	0.2%
Surface area	1,928 ha
Volume	16.08 Mm^3
Shoreline length	39 km
Shore development	2.69
Island area	100 ha
Bottom roughness	9.05
Volume development	0.67

Pan, located in the northeastern area of the wetland. The effective wind exposition (fetch) can oscillate 1–2 km NW–SE in the Tablazo area (Fig. 1.3; Álvarez-Cobelas et al. 1996); however, recent emergent macrophyte changes, heterogeneity and macrophyte management practices make an accurate estimation difficult.

Wetland landscape uniqueness and richness are provided from two main hydrological and hydrochemical processes: TDNP provided an area where saline sulfate-rich surface waters flowing from the Gigüela River were mixed with freshwater coming from the groundwater sources (Coronado et al. 1974; see Chapter 6); besides, the distinctive seasonality of water sources increased wetland complexity. Therefore, aquatic emergent macrophytes appeared with European cut-sedge (*Cladium mariscus*) as the dominant population, accompanied by reed (*Phragmites australis*) and cattail (*Typha domingensis*) restricted to littoral areas (Cirujano 1996; see Chapter 8). TDNP represented the more important cut-sedge cover in Western Europe (Álvarez-Cobelas and Cirujano 1996) but, at present, this macrophyte is in regression due to wetland degradation (Álvarez-Cobelas et al. 2001). Wetland richness was also supplied by numerous vertebrate and invertebrate taxa, which regrettably disappeared or are undergoing a severe number reduction (Álvarez-Cobelas and Cirujano 1996).

International recognition of TDNP and its protected status appeared as a consequence of the habitat importance for waterfowl. In fact, the first written reference known about TDNP comes from "Libro de Caza" (*The hunting book*; 1325) by the Infante Don Juan Manuel (1282–1345) which described the area as "*a propitious place for every hunting activity*". Subsequently, following the "*Relaciones Topográficas*" by King Felipe II (a detailed statistical description of Spanish towns carried out during the sixteenth century), protection was ordered for the TDNP area, in order to protect its avifauna richness for hunting purposes (Sarria 1986). Thus, from the seventeenth to the twentieth century, the exceptional waterfowl richness of TDNP attracted important politicians and aristocrats devoted to hunting (Cobelas et al. 1996). Most representative and profuse waterfowl in TDNP are the mallard (*Anas platyrhynchos*), the common teal (*Anas crecca*) and the red-crested pochard (*Netta rufina*), the emblematic waterfowl in La Mancha Húmeda wetlands (see Chapter 8).

1.3 The Upper Guadiana Basin and the UNESCO's La Mancha Húmeda Biosphere Reserve

The Upper Guadiana basin comprises the TDNP's natural drainage area (Fig. 1.1) and extends over 15,000 km^2, covering the provinces of Ciudad Real (6,640 km^2), Cuenca (5,012 km^2), Toledo (2,460 km^2) and Albacete (1,898 km^2). The basin is located in the Submeseta Meridional Castellana (Castilian Southern Plateau) and limits to the West by the Toledo Mountains and the Tagus river basin, to the North by the Iberian mountain range, to the South by the Campo de Montiel region and the Sierra Morena and to the East by the Júcar river basin (Fig. 1.1). Ground elevations range from 550 to 1,200 m. The climate of the basin is semi-arid, with an average annual rainfall of 300–500 mm and an average annual temperature ranging 14–15°C (Pérez-González and Sanz-Donaire 1998). Annual potential evapotranspiration exceeds rainfall, reaching 800–900 mm. Spatial and temporal distribution of the rainfall is very heterogeneous at the basin scale (Pérez-González and Sanz-Donaire 1998). The fluvial net is ephemeral and closely linked to the aquifers. The main river at present is the Gigüela which flows NE–SW. Other minor rivers are Záncara, Córcoles, Azuer, Riánsares and Rus (Fig. 1.1). At the river basin scale, water is managed by the Guadiana Water Authority (Confederación Hidrográfica del Guadiana, http://www.chguadiana.es/), a public agency depending on the the Spanish government through the Ministry of the Environment. This basin management agency interacts with autonomous and local governments as well as with water user associations and, theoretically, has the last say in any water-related decisions at the basin scale.

Geology of the Upper Guadiana basin is basically composed of Jurassic, Cretaceous and Tertiary limestone underlain by a gneiss basement that crops out along the southern and eastern basin boundaries (Portero and Ramírez 1988). The most important geomorphologic landmarks are both those shaped by carbonate dissolution and collapse (sinkholes) and those related to floodplain wetland or small lagoon processes (alluvial fans, floodplain wetland, peatbogs and salts; Pérez-González 1996). Main soil types, according to FAO (1998), belong to the cambisol group, although regosol and, at the southeast, luvisol and podzol are found in the basin (Horra 1996).

Five aquifers are found along the Upper Guadiana basin (Fig. 1.4), two of them – 04.04 and 04.06 aquifer systems – being considered large groundwater reserves which feed lowland river areas and depressional landscapes that develop extended marshlands. The aquifer system 04.04 extends throughout 5,500 km^2 and includes most wetland ecosystems of the MHBR. This groundwater system is comprised by Miocene and Jurassic limestone and marls and can be subdivided into two aquifers, the upper one being a heterogeneous unconfined aquifer, while the lower one is confined. The Campo de Montiel aquifer or the 04.06 Hydrogeologic Unit is an unconfined Jurassic limestone system occupying an extension of 2,700 km^2. Aquifer geometry and hydrogeological parameters are yet to be accurately determined, although there seems to be little doubt about the mechanisms of groundwater flow (Cruces and Martínez-Cortina 2000). While thickness of each aquifer system oscillates between 30 and 300 m, transmissivity (the rate at which groundwater can flow through an aquifer section of unit width under a unit hydraulic gradient) and storage coefficient (the volume of water released from storage per unit surface area of aquifer per unit decline in hydraulic head) of the 04.04, is higher than those of 04.06 aquifer (transmissivity: 500–20,000 and 50–1,500 m^2 day^{-1}; storage coefficients: 5% and 2–5%, in the 04.04 and 04.06 aquifers, respectively) (Cruces et al. 1997). Therefore, the former controls the hydrogeology of the region and is considered as the most important aquifer of the Upper Guadiana basin. Both aquifers are ground connected but its complex hydraulic

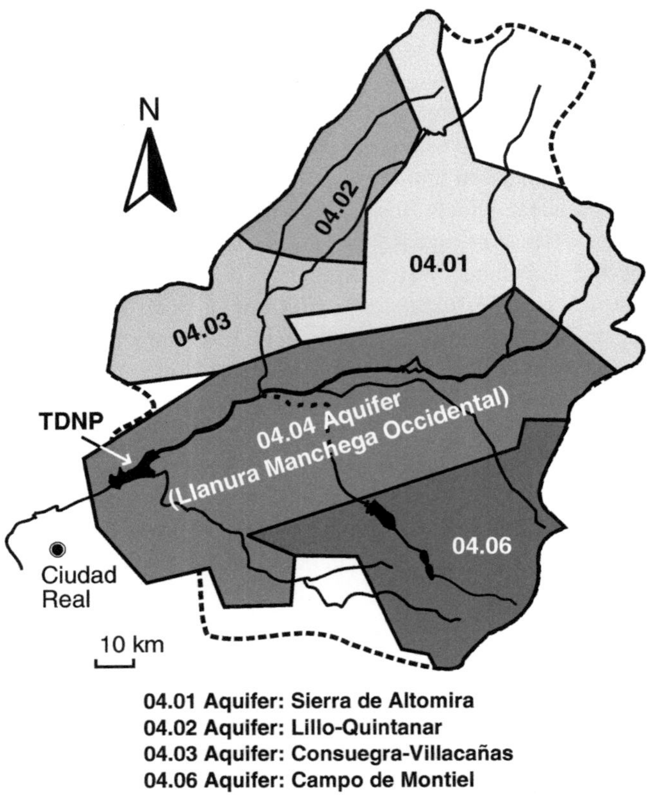

Fig. 1.4 Aquifers at the Upper Guadiana basin

transference is little known yet. There is evidence that 04.06 aquifer surplus discharges both through seepage and groundwater flow into the 04.04 the Aquifer area (around 50 Mm^3 $year^{-1}$ according to some authors (Llamas and Martínez-Santos 2005); see Chapter 2). Under natural conditions, groundwater flow is observed to follow a westward trend, from the recharge areas (Campo de Montiel plateau) to the main discharge zone (Las Tablas de Daimiel National Park and surroundings; IGME 1989; Fig. 1.5 in 1974).

Wetland abundance along the Upper Guadiana Basin was so high that, in 1980, UNESCO recognized its environmental importance and value, designating La Mancha Húmeda region as a Biosphere Reserve (Man and the Biosphere Programme) in order to preserve its highly valued and strongly threatened wetlands. By this date, more than 50% of the wetlands covering MHBR were already lost. MHBR includes more than 100 wetlands (from freshwater to saline) constituting the most important wetland district in Spain. The inner area of MHBR is occupied by the "Llanura Manchega". Wetland and lagoon occurrence and ecosystem functioning were intrinsically linked to the 04.04 Aquifer groundwater discharges. Nowadays, most wetlands

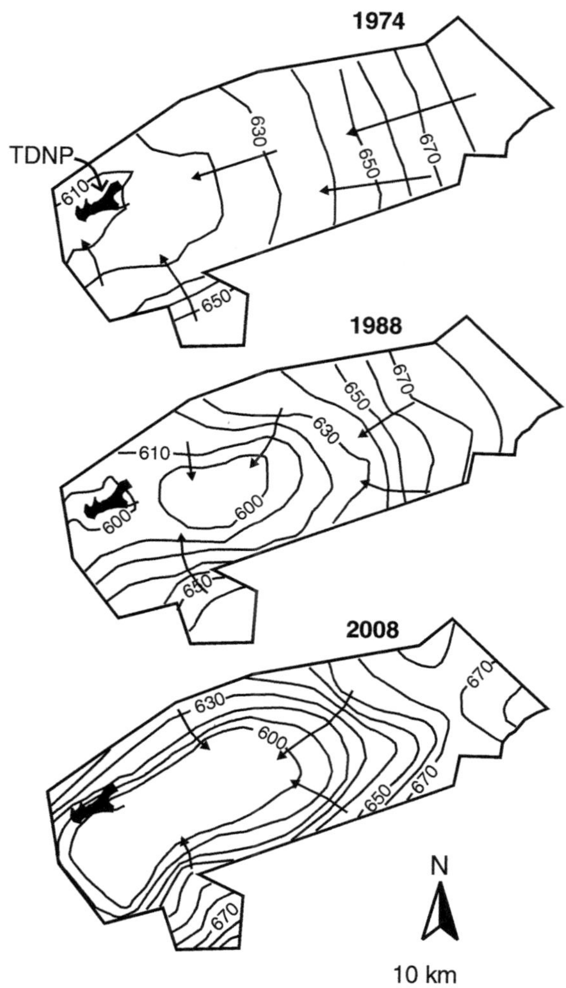

Fig. 1.5 Evolution of the 04.04 Aquifer (Llanura Manchega aquifer) piezometric levels from 1974 to the present day (Data source from SGDGOH 1989; IGME 2008)

at the MHBR are temporary and very fluctuating wetlands but before the 1950s most of them were permanent (Álvarez-Cobelas et al. 2001).

The designation of MHBR as a UNESCO Biosphere Reserve is only one among the many international acknowledgements of its wetlands. Five wetlands are listed in the Ramsar Convention and one of them, Las Tablas de Daimiel, also represents the only inland wetland of the Spanish Network of National Parks. MHBR contributes significantly to the European biodiversity (Florín and Montes 1999). Furthermore, these wetlands are also important at the community level, where habitats are considered to be of interest within the European Union (Directive 97/62/EC): inland salt meadows, Mediterranean halophilous bush formations (*Sacocornetea fruticosi*); Mediterranean salt steppes (*Limonietalia*), oligotrophic calcium-carbonate waters with a benthic vegetation of *Chara* spp., Mediterranean temporary ponds, and calcium-carbonate wetlands of *Cladium mariscus*. Other valuable communities not officially acknowledged include microbial mats of the cyanobacteria *Microcoleus chtonoplastes*, submerged macrophytes dominated by the Potamogetonaceae *Ruppia drepanensis*, crustacean communities of *Arctodiaptometum saline* and the communities of riparian carabids (Florín and Montes 1999).

1.4 Groundwater Overexploitation and the Wetland Complex

The importance of groundwater discharges to sustain wetland hydrology and ecology at the Upper Guadiana basin and, particularly, to TDNP until the 1970s has already been mentioned. Besides supporting most wetland occurrences at the Upper Guadiana basin, 04.04 Aquifer also sustains most human and economic activity in the region. Historically, the 04.04 Aquifer groundwater reservoir has been considered to store a volume of water up to 100 times its annual recharge (Sahuquillo-Herráiz et al. 1982). However, from the 1970s, groundwater pumping (mostly for irrigation practices) changed this situation dramatically. During the period between 1974 and 1989 the land devoted to irrigation increased by 324% and the groundwater extractions rose from 152 to 568 Mm^3 $year^{-1}$ (IGME 2004; Fig. 1.6). In the 1970s groundwater reserve at the 04.04 Aquifer was deemed to be 11,000–12,000 Mm^3 (IGME 2004). Groundwater recharges measured in different areas of the aquifer have been estimated as ranging from 10 to 625 Mm^3 $year^{-1}$ (Cruces et al. 1997), although numerous inconsistencies, including the location of the experimental plots and methodological procedures, can be found, preventing extrapolation to the entire aquifer. While frequency and extent of droughts in this region complicate the definition of an averaged hydrological year, most authors agree that the mean annual recharge oscillates between 150 and 350 Mm^3 $year^{-1}$ (Cruces et al. 1997; Acreman et al. 2000; IGME 2004, 2008). At this exploitation regime, groundwater extraction exceeds

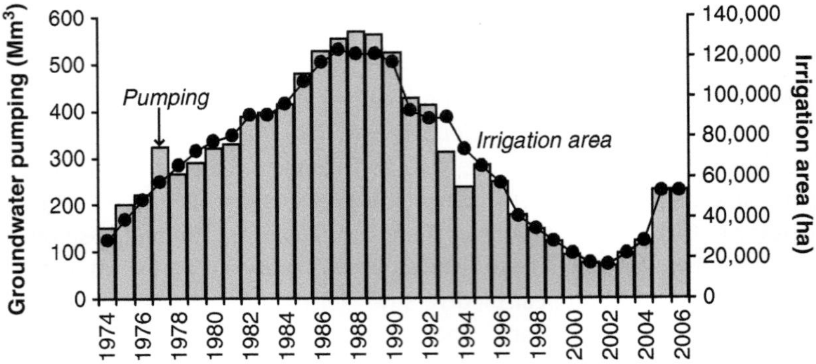

Fig. 1.6 Land devoted to irrigation and groundwater pumping at the Llanura Manchega aquifer (04.04 Aquifer) for the period 1974–2006

groundwater recharge and water reserve of the aquifer dropped severely until 1987 (from 5 to 30 m; Navarro et al. 1993), when the aquifer was declared temporarily overexploited (Box 1.1). Consequences of the piezometric level decline are that the aquifer was detached from wetlands in many areas, natural discharge areas disappeared, such as the Ojos del Guadiana spring (dry from 1983), flow dynamics changed so that groundwater does no longer flow westward, but rather towards the centre of the aquifer (where the main pumping cones are; Fig. 1.5). Furthermore, a dramatic drought was experienced from 1991 to 1995, perhaps the most significant ever in Spain during the twentieth century, exacerbating the situation all the more. In the period between 1980 and 1996, official data from the Guadiana River Water Authority assumed a groundwater loss of around 3,750 Mm^3 (IGME 2008), different to the 6,000 Mm^3 estimated by Cruces et al. (1997). The surface devoted to irrigation decreased in the aquifer area after pumping restrictions were imposed on farmers, followed as of 1992 by the application of the European Agro-Environmental Program on water demands for irrigation. This program also served as a vehicle to compensate farmers for income losses associated with voluntarily cutting down on water use (Viladomiu and Rossell 1997). These measures were not enough to recover the groundwater table to the 1970 levels and the aquifer was definitely declared

Box 1.1 Aquifer overexploitation

Groundwater overexploitation and aquifer overexploitation are terms that are common in water resources management. Hydrologists, managers and journalists use them when talking about stressed aquifers or groundwater conflict. Overexploitation may be defined as the situation in which, for some years, average aquifer abstraction rate is greater than, or close to the average recharge rate. Intensive exploitation of aquifers can give rise to overexploitation problems. Aquifer overexploitation depends on the balance between demand and renewable resources. In semi-arid regions, the absence of high rainfall and the existence of ephemeral rivers require an increased use of groundwater resources for maintaining specific socioeconomic activities (e.g. agriculture). This leads to overexploitation, which commonly arises from excessive abstraction for irrigation. The resulting increase in productivity and change in land use can establish a cycle of unsustainable socio-economic development within an irrigated region. Additional resources are exploited to satisfy the increased demand from the population and agriculture, exacerbating the already fragile environment by reducing groundwater levels and, in some circumstances, accelerating the desertification processes. Most cases, overexploitation is mainly a consequence of the fact that groundwater resources have historically provided a low-cost, high quality source for public water supply. Examples of aquifer overexploitation can be found in southeastern and southwestern USA, central and northern Mexico, in the arid Middle East and North African countries.

overexploited in 1994. From this date, the irrigation area at the watershed decreased until 2002, but it was not responsible for the slight aquifer recuperation registered between 1995 and 1999 (1,750 Mm^3), which was due to an uncommonly humid period registered in this region in 1997–1998 (Fig. 1.6). This aquifer recovery was used then by the farmers, to pressure policy makers for increased groundwater extractions for irrigation; once again, this meant an increase of groundwater pumping from 2002 and a descent of the aquifer groundwater reserve of around 800 Mm^3 from 1999 to 2004 (IGME 2008; Fig. 1.6). IGME (2004) stated that the 04.04 Aquifer water table levels dropped at an approximate average rate of 1–2 m $year^{-1}$ over the period 1999–2004. Again a slight aquifer recovery was computed by IGME (2008) for the humid period 2004–2005, estimated in a surplus of 300 Mm^3. From 2005, using the above mentioned lower annual recharge of 250 Mm^3 $year^{-1}$, groundwater extractions have been set at around 230 Mm^3 $year^{-1}$, although a continuous drought period evidenced that groundwater exploitation continues consuming the aquifer. Water table monitoring suggests that pumping still exceeds the aquifer renewable resources, dropping piezometric levels –3.32 m and consuming 500 Mm^3 for the period 2003–2008 (IGME 2008). The high number of illegal wells (58% of total wells) not taken into account, the inaccurate estimates of the remotely-based irrigation surface and the unrealistic water consumption per crop applied are significant uncertainties that obstruct management and restoration of the 04.04 Aquifer. After around 40 years of pumping irrigation being triggered in the Upper Guadiana basin, and after almost 25 years of different management strategies being implemented to reverse the overexploitation of the 04.04 Aquifer, the situation has not yet changed significantly, jeopardizing the recovery of TDNP.

1.5 Socioeconomic Aspects in the Upper Guadiana Basin

In 2005, about 570,000 people lived in the Upper Guadiana basin, with population growing at a rate close to 9% in the last 15 years (CHG 2009). However, 70% of Upper Guadiana basin municipalities, especially small villages, showed a negative demographic trend during this period. Average population density is 29 inhabitants km^{-2}, which is significantly below Spain's 87 inhabitants km^{-2}. The Upper Guadiana basin comprises 140 municipalities, four of which are home to 20,000–30,000 people, and seven to 10,000–20,000 (Fig. 1.1). Large urban agglomerations are absent in a region where most people live in towns with population density below 150 inhabitants km^{-2}. In view of these parameters, and according to OECD criteria, the Upper Guadiana basin can be classified as a rural area (OECD 1996; Viladomiu and Rossell 2002).

The Upper Guadiana basin presents a fairly young population, 21% of which is under the age of 16. This figure is comparable to the country's average, and slightly higher than Castilla-La Mancha's regional average. The number of people over 65 represents about 17% of the population (significantly lower than other rural areas of Spain). Population growth has been sustained in the last decades, although it slowed down in the 1980s and the first half of the 1990s (Martínez-Cortina 2002; Olmedo 2002).

The 04.04 Aquifer shelters about two thirds of the basin population (around 300,000 inhabitants; 35 inhabitants km^{-2}), and encompasses all of the more populated villages. This area is considered as one of the most dynamic areas of Castilla-La Mancha in terms of population and economy.

By 1997, per capita income in the Upper Guadiana basin was 85% of Spain's average, and 65% of the European Union (EU) average (Eurostat 1997). Thus, this region was considered as under objective one of the EU's structural funds (promoting the development of regions where per head Gross Domestic Product (GDP) is below 75% of the EU average) and has been until now in the receiving end of European subsidies.

In 1991, agriculture accounted for over 21% of employment in the basin, a rate similar to the industry and building sectors (Martínez-Cortina 2002). This percentage is significantly higher in the area of the 04.04 Aquifer where, in the year 2000, agriculture made up for approximately 38% of employment (Table 1.2) (up 60% in some municipalities as Las Labores, Arenas de San Juan, Santa María de los Llanos, Ossa de Montiel, Las Mesas, Carrizosa and Las Pedroñeras; Olmedo 2002). Agricultural employment has nevertheless been in constant decrease in recent times in the whole country, while employment has increased in the service sector (Table 1.2). Notwithstanding, manufacturing is limited in this region where most industry employment is found in building activities. Most manufacturing activities are linked to agricultural practices through food processing (wineries, meat and cheese; Martínez-Cortina 2002). Although the clothing industry and food manufacturing employment are the same in the basin, the latter activity creates more gross wealth (CHG 2009). Tourism industry still presents a low degree of development in this area, despite Don Quixote's excellent legacy.

Land uses in the Upper Guadiana basin are linked to agriculture. The useful farming surface in the basin accounts for 81% of the total area (Tarjuelo 1999). Around 73% of the basin surface is used for agricultural practices, with 65% of it devoted to herbaceous crops, 35% to vineyards and 5% to olives (Tarjuelo 1999). Forest areas represent less than 20% of the total basin surface and are restricted to the Toledo Mountains and to the east, in the Campo de Montiel. Most forests (68%) are privately owned, with a smaller surface belonging to community and public forests (6% and 7%, respectively). Forests are basically composed of Holm oaks which have been frequently replaced by the Stone pine (CHG 2008).

Table 1.2 Comparison of employment in the economic sectors between Spain, Castilla-La Mancha and the 04.04 Aquifer area. Data cover the period between 1991 and 2000 (Martínez-Cortina 2002; Olmedo 2002)

	Spain	Castilla-La Mancha	04.04 Aquifer
	1991/2000 (%)	1991/2000 (%)	1991/2000 (%)
Agriculture	10/7	16/10	44/38
Industry[a]	31/31	35/35	24/26
Services	59/62	49/55	32/36

[a] The building sector has been included within industry, accounting for 11%, 15% and 14% of the total employment respectively, in 2000. 1991 data were not available

Groundwater resources at the basin are also particularly important from the social point of view, catering to most of the urban water supply (serving 75% of the population; CHG 2008). For the whole basin, joint water demand for urban supply and industrial uses is of approximately 51 Mm^3 $year^{-1}$ (43 and 8, respectively; CHG 2008). Gascó et al. (2004) estimated that groundwater withdrawals for urban supply currently amount to only 2–4% of those for irrigation. Urban consumption showed a slow growth, with an annual rate of close to 1% (Martínez-Cortina 2002; see also Chapter 3).

1.6 A Chronological Summary of TDNP Impacts and Its Degradation

From 166 to 1400 BC there is archeological evidence of human influence in this wetland, through buildings named "motillas" (artificial hills constructed in a plain, used as settlements as, for instance, the archeological site "Motilla de las Cañas" inside TDNP). From then on, human presence has been constant in TDNP but its influence over the wetland has intensified in the last decades. Historical information compiled by Álvarez-Cobelas and Cirujano (1996) proved that this extended wetland territory, crossed and settled by numerous civilizations during thirty-six centuries, has been reduced to one third of its surface in a short period (1965–1985) due to desiccation, groundwater overexploitation and water pollution.

Natural wetland functioning was first altered in 1956, when the Law of July 17th on "*Saneamiento y Colonización de los Terrenos Pantanosos próximos a los márgenes de los ríos Guadiana, Gigüela y Záncara y afluentes de éstos últimos*" (Sanitation and Colonization of swampy areas close to the Guadiana, Gigüela and Záncara riversides and their tributaries). This Law was enacted based on the premise that recurrent flooding increased insalubrities, as well as to satisfy requirements to increase farming land to foster growth in this economically undeveloped area. Because uncultivated lands were considered rather as poor terrains, then the aim of this Law was to convert deep wetland areas to farmlands. River canalization and wetland draining increased quickly from 1967, adversely affecting wetland inundation patterns. In fact, the central southern area of the wetland (Las Cañas) was used for rice crops during some years, until the late 1970s. It stimulated an active social protest against the disappearance of Las Tablas, which forced the Government to stop desiccation works, after announcing that 1,000 ha of the wetland would be protected. In 1973, those 1,000 ha were declared as National Park by means of the Ordinance 1874/1973 (BOE 181 of June 30, 1973) in order "to conserve one of the most valuated ecosystems in Spain and the most representative wetland of "La Mancha Húmeda". The purpose of the conservation measures adopted was to avoid the increased wetland degradation by trying to maintain inundation patterns.

From 1974, this region experienced a huge agrarian transformation, when traditional agriculture with rain-fed lands (wheat, vineyard and olive) were replaced by herbaceous crops (maize, alfalfa, beet, melon, etc.) with more water requirements; this only could be obtained through an increase of groundwater pumping.

The Public Administration promoted and subsidized the change to this intensive agriculture, causing a quick raise of groundwater consumption, leading to aquifer overexploitation. Given that the adopted conservational measures were only applicable at the wetland scale, groundwater and river discharges declined and disappeared 10 years later, with the last known groundwater discharge occurring in 1986.

In the mid-1970s, TDNP degradation was already obvious, as warned by Sáez-Royuela (1977) who pointed out the need to increase water discharges into the wetland, controlling water quality and eutrophication. Thus, in 1980, the National Park was again legally classified (Law 25/1980, BOE 110 of May 7, 1980) providing a special legal regime devoted to protect the ecosystem, including conservation measurements of ground and surface waters as wetland supporters. Nowadays, the National Park area covers 1,928 ha, including the Las Cañas area as well, since it was also established as a buffer area (the protected zone includes 5,410 ha). Aiming to retain surface waters flowing by draining channels downstream of TDNP, a dam (Puente Navarro dam) was built at the end of the wetland. This measure should allow the restoration of the severely damaged area of Las Cañas. In 1986, a heavy intentional fire burnt some 185 ha of cut-sedge vegetation in the central area of the National Park (Alvarez-Cobelas et al. 2008).

Since 1986, due to the endless reduction of wetland inundation, the Public Administration established a Hydrologic Remediation Plan (HRP) at TDNP, in order to conserve its ecological values, in critical decline at that time. HRP estimated that, to preserve wetland ecological integrity, an inundation of 1,800 ha should be achieved after the humid season (Spring) and 600 ha after the dry season (Summer), through 18 Mm^3 $year^{-1}$ from external water sources (EPTISA 1986). The measures adopted by HRP included (i) groundwater pumping inside the wetland, (ii) water diversions from the Tagus basin, (iii) building an inner dam (Morenillo dam) in order to ensure inundation after the humid season in the shallowest part (NE) of the wetland, and (iv) recovering the 04.04 Aquifer groundwater level, declaring it as overexploited and limiting pumping, under the shelter of the European Agro-Environmental Program (UE 2078/92) by means of compensatory incomes for voluntarily cutting down on water use.

Nowadays, TDNP degradation continues, mainly due to water input deficit and water quality impairment. Wildlife and flora transformations have been significant during the last 40 years because wetland functioning changed severely since then. In 2008, a special Plan to restore aquatic ecosystems in the Upper Guadiana basin (known as the PEAG, Upper Guadiana Special Plan) was launched, intending to recover groundwater levels by 2027, thus enhancing the ecological performance of aquatic ecosystems which are mostly groundwater-dependent. As a part of this Plan, the Guadiana Water Authority promoted a smaller Plan to restore Las Tablas de Daimiel National Park and, as evidence that the PEAG is being useful, the National Park is now under restoring. The Plan for Las Tablas de Daimiel is called REGATA, which is the acronym for Gradual Restoration of Las Tablas. The REGATA Plan intends to act at two spatial levels, namely, that of the wetland and that of the Upper Guadiana catchment, both impinging on water quantity and the quality, to guarantee enough water of good quality for the wetland. Some actions within the Park have started in 2009, such as the disposal of sediment and decaying vegetation, but more actions should be implemented both in the wetland and in the catchment in the years to come if the wetland is to stay in good ecological health, as was the case with the measures pursued in the 1950s.

References

Acreman M, Almagro J, Alvarez J, Bouraoui F, Bradford R, Bromley J, Croke B, Crooks S, Cruces J, Dolz J, Dunbar M, Estrela T, Fernandez-Carrasco P, Fornes J, Gustard G, Haverkamp R, Hernández-Mora N, Llamas R, Martinez CL, Papamasorakis J, Ragab R, Sánchez M, Vardavas, I, Webb T (2000) Groundwater and River Resources Programme on a European Scale (GRAPES). Technical Report to the European Union ENV4 – CT95-0186. Institute of Hydrology, Wallingford

Allanson BR, Hart RC, O'Keeffe JH, Robarts RD (1990) Inland waters of southern Africa. Kluwer, Dordrecht

Álvarez-Cobelas M, Cirujano S (eds) (1996) Las Tablas de Daimiel: Ecología Acuática y sociedad. Ministerio de Medio Ambiente, Madrid

Álvarez-Cobelas M, Verdugo M, Cirujano S (1996) Geografía y morfometría. In: Álvarez-Cobelas M, Cirujano S (eds) Las Tablas de Daimiel: Ecología Acuática y sociedad. Ministerio de Medio Ambiente, Madrid

Álvarez-Cobelas M, Cirujano S, Sánchez-Carrillo S (2001) Hydrological and botanical man-made changes in the Spanish wetland of Las Tablas de Daimiel. Biol Conserv 97:89–97

Álvarez-Cobelas M, Sánchez-Carrillo S, Cirujano S, Angeler DG (2008) Long-term changes in spatial patterns of emergent vegetation in a Mediterranean floodplain: natural versus anthropogenic constraints. Plant Ecol 194:257–271

Bradley C (1997) The hydrological basis for conservation of floodplain wetlands: implications of work at Narborouth, UK. Aquat Conserv Mar Freshw Ecosyst 7:41–62

Carignan R, Neiff JJ (1992) Nutrient dynamics in the floodplain ponds of the Paraná (Argentina). Biogeochemistry 17:85–121

Casado S, Montes C (1995) Guía de los Lagos y Humedales Españoles. Reyero Ediciones, Madrid

CHG (2008) Plan Especial del Alto Guadiana (PEAG), Memoria Técnica: Estudio socio-económico. Guadiana River Basin Authority. Ministry of Environment and Rural and Marine Affairs. http://www.chguadiana.es/?url=planificaci%F3n+planes+hidrol%F3gicos+vigentes+plan+especial+del+alto+guadiana+%28peag%29&corp=chguadiana&lang=es&mode=view. Accessed 21 Oct 2009

CHG (2009) Summary of population. Guadiana River Basin Authority. Official website http://www.chguadiana.es. Accessed 21 Oct 2009

Cirujano S (1996) Bentos vegetal: flora y vegetación superior. In: Álvarez-Cobelas M, Cirujano S (eds) Las Tablas de Daimiel: Ecología Acuática y sociedad. Ministerio de Medio Ambiente, Madrid

Cirujano S, Álvarez-Cobelas M, Sánchez-Carrillo S, Angeler DG, García-Murillo P (2009) Wetland management in Spain: a history of controversial conservation. In: Herrera JR (ed) International wetlands: Ecology, conservation & restoration. Nova Science, New York

Cobelas A, Sánchez-Soler MJ, Carrasco M, García-Consuegra B, Escuderos-Hernández J, Álvarez-Cobelas M (1996) Aspectos históricos. In: Álvarez-Cobelas M, Cirujano S (eds) Las Tablas de Daimiel: Ecología Acuática y sociedad. Ministerio de Medio Ambiente, Madrid

Coronado R, del Portillo F, Sáez-Royuela R (1974) Guía del Parque Nacional de Las Tablas de Daimiel. ICONA, Madrid

Cruces J, Martínez-Cortina L (2000). La Mancha Húmeda: explotación intensiva de las aguas subterráneas en la cuenca alta del río Guadiana. Papeles del Proyecto Aguas Subterráneas, Serie A(3). Fundación Marcelino Botín, Madrid

Cruces J, Casado ME, Llamas MR, de la Hera A, Martínez-Cortina L (1997) El desarrollo sostenible de la Cuenca Alta del Río Guadiana: aspectos hidrológicos. Revista Obras Públicas 3362:7–18
de la Horra JL (1996) El medio edáfico. In: Álvarez-Cobelas M, Cirujano S (eds) Las Tablas de Daimiel: Ecología Acuática y sociedad. Ministerio de Medio Ambiente, Madrid, pp 35–46
DGCN (1998) Estrategia española para la conservación y el uso sostenible de la diversidad biológica. Dirección General de Conservación de la Naturaleza, Ministerio de Medio Ambiente, Madrid. http://194.224.130.185/secciones/biodiversidad/banco_datos/info_disponible/index_estrategia_espaniola.htm. Accessed 15 July 2009
EPTISA (1986) Estudio de viabilidad de un Plan de Regeneración Hídrica en el Parque Nacional Las Tablas de Daimiel (Ciudad Real). Internal Technical Report for ICONA, Madrid
EUROSTAT (1997) Regions: statistical yearbook. Statistical Office of the European Communities, Luxembourg
FAO (1998) World reference base for soil resources. Food and Agriculture Organization of the United Nations, Rome
Finlayson CM, Spiers AG (eds) (1999) Global review of wetland resources and priorities for inventory. Supervising Scientist Report No. 144, Canberra
Florín M, Montes C (1999) Functional analysis and restoration of Mediterranean lagunas in the Mancha Húmeda Biosphere Reserve (Central Spain). Aquat Conserv Mar Freshw Ecosyst 9:97–109
Gao X, Giorgi F (2008) Increased aridity in the Mediterranean region under greenhouse gas forcing estimated from high resolution simulations with a regional climate model. Global Planet Change 62:195–209
Gascó JM, López-Sanz G, Naredo JM (2004) Informe sobre los problemas del agua en La Mancha y sus posibles tratamientos, con un dictamen sobre el Proyecto de Trasvase Tajo-La Mancha llamado Sistema de abastecimiento de agua potable a las poblaciones de la llanura manchega desde el acueducto Tajo-Segura. Fundación Nueva Cultura, Zaragoza (http://www.fnca.eu/fnca/docu/docu85.pdf)
Hamilton SK, Lewis WM (1990) Basin morphology in relation to chemical and ecological characteristics of lakes on the Orinoco river floodplain. Arch Hydrobiol 119:393–425
Heery S (1993) The Shannon floodlands. Tír Eolas, Galway
Heras G, Prieto E, Guerrero T, Oñate JM, Ontalba A, Cortés M, Lorenzo A, Marcilla JJ, Palomo JL, Tudanca M, Carrión J (1971) Recursos y aprovechamientos hidráulicos. Consejo Económico Sindical Interprovincial de La Mancha, Ciudad Real
IGME (1989) Sistema acuífero nª 23 – Mancha Occidental: síntesis hidrogeológica. Serie: Manuales de Utilización de Acuíferos. Instituto Tecnológico y Geominero de España, Dirección de Aguas Subterráneas, Madrid
IGME (2004) Informe sobre evolución piezométrica de la Unidad hidrogeólogica 04.04, Mancha occidental, Año 2004. Instituto Geológico y Minero de España, Ministerio de Ciencia e Innovación, Madrid
IGME (2008) Informe sobre evolución piezométrica de la Unidad hidrogeólogica 04.04, Mancha occidental, Año 2008. Instituto Geológico y Minero de España, Ministerio de Ciencia e Innovación, Madrid
John DM (1986) The inland waters of tropical Africa. Arch Hydrobiol 23:1–236
Junk WK (1983) Ecology of the várzea, floodplain of Amazonian whitewater rivers. In: Siole H (ed) The Amazon. Dr W. Junk Publishers, Dordrecht
Köhler J (1993) Growth, production and losses of phytoplankton in the lowland river Spree. I Population dynamics. J Plankton Res 15:335–349
Llamas MR, Martínez-Santos P (2005) NeWater WB3 Report (WP3.4 Guadiana Basin), Baseline Condition Report (A) Upper Guadiana Basin. http://newater.topshare.com/public-files/2037_22382554010_200804230920859_83602_Guadiana_Baseline.pdf. Accessed 21 Oct 2009
Martínez-Cortina L (2002) Marco hidrológico de la Cuenca Alta del Guadiana. In: Coleto C, Martinez-Cortina L, Llamas MR (eds) Conflictos entre el desarrollo de las aguas subterráneas

y la conservación de los humedales. Fundación Marcelino Botín-Ediciones Mundiprensa, Madrid

Navarro A, Fernández-Uría A, Doblas JG (1993). Las aguas subterráneas en España. Instituto Geológico y Minero de España, Madrid, Spain. http://aguas.igme.es/igme/publica/libro20/lib20.htm. Accessed 23 Oct 2009

OECD (1996) Territorial indicators of employment. Focusing on rural development. Paris

Olmedo A (2002) La participación ciudadana y las comunidades de usuarios en la gestión de las aguas subterráneas y de los humedales de Castilla-La Mancha. In: Coleto C, Martinez-Cortina L, Llamas MR (eds) Conflictos entre el desarrollo de las aguas subterráneas y la conservación de los humedales. Fundación Marcelino Botín-Ediciones Mundiprensa, Madrid

Pérez-González A (1996) Marco geológico y geomorfológico. In: Álvarez-Cobelas M, Cirujano S (eds) Las Tablas de Daimiel: Ecología Acuática y sociedad. Ministerio de Medio Ambiente, Madrid

Pérez-González ME, Sanz-Donaire JJ (1998) Clima y microclima de La Mancha Húmeda. Anales Geografía Universidad Complutense 18:239–256

Portero JM, Ramírez JI (1988) Memoria y hoja geológica a escala 1:50.000 de Daimiel (nº 760). Instituto Geológico y Minero de España, Madrid

Richardson CJ, Reiss P, Hussain NA, Alwash AJ, Pool DJ (2005) The restoration potential of the Mesopotamian marshes of Iraq. Science 307:1307–1311

Sáez-Royuela R (1977) Contribución al estudio ecológico de Las Tablas de Daimiel III. Las aguas (1974–1975). Anales INIA. Serie Recursos Naturales 3:101–149

Sahuquillo-Herráiz A, López-García J, López-Camacho B (1982) Transient simulation of an aquifer connected to a wetland. Math Comput Simulat 24:161–172

Sarria A (1986) Transcripción del original de la Biblioteca del Real Monasterio de San Lorenzo del Escorial de las Relaciones topográficas mandadas hacer por Felipe II, año 1575. Ayuntamiento de Daimiel, Daimiel

SGDGOH (1989) Estudio para la ordenación de las extracciones y niveles piezométricos en el acuífero de La Mancha Occidental. Estudio 12/89. Ministerio de Obras Púrlicas y Transporte, Madrid

Tarjuelo JM (1999) Agricultura. In: Diaz-Pineda M (ed) Comisión de Expertos sobre le Plan de Ordenación de los Recursos naturales y Desarrollo Sostenible del Alto Guadiana. Dictamen, prospección, diagnóstico y propuesta de actuación. Technical Report. Organismo Autónomo de Parques Nacionales and Conserjería de Agricultura y Medio ambiente, Castilla-La Mancha, Madrid

Viladomiu L, Rossell J (1997) Gestión del agua y política agroambiental: el Programa de Compensación de Rentas por reducción de regadíos en la Mancha Occidental y Campo de Montiel. Revista Española Economía Agraria 179:331–350

Viladomiu L, Rossell J (2002) Intensificación agraria, agua y humedales en la cuenca alta del Guadiana. In: Coleto C, Martinez-Cortina L, Llamas MR (eds) Conflictos entre el desarrollo de las aguas subterráneas y la conservación de los humedales. Fundación Marcelino Botín-Ediciones Mundiprensa, Madrid

Part II
Abiotic Environment and Historical Reconstructions

Chapter 2
Paleoenvironmental Reconstruction of Las Tablas de Daimiel and Its Evolution During the Quaternary Period

B. Ruiz-Zapata, M.J. Gil-García, and I. de Bustamante

Abstract Pollen analyses demonstrated that changes experienced in Las Tablas de Daimiel (TDNP) area throughout the Quaternary took place mainly through climate-induced landscape transformations as well as by human activities occurring in the region during the last 1,000 years. From a climatic point of view, the regional Pleistocene landscape was dominated by Mediterranean forest (*Quercus*, mesophilus taxa and *Pinus*) during warm and humid periods, while open landscapes, depicted by xeric grass and steppes, were abundant in dry and cold events (Late Maximun Glaciar and Younger Dryas). The Holocene period showed an increased trend towards the Mediterranean forest. In this historical framework, TDNP evolved from a lacustrine environment to a fluctuating freshwater marsh, through to a braided river during the last 250,000 years. The use of a high resolution pollen record, in combination with geochemical data in charophyte and detrital sediment layers allowed wetland reconstruction during the last 3,000 years, introducing a new climatic dataset for the Late Holocene. Despite the lack of any direct evidence of human action, there were some deforestation episodes during the Spanish Re-conquest of Iberian Muslim territories (Middle Age) which did not to correlate with the climatic signal. It was clearly observed, since ecological changes were controlled by climatic variability at the beginning of the sequence, while the later intense human-made disturbances abruptly interfered with the natural recovery mechanism of the ecosystem.

2.1 Introduction

Paleoclimatic reconstructions from marine environments (Ruddiman and McIntyre 1981; Moreno et al. 2002, 2005; Huntley et al. 2003; Duplessy et al. 2005), from ice cores (Johnsen et al. 1992; Dansgaard et al. 1993; Bond et al. 1997; Bradley

B. Ruiz-Zapata (✉)
Department of Geology, University of Alcalá de Henares, Science Building, Ctra. A-II km 33.600, E-28871 Alcalá de Henares, Spain
e-mail: blanca.ruiz@uah.es

S. Sánchez-Carrillo and D.G. Angeler (eds.), *Ecology of Threatened Semi-Arid Wetlands: Long-Term Research in Las Tablas de Daimiel*, DOI 10.1007/978-90-481-9181-9_2,

1999; Broecker 2000) and from continental environments (Frenzel et al. 1992; Allen et al. 1999, 2000) demonstrate strong climatic oscillations during the Quaternary period. These oscillations are the result of the break, due to natural reasons, of the system's relative balance across complicated and only partially known mechanisms. However, long record sequences provided by marine environments and ice cores come mainly from high latitudes, complicating extrapolations to understand climate evolution in the Mediterranean region during the Quaternary. On the other hand, the continental sequences of Mediterranean Europe are not constant, except in few cases (e.g. Tzedakis et al. 1997, 2003, 2004; Allen et al. 2000); therefore a certain degree of chronological uncertainty prevails (Tzedakis et al. 2006). In the case of the Iberian Peninsula, although there are few terrestrial sequences covering the entire last glacial period and previous glacial periods (e.g. Pons and Reille 1988; Montserrat 1992; Burjachs and Juliá 1994; Ruiz-Zapata et al. 1996, 2002; Sánchez-Goñi et al. 1999, 2000, 2002; Schulte et al. 2008; Gómez-Orellana et al. 2007), and most of them are located in high mountain ecosystems or related to archaeological sites (Dupré 1988; Carrión et al. 1998; Ruiz Zapata et al. 2005, 2008), reconstructions revealed the unequal distribution of Quaternary climate oscillations. Paleorecords showed numerous climate fluctuations in the North, Centre and Southeast areas of the Iberian Peninsula (Julià et al. 1994; Peñalba et al. 1997; Martín Arroyo et al. 1999; Valero-Garcés et al. 2000, 2008; Carrión 2002; Gil García et al. 2002; Gómez-Orellana et al. 2007; Luzón et al. 2007; González-Sampériz et al. 2008) whilst suggesting very slight changes in the Southern low altitude areas. These findings could be verified in the two recent paleorecords performed in the Central area of the Iberian Peninsula, Fuentillejo (Ruiz Zapata et al. 2008; Vegas et al. 2006, 2009) and Las Tablas de Daimiel (Valdeolmillos 2004).

We acknowledge that the current aridity experienced in the middle South area of Spain seems to have originated during the Holocene, as a response to the Azores anticyclone displacement toward the North. This is due to blocking the arrival of areas of low pressures from the West. On the contrary the arid phases experienced in the Pleistocene are associated with the latitudinal decrease of the polar front (Bailarón 1997). Independently of their origin, both climatic fluctuations have been recorded with distinctive intensity in the Las Tablas de Daimiel area, inducing changes on wetland ecology and modifying plant community compositions.

Paleoecological reconstructions can be approached from the perspective of mineralogical, limnological, petrographic, palynologic and isotopic analysis (Alonso-Zarza et al. 2004, 2006; Valdeolmillos 2004). Because the formation of a wetland belongs to the conjunction of several climate-controlled parameters, pollen appears as one of main sources to detect wetland ecological changes throughout time. The study of the content of the pollinic palynomorphs of sediments (Faegri et al. 1989; Moore et al. 1991) constitutes one of the principal instruments in the reconstruction of vegetation landscape (Birks and Birks 1980). This helps reconstructing environmental evolution and/or climatic change, providing a useful guide to eventual future restoration of the environment (Birks et al. 1988; Carrión 2001).

It is generally assumed that detected changes in plant community structure in paleoecological reconstruction can be understood as a response to the prevailing climatic conditions during certain periods of the Quaternary period. Nevertheless, from the Middle Holocene, the anthropogenic activity (Behre 1988; Moore et al. 1991; Carrión et al. 2000; Carrión 2001; Costa Tenorio et al. 1990; García Antón et al. 2002; López Sáez et al. 2003; González-Sampériz 2001, 2004) begins to be a determinant factor in vegetation development. These facts make the knowledge of the Holocene paleoclimatology and paleoecology hard to interpret (Carrión et al. 2000), since it is not always possible to differentiate phenomena which are produced by humans or natural factors. But they can be useful to understand the consequences that environmental changes have had on the socio-economic settings of prehistoric and protohistoric societies (López Sáez et al. 2003). This helps determining whether synchronies exist between climatic and social changes, as well as to evaluate the degree and category (agriculture, livestock, deforestation, etc.) of the human activities for every cultural period (Girard 1973; Dupré 1988; Berglund 1991; Galop 1998; López Sáez et al. 2000, 2003; Barbier et al. 2001).

The pollinic analyses of the sequences available up to this moment at Las Tablas de Daimiel have allowed for the establishment of an evolution of the vegetation from the Middle-Pleistocene up to the present day. In addition, the manner in which the climate and/or the anthropogenic activity in the Holocene have determined the changes of the vegetation landscape in the wetland and in its surrounding areas could be deduced.

Because sediment records are somehow incomplete, be it for the existence of hiatuses and/or for the temporary range they register, the study of the climatic variability in the area must be done by means of a regional review of the available records. In spite of that, the evolution reconstructed from these sequences is coherent with the information valid for the rest of the Iberian Peninsula.

2.2 Material and Methods

Paleoecological and paleoclimatic reconstructions of Las Tablas de Daimiel (TDNP hereafter) were fundamentally based on the identification of the pollinic contents from seven paleorecords located inside of the wetland and its surroundings (Fig. 2.1): the Gigüela profile 4.2 (Gil García et al. 2007), and the CC-17 (Dorado Valiño et al. 1999, 2001, 2002), TD (Valdeolmillos et al. 2003), LT (Valdeolmillos 2004; Dorado Valiño et al. 2004; Alonso-Zarza et al. 2006), DTD (Valdeolmillos et al. 2002) sequences. In order to complete the temporal sequence (Fig. 2.2) as well as to determine if the detected climatic events extended to local and / or regional ranges, the eolian sequences MO and TASG located in the context of La Llanura Manchega (Ruiz Zapata et al. 2000) were included. The above-mentioned events have been temporarily located from the numerical indexed dating; for those

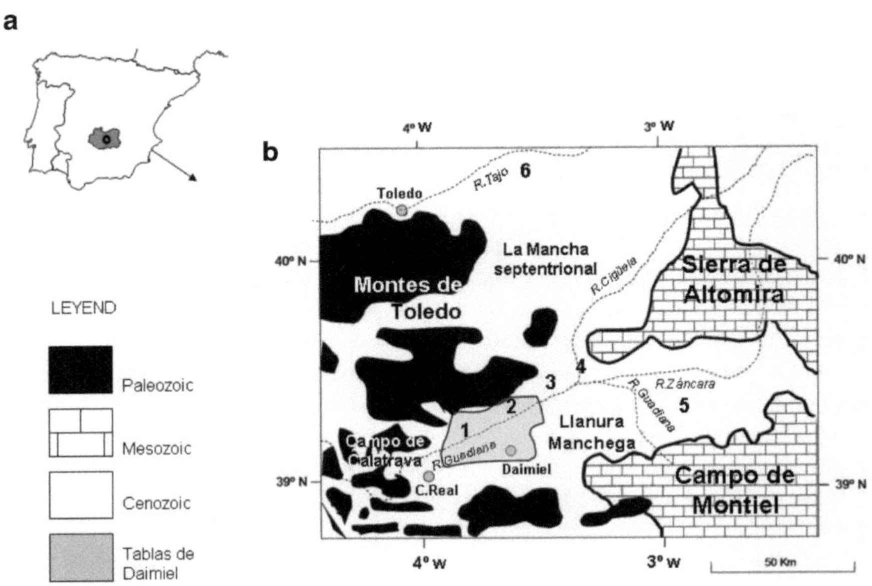

Fig. 2.1 Location of (**a**) La Mancha region in the Iberian Peninsula (**b**) and closer view showing studied cores (1: CC-17, 2: LT and TD, 3: DTD, 4: Gigüela 4.2, 5: TASG, 6: MO) in their geological setting

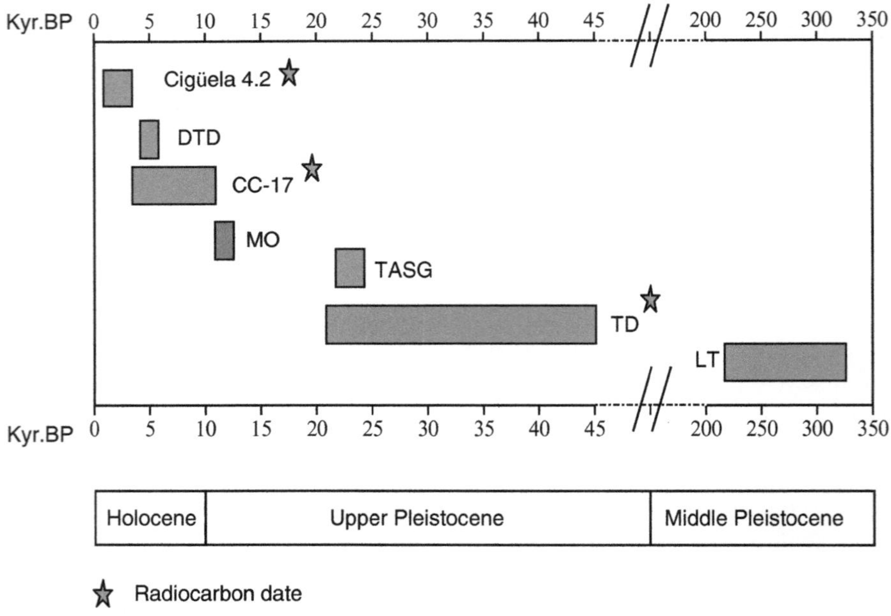

Fig. 2.2 Chronology of the paleo-sequences studied in Las Tablas de Daimiel area

levels without numerical dating, the adopted chronology is estimated by the authors.

The core sampling methodologies were detailed in the previously cited studies. Briefly, the Gigüela sequence 4.2 was obtained by means of a vibracorer drill, CC-17 sequence with an Eijkelkamp mechanical drill and TD and LT sequences, with a constant witness mechanical drilling. DTD, MO and TASG sequences were directly taken on the vertical profiles which emerged on the surface.

Different degrees of resolution have been used for the pollinic samplings of every sequence, depending on the aims pursued in each paleo-record. Thus, when the goal was to identify vegetation throughout a climatic sequence, as in TD and LT, the interval of the analyzed samples (194) was selected by changes in lithology. In the case of the sequences DTD, MO, TASG and CC-17, lithological homogeneity advised a sampling every 10 cm. The Gigüela 4.2 profile was sampled with a constant resolution of 0.7 cm, in order to discern between climatic and/or anthropogenic activity. Chemical treatments for pollen grain extraction were applied following standard protocols (Coûteaux 1977; Faegri et al. 1989; Moore et al. 1991). The obtained residues were submitted to enrichment technologies by means of the pollen concentration by blooming in Thoulet dense liquor (Girard and Renault-Miskovsky 1969) and, finally, preserved in glycerine.

Reading and counting the pollinic contents was carried out following the method proposed by Cambón (1981). The number of grains counted has always been higher than 300, except in poor samples, as those of MO and ASG sequences, where pollen counted only reached 215 grains. For the determination of the pollinic types, the reference collection at the University of Alcala and the Atlas of Pollen by Valdés et al. (1987), Moore et al. (1991) and Reille (1992) was used.

Pollinic taxa were quantified through pollinic diagrams using the TILIA® and TILIA-GRAPH® (© Eric C. Grimm 1987) computer packages. Taxa were classified into arboreal, shrubby and herbaceous types. Frequencies were computed from sume base, excluding the pollen and spores from aquatic vegetation. These percentages were calculated as the total sum of the content in palynomorph of every sample.

2.3 Overview of Vegetation Composition in La Mancha Region During Quaternary Period: Palaeo-Sequences Description

Table 2.1 shows the radiocarbon dating of the studied sequences. Most recent information was provided by the Gigüela profile 4.2 sequence (2,700 years before the present day (BP) while TD and LT paleorecords extended to the Upper and Middle Pleistocene (45,000 and 325,000 years BP). Therefore, chronology of the sequences demonstrated nearly complete records until the first third of the Upper Pleistocene (Fig 2.2). There was a long hiatus until halfway the Middle Pleistocene which is covered partially by the LT sequence (Fig. 2.2). Both the dating and the chronological sequence served then as the starting point to establish the evolutionary frame of the vegetation during the Quaternary.

Table 2.1 Radiocarbon dating of the cores taken at Las Tablas de Daimiel National Park and the surrounding area

Core	Sample depth (cm)	Conventional ^{14}C age (year BP)	Calibrated results (2σ, 95%)	Laboratory reference
Gigüela 4.2	560	521 ± 37	AD 1,389–1,445	GdA-308
	730	1,098 ± 39	AD 877–1,020	GdA-309
	990	2,699 ± 53	939–795 BC	GdA-306
CC-17	346–356	6,150 ± 60	5,240–4,925 BC	Beta-79215
	635–645	9,890 ± 180	>8,590 BC	Beta-79216
	630	21,120 ± 60		Beta-135637
LT-TD	810	25,160 ± 100		Beta-135639
	1,340	>44,940*		Beta-132973
	1,342		U/Th	Valdeolmillos (2004)
	1,480–2,180	180,000 OIS 7	Racemization	Dorado (2004)
MO	270	11,000	IRSL	Rendell et al. (1994, 1996)
TASG	200	22,000–23,000	OSL/TL	Rendell et al. (1994, 1996)

*radiocarbon dead

2.3.1 Landscape Evolution from 325,000 to 21,000 Years BP: LT, TD, TASG-1 and FUENT-1 Profiles

LT and TD profiles were taken inside TDNP in the central part of the wetland close to the right side of the Guadiana riverbank. From a sedimentological point of view (Alonso Zarza et al. 2004, 2006), both sequences were differentiated into three units, reflecting sediment compositions pertaining to a very shallow fluvio-lacustrine system. The lowest unit (TD-A) appeared in the last 17 m and it represented a shallow lake system. The dating compiled for this interval (from 14.80 to 21.80 m) indicates an age previous to the OIS 7 (Dorado Valiño et al. 2004; Alonso-Zarza et al. 2006). The intermediate section (LT-B) registered the settling of a freshwater swamp with abundant vegetation, and it was dated to be around 180,000 years by Valdeolmillos (2004) and Alonso-Zarza et al. (2006) (Table 2.1). The highest section (LT-C and TD sequences, Valdeolmillos et al. 2003) revealed the presence of a marshy system where fluvial discharges alternated with some desiccation phases, corresponding to an age of from 20 to 45,000 years BP (Table 2.1). Pollen analyses of the section LT-A (Valdeolmillos 2005) displayed the presence of Pinaceae, Cupressaceae, Asteraceae and Poaceae which indicated the change from dry conditions to wet ones. In the intermediate section or LT-B, there was an increase of both tree (*Pinus*, *Quercus*, deciduous trees) and aquatic plant diversities, denoting a temperate climate with increasing humidity. Although in LT-C layers without pollen content were recorded most frequently, a landscape change was detected, from mesophile and Mediterranean forests to Betulaceae forest, which were replaced by Pinaceae and, finally, by Cupressaceae, culminating in the development of open landscapes dominated by Poaceae and Chenopodiaceae (Valdeolmillos 2004; Valdeolmillos et al. 2003). A colder and arid environmental condition was reconstructed which appeared typical of the Last Glacial Maximum. The sterile layers registered at the top of the impaired vegetation period (Valdeolmillos 2004) have been correlated with the Heinrich events (Heinrich 1988), H4 about 36,000 years BP (between 10.80 and 11.20 m) and the event H3 about 28,000 years (7.40 and 7.80 m) (Fig. 2.4).

These cold and arid conditions detected in LT-C and TD (Valdeolmillos 2004) seem to be widespread at regional landscape level, and they were also recorded in the FUENT-1 sequence (Vegas et al. 2009), located in the neighbouring volcanic area of Campos de Calatrava (Fig. 2.1). The first 20 m of the FUENT-1 sequence developed during the isotopic marine OIS 1–2–3, corresponding to a succession of arid phases across two steppe associations: one warm and dry steppe composed by Chenopodiaceae, Asteraceae and Ephedraceae and another cold and dry steppe, dominated by *Juniperus* (Cupressaceae). These cold episodes were correlated with Heinrich's climate fluctuations H2 (24,800–23,000 years BP) and H1 (17,500–16,000 years BP) as well as with the latter Younger Dryas cold period (12,500–11,500 years BP) (Vegas et al. 2009). After that, the warm steppe increased, although this did not imply an increase in the humid conditions.

While these climate conditions appeared to be generalized in the area, the eolian sequence TASG-1 placed to the East of the Llanura Manchega (Fig. 2.1) and dated in the Last Glacial Maximum (22–23,000 years BP), revealed that the arid phases showed different drought intensities. Likewise, this sequence demonstrated a spatial heterogeneity on the region as a result of the presence of shelter areas where vegetation associated to humid environments was able to grow during arid periods (Rendell et al. 1994). The pallinologic analysis of this sequence (Fig. 2.3) displayed an open vegetation landscape, settled by Asteraceae tubiliflorae, Asteraceae liguliflorae, Chenopodiaceae, and Poaceae. Although tree vegetation did not appear as predominant, its onset defined four distinctive climate phases: 1 and 3 characterized by dry conditions, as opposite to 2 and 4, which recorded the presence of *Quercus*, *Betula*, *Corylus*, *Fagus* and *Castanea* (Ruiz Zapata et al. 2000).

2.3.2 Landscape Evolution from 11,000 to 4,000 Years BP: MO and CC-17 Sequences

The most ancient paleo-record at La Mancha region during the Holocene is the MO sequence, which is located in Villarrubia de Santiago (Toledo), about 75 km North of TDNP (Fig. 2.1). This record was taken in a sandy formation with massive structure and "loess-like" type facies, dated by IRSL as 11,000 years BP (Table 2.1, Rendell et al. 1994) and corresponding to the Younger Dryas (Fig. 2.2). Along its 2.70 m of thickness it is possible to observe the progressive loss of arboreal vegetation, constituted by *Pinus*, which gave way to *Juniperus* and finally to an open landscape of Asteraceae and Chenopodiaceae. It appeared to be an indication of increased aridity in the area.

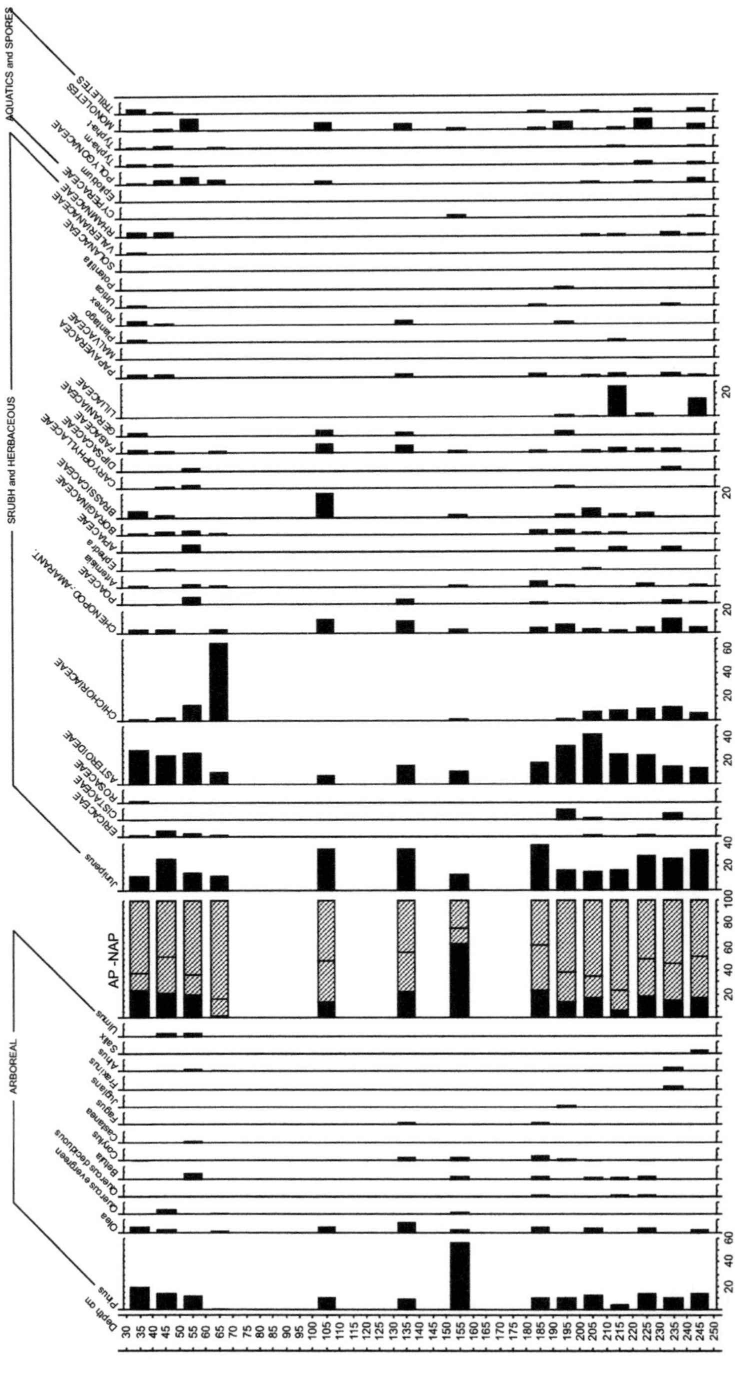

Fig. 2.3 Pollen diagram of the eolian sequence TASG-1 (Modified from Ruiz Zapata et al. 2002)

The next sequence in the temporal paleo-record is the CC-17 (Dorado Valiño et al. 2004; Fig. 2.4), a peat deposit formed in a meander of the Guadiana River, at approximately 4 km downstream from TDNP (Carrión de Calatrava, Fig. 2.1), with a radiometrically dated age of around 10,000 years BP (Table 2.1). Eight pollinic zones were defined (Fig. 2.4, Dorado Valiño et al. 2002) along its 6.81 m of length. Palinologically, the lowest and oldest one (A) was characterized by the scanty abundance of trees – almost exclusively defined by *Pinus* and to a lesser extent by Cupressaceae – which, together with *Artemisia* and Chenopodiaceae, defined conditions of cold and arid nature. The next layer (B) accumulated *ca.* 9,890 ± 180 years BP, represented a time interval where the arboreal set was more profuse (both types of *Quercus*, *Olea* and *Pinus*). It corresponds to an increase of temperature and humidity and was supported by the development of aquatic taxa. Chenopodiaceae and Asteraceae were expanded in layer C, while the arboreal strata diminished, marking a slightly arid phase. In layer D, dated at the top sequence in 6,150 ± 60 years BP (Table 2.1), the arboreal elements recovered, basically to *Quercus* and *Olea* trees, together with aquatic plants, similar to what has been documented in the area of Castillo de Calatrava (García Antón et al. 1986) that is dated to 6,240 ± 190 years BP. These facts defined an increase of temperature and humidity that might correspond to the Holocene climatic optimum, a warm period roughly present in the interval between 9,000 and 5,000 years BP. In layer E, the environment returned again to aridity through the decrease of the arboreal pollen and the development of Cupressaceae, Asteraceae and Chenopodiaceae. This pattern was also observed in the DTD sequence (Fig. 2.4) (Valdeolmillos et al. 2002). During deposition of layer F, warm and dry Mediterranean conditions were settled; evergreen *Quercus* and, to a lesser extent, deciduous *Quercus* increased and the constant curve of *Olea* began. This is also observed in the Daimiel II profile dated in 3,190 ± 70 years BP (Menéndez and Florchütz 1968).

2.3.3 Landscape Evolution from 4,000 Years BP to the Present: CC-17 and Gigüela 4.2 Paleo-records

The two uppermost layers of the CC-17 sequences define the environmental conditions of the area in the Holocene (Dorado Valiño et al. 2002, Fig. 2.4). Layer G again represented an arid episode associated with the fall of the Mediterranean taxa and the development of Cupressaceae and *Artemisia*. The uppermost layer (H) registered an increase of the arboreal cover due to *Quercus* and *Artemisia* spread but as in the nearby area of Castillo de Calatrava, which dated in 1,730 ± 80 years BP (García-Antón et al. 1986), the defined conditions were markedly Mediterranean. In the Holocene of the FUENT-1 sequence (4,000 years BP), vegetation was detected that is clearly of Mediterranean nature (Ruiz Zapata et al. 2008; Vega et al. 2009).

The arid episodes described for the Holocene in the central Iberian Peninsula area (i.e. García-Hidalgo et al. 2007) were not registered in the CC-17 sequence, due either to the lack of sedimentation or to the absence of any response from the

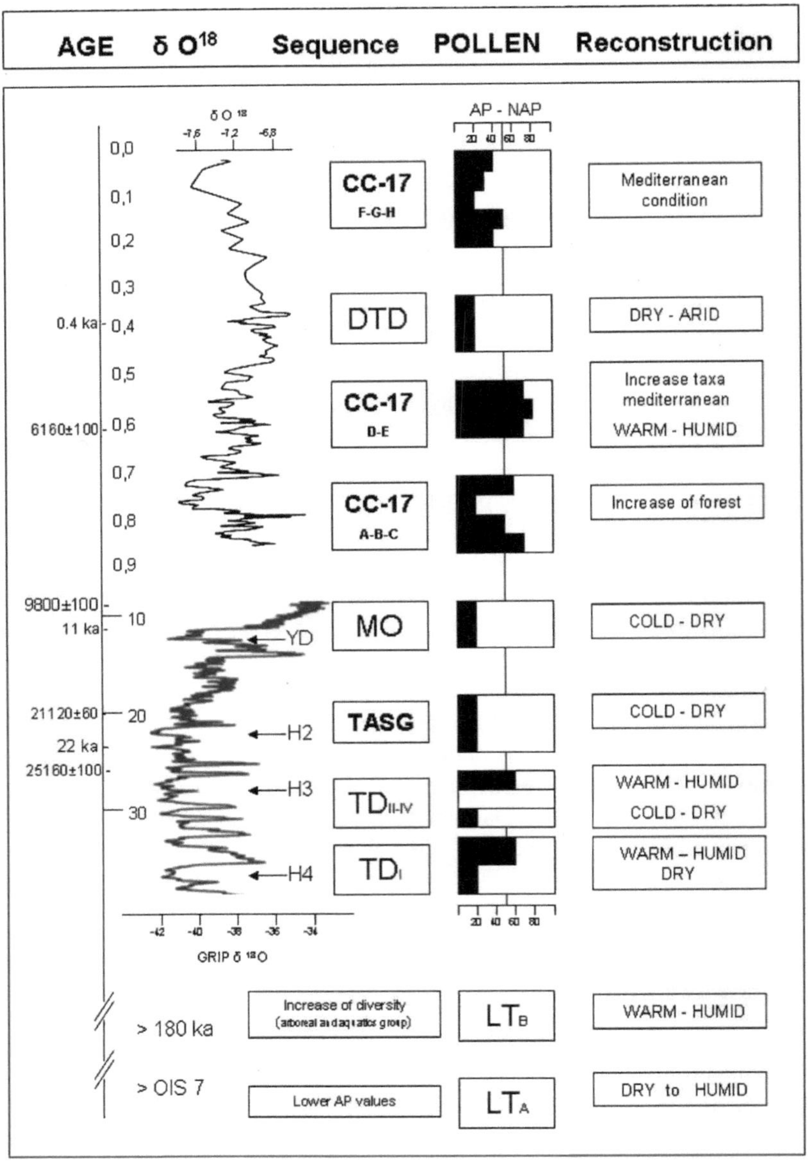

Fig. 2.4 Correlation between the main pollen sequences (AP/NAP) at Las Tablas de Daimiel National Park area and the GISP-2 isotope record (Taken from Dansgaard et al. 1993). Heinrich events are noted as H4, H3 and H2. YD is the Younger Dryas

plant cover during these episodes. Although the CC-17 sequence showed great similarities with the Castillo de Calatrava record (García Antón et al. 1986), the former did not to detected any human influence associated with the cereal development later than 1,730 ± 80 years BP, in spite of settlements in the area and the agrarian practices which have been verified from the Age of the Bronze (Molina et al. 2005), as well as during the Roman occupation (Ramos 1988).

The last sequence representing the most recent records in TDNP is Gigüela 4.2 (Fig. 2.5), which was obtained from the right side of the Gigüela river, inside the Tablas de Daimiel National Park (Fig. 2.1). At the beginning of the sequence (Area 0; about 2,800–2,100 years BP or <150 BC) open field landscapes were present, composed by Asteraceae (liguliflorae and tubuliflorae) and Poaceae, followed by *Artemisia*, Brassicaceae and Chenopodiaceae-Amaranthaceae, with scarce trees (*Pinus*) and shrubs (*Calluna*). This information, together with the lower C and N contents, revealed the existence of slightly arid and cold conditions, typical of the Sub-Atlantic cold and arid phase (Late Iron Cold Period). This phase also has been detected in the Castillo de Calatrava (García Antón et al. 1986) and Daimiel II (Menéndez Amor and Florchütz 1968), where *Artemisia* and Chenopodiaceae-Amaranthaceae increased simultaneously. This arid phase has been identified in many other places of the Iberian Peninsula such as in the North-West (Desprat et al. 2003) and in the North-East (Gutierrez-Elorza and Peña Monné 1998). In the *A* area of the sequence (2,100–1,680 years BP, 150 BC–AD 270) the arboreal recovery, the greater diversity of aquatic taxa and the decrease of *Calluna*, together with the increase of organic C, suggested a more humid period and a seasonal raise of the water-level in the wetland. The simultaneous increase experienced by the *Quercus* evergreen and the setback of *Pinus* and *Artemisia* revealed slightly higher temperatures. This thermal improvement was also identified in the North-West of the Iberian Peninsula (Desprat et al. 2003) and it belongs to the warm Roman episode. In the beginning of *B* area (1,600–1,000 years BP, 270–AD 950) a generalized increase of the wetland is apparent, allowing for a shallow and lentic water ecosystem where Charophyceae, as well as Poaceae, Cyperaceae, *Typha*, Ranunculaceae, *Polygonum* and *Potamogeton*, developed. However, the increase of Chenopodiaceae-Amaranthaceae (revealing the existence of saline soils in the surroundings), together with a dieback of the tree taxa, specially evergreen *Quercus,* and a slight increase of *Artemisia,* evidenced a climatic deterioration related to slightly lower temperatures, as assigned to the Dark Ages by Desprat et al. (2003). Riera et al. (2004) also identified similar conditions, due to an increase of salinity and aridity, in lakes of the NE of Spain for the same period. Data of the *C* area of the Gigüela 4.2 sequence (1,000–860 years BP, 950–AD 1,090) reveal an important anomaly where arboreal and shrubs taxa tended to disappear in the lower zone of the area, recovering later and declining again towards the top. These dramatic changes have been explained in other Spanish regions as related to human disturbance (Riera et al. 2004) and to climate change (Desprat et al. 2003). In this instance, human disturbance can be connected to the Muslim dominance during the eighth century, which introduced water mills and exploited the pastures, thereby explaining the low values of arboreal and shrubs pollen. Likewise, the recurring fights until Christians reconquested the territory in the eleventh century, might also partially explain the deforestation at the end of this cycle. Despite anomalous vegetation records, aquatic taxa diversity increased during this period, due to a significant rise of water depth in the wetland. The abundance of riparian taxa (Poaceae, Cyperaceae and *Tamarix*)

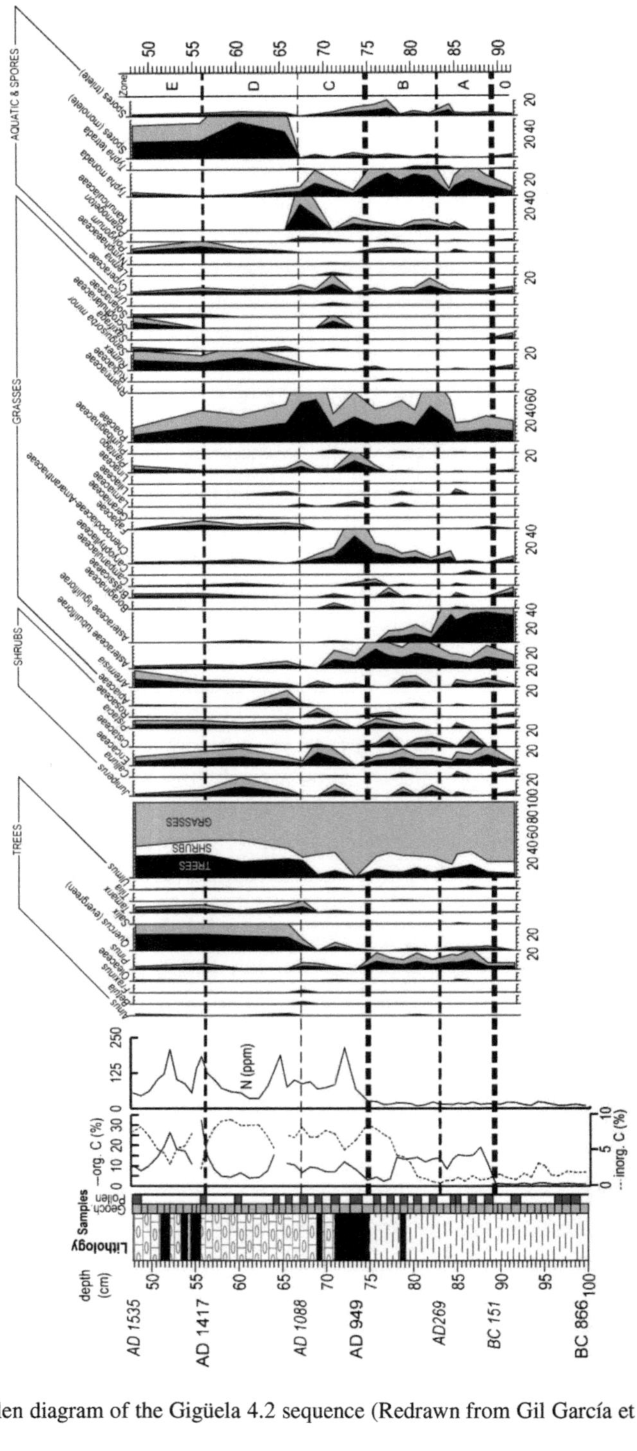

Fig. 2.5 Pollen diagram of the Gigüela 4.2 sequence (Redrawn from Gil García et al. 2007)

together with the presence of *Betula* and *Fraxinus* and a decrease of Asteraceae and Chenopodiaceae-Amaranthaceae, also suggest a raise in wetland water level, related to a gradual change towards a wet environment. The detected increase of evergreen *Quercus* versus the slight setback of *Pinus* and *Artemisia* reveal that temperature increases joined humid conditions, to shape the climate during this period.

The *D* area (860–530 years BP, 1,090–AD 1,400), showed a notable increase of the perennial and evergreen *Quercus*, accompanied by a rise of some taxa of pasture and nitrophilous herbs, revealing the existence of landscapes very similar to the "meadow". Significant changes were also observed in the aquatic environment, such as the increase of the emergent vegetation (Cyperaceae, *Typha* and Poaceae), the sharp increase of monolete spores and the low diversity of the submerged aquatic taxa. These would be consistent with the existence of frequent periods of eutrophication. These changes would also agree with an increase of the temperature and humidity, as revealed by the increase of the evergreen *Quercus*. Comparable changes were described by Desprat et al. (2003), Julià et al. (1998) and Riera et al. (2004) in NW, central and NE of Spain, and were assigned to the "Warm Medieval Period".

Finally, during the *E* area of the Gigüela 4.2 sequence (after AD 1,400), an increase of *Pinus* and *Artemisia* was observed, both indications of a temperature descent. A dominance of the monolete spores appeared again in the aquatic environment, depicting frequent periods of eutrophication. Finally, fluctuations observed in the organic carbon curve demonstrated an unstable climate where cold and warm periods alternated. These characteristics suggest that this period represents the beginning of the "Little Ice Age" also identified by Desprat et al. (2003), Julià et al. (1998), Gutierrez-Elorza and Peña Monné (1998) and Riera et al. (2004) in other Spanish sequences.

2.4 Summary of Climatic and Wetland Environmental Evolution During the Quaternary Period

The previously described pollinic sequences (Fig. 2.2) collect regional information from Middle Pleistocene up to the present. From the palynological point of view, all the sequences studied showed a great homogeneity, with a constant presence, although not very abundant, of Pinaceae, Cupressaceae, and herbaceous taxa, fundamentally of xeric nature (Asteraceae, Chenopodiaceae and *Artemisia*) (Fig. 2.4). Evergreen *Quercus* and, occasionally, deciduous *Quercus* were detected, which defines the Mediterranean landscape of the area along the Quaternary period. Riparian taxa such as *Alnus* and *Salix* appeared in a constant pattern, although in low proportion, as did the aquatic taxa (Cyperaceae, Ranunculaceae, *Potamogeton*, *Typha monada* and *tetrada*, Nymphaceae). Abundant fluctuations of both groups can be related to water availability. In this sense, it was relatively frequent to observe presences of mesophilous taxa (*Betula*, *Corylus* and *Castanea*) that alternated with xeric herbaceous taxa, which indicates climate fluctuations from temperate and

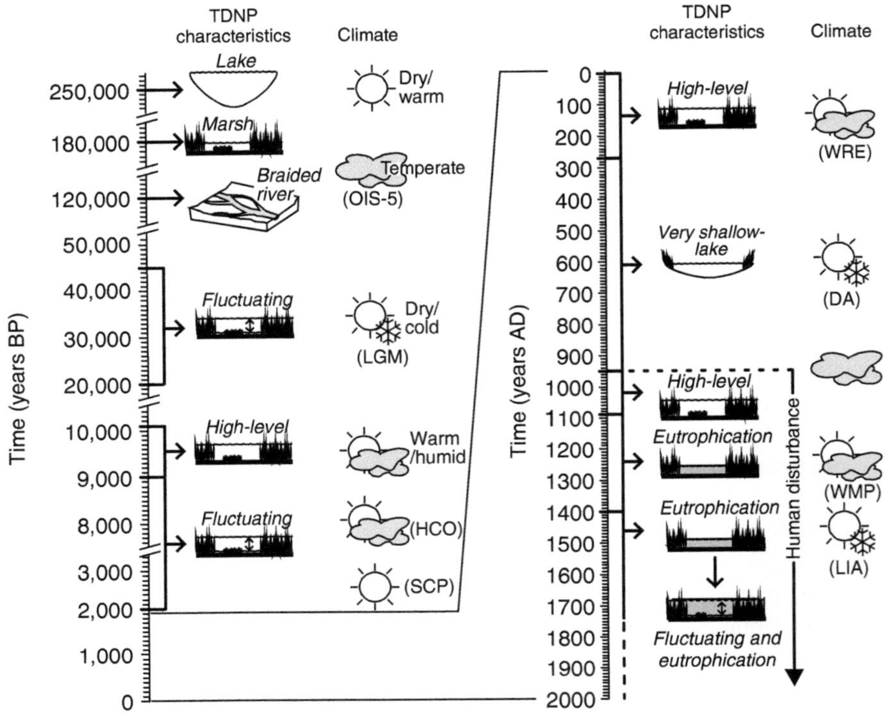

Fig. 2.6 Summary of the temporal evolution of Las Tablas de Daimiel and climatic variability during the Quaternary period. (LGM: Last Glacial Maximum, HCO: Holocene Climate Optimum, SCP: Sub-Atlantic Cold Period, WRE: The Warm Roman Episode, DA: The Dark Ages, WMP: Warm Medieval Period, and LIA: Little Ice Age)

humid to dry and cold conditions. It was in this regional framework when the ecosystem evolution of TDNP during the Quaternary took place, where fossil records prove its occurrence more than 200,000 years ago. There is evidence of human disturbance in the ecosystem from the year 1250; however, it is only recently that these disturbances are severely changing the ecosystem. The main ecosystem transformations in the regional climatic context from more than 200 ky BP can be summarized as (Fig. 2.6):

2.4.1 Middle Pleistocene

Before 200 ky BP, the wetland appeared as a lacustrine environment with slight fluvial discharges. Regional landscape was composed by mosaics of small coniferous forests (*Pinus*), Cupressaceae fields, mesophilous (*Salix*) and thermophilous (*Quercus*) forests and steppes with *Artemisia* and Poaceae. Climate tended to be dry.

≈180 ky BP. TDNP changed to a freshwater marsh with abundant and diverse aquatic vegetation. Also, tree composition increased its diversity during this period. Climate appeared as temperate and humid, although fluctuating.

2.4.2 *Upper Pleistocene*

About 120 ky BP, the wetland was a braided fluvial system. *Pinus* and *Quercus* forest expanded as well as consolidated *Castanea*, *Corylus* and *Betula* trees. Climate was then temperate and humid, corresponding with some interval of the OIS 5.

20–45 ky BP. The wetland transformed again into a marshy ecosystem where fluvial discharges alternated with desiccation cycles. The landscape registered a progressive loss of forests, while xeric vegetation expanded. Climate became cold and dry, corresponding to the Last Glacial Maximum.

2.4.3 *Holocene*

During the Early Holocene (11–9 ky BP.) TDNP continued to be a marsh although the increase of aquatic taxa indicated a period of high water levels. The landscape showed a recovery of evergreen and deciduous *Quercus* forests as a result of a more humid and warm climate. This environmental pattern seems to be found across the Iberian Peninsula, as evidenced in other sequences.

9–1 ky BP. The wetland fluctuated between dry and humid periods as evidenced by the changes in aquatic plant diversity. As a result of climatic variations, the landscape also oscillated between forest reduction and recovery. The increase of temperature and wetness recorded during 9–5 ky BP corresponded to the Holocene Climate Optimum. At the end of this cycle, climate returned to warm and arid conditions similar to those prevailing nowadays.

Above 150 BC. The Sub-Atlantic Cold Period started at this time, characterized by cold and dry conditions. The landscape featured large steppes with scarce trees and shrubs. Wetlands did not change significantly from the previous period.

150 BC–AD 270. This cycle corresponded to The Warm Roman Episode, a warmer and wetter climate period characterized by an arboreal recovery in the landscape. The wetland experienced a significant increase of seasonal water-level, resulting in the development of numerous aquatic taxa. During this period, several Spanish rivers had frequent floods (Macklin et al. 2006).

AD 270–950. At the beginning of this period, a generalized increase of water level appeared in TDNP, which was transformed into a shallow and lentic water with abundant Charophyceae, Poaceae, Cyperaceae, *Typha*, Ranunculaceae, *Polygonum* and *Potamogeton*. However, the climate suffered deterioration, changing to the colder and drier conditions corresponding to The Dark Ages. Arboreal and shrub taxa tended to be lost in the landscape, while *Artemisia* appeared to be dominant. Soil salinity also increased in the area, as revealed by the increase of Chenopodiaceae-Amaranthaceae. After this climatic anomaly, an improvement of temperature and wetness was observed, allowing a slight recovery of the arboreal and aquatic taxa (mainly *Potamogeton*). These changes spread to the entire Iberian Peninsula.

AD 950–1,090. A dramatic decline of arboreal and shrub taxa took place during this period, and may be related to either human disturbance connected to Muslim pasture exploitations, Muslim–Christian conflicts or to a climatic change. On the contrary, the wetland increased water retention, as revealed by the increase of aquatic taxa during this cycle. This improvement can be probably explained by Muslims introducing water mills in the floodplain 1,090–1,400 years AD. During this period the climate was warmer and wetter corresponding to the Warm Medieval Period. The landscape was mainly composed of grasslands. Wetland eutrophication episodes started to be frequent and TDNP appeared invaded by emergent macrophytes (Cyperaceae, *Typha* and Poaceae), while submerged taxa were scarce.

AD >1,400. The Little Ice Age was characterized by a cooling and drying of climatic trend with dramatic landscape results: forest and shrubs were severely degraded. In the TDNP surroundings, the climate was highly variable, alternating cold – warm periods with other dry – humid ones. These climate fluctuations impinged on the wetland ecological status, which continued suffering eutrophication episodes.

References

Allen JRM, Brandt U, Brauer A, Hubberten HW, Huntley B, Keller J, Kraml M, Mackensen A, Mingram J, Negendank JFW, Nowaczyk NR, Oberhänsli H, Watts WA, Wulf S, Zolitschka B (1999) Rapid environmental changes in southern Europe during the last glacial period. Nature 400:740–743

Allen JRM, Watts WA, Huntley B (2000) Weichselian palynostratigraphy, palaeovegetation and palaeoenvirnmental the record from Lago Grande di Monticchio, Southern Italy. Quat Int 73(74):91–101

Alonso Zarza AM, Dorado Valiño M, Valdeolmillos Rodríguez A, Ruiz Zapata MB (2004) Caracterización de los sistemas palustres de Las Tablas de Daimiel durante el Cuaternario: textura y composición de sus barros micríticos. Geo-Temas 6:13–16

Alonso-Zarza A, Dorado Valiño M, Valdeolmillos Rodriguez A, Ruiz Zapata MB (2006) A recent analogue for palustrin carbonate environments: the Quaternary deposits of Las Tablas de Daimiel wetlands, Ciudad Real Spain. Geol Soc Am Special Paper 416:153–168

Bailarón L (1997) El clima mediterráneo y sus características en el contexto de la circulación general atmosférica. In: Ibañez JJ, Valero BL, Machado C (eds) El paisaje mediterráneo a través del espacio y del tiempo. Implicaciones en la desertificación. Geoforma, Logroño

Barbier D, Burnouf J, Visset L (2001) Les diagrammes société/végétation: un outil de dialogue interdisciplinaire pour la compréhension des interactions homme/milieu. Quaternaire 12(1–2):103–108

Behre KE (1988) The role of man in Europe vegetation history. In: Huntley B, Webb T (eds) Vegetation history. Kluwer, Dordrecht

Berglund BE (ed) (1991) The cultural landscape during 6000 years in southern Sweden – the Ystad Project. Ecol Bull 41:1–495

Birks HJB, Birks HH (1980) Quaternary palaeoecology. Edward Arnold, London

Birks HH, Birks HJB, Kaland PE, Moe D (eds) (1988) The cultural landscape – past, present and future. Cambridge University Press, Cambridge

Bond G, Showers W, Cheseby M, Lotti R, Almasi P, de Menocal P, Priore P, Cullen H, Hajdas I, Bonani G (1997) A pervasive millennial-scale cycle in North Atlantic Holocene and glacial climates. Sci Am 278:1257–1266

Bradley P (1999) Paleoclimatology. Reconstructing climates of the Quaternary. Academic, San Diego

Broecker WS (2000) Abrupt climate change: causal constraints provided by the paleoclimate record. Earth Sci Rev 51:137–154

Burjachs F, Juliá R (1994) Abrupt climatic changes during the Last Glaciation based on pollen analysis of the Abric Romani, Catalonia, Spain. Quat Res 42:308–315

Cambón G (1981). Relations entre le contenu pollinique de l'atmosphère et le couvert végétal mediterrannee occidentale a Montpellier (France), Valencia (Espagne) et Oran (Algérie). Ph.D. Univ. Sci. Tech. Languedoc

Carrión JS (2001) Condicionantes de la respuesta vegetal al cambio climático. Una perspectiva paleobiológica. Ac Bot Malacitana 26:157–176

Carrión JS (2002) Patterns and processes of late Quaternary environmental change in mountane region of southwestern Europe. Quat Sci Rev 21:2047–2066

Carrión JS, Munuera M, Navarro C (1998) The palaeoenvironment of Carihuela Cave (Granada, Spain): a reconstruction on the basis of palynological investigations of cave sediments. Rev Palaeobot Palynol 99:317–340

Carrión J, Giner M, Navarro C, Sáez F (2000) Paleoclimas e historia de la vegetación cuaternaria en España a través del análisis polínico viejas falacias y nuevos paradigmas. Complutum 11:115–142

Costa Tenorio M, García Antón M, Morla Juaristi C, Sainz Ollero H (1990) La evolución de los bosques de la Península Ibérica: una interpretación basada en datos paleobiogeográficos. Ecología 1:31–58

Coûteaux M (1977) À propos de l'interprétation des analyses polliniques de sédiments minéraux, principalement archéologiques. In Laville H, Renault-Miskovsky J (eds) Approche écologique de l'homme fossile. Supplément du Bull Soc Fr Stu Quat 47:259–276

Dansgaard W, Johnsen SJ, Clausen HB, Gundestrup Dabl-Jensen D, N-S HCU, Hvidberg CS, Steffensen JP, Svembjömsdottir AE, Jouzel J, Bond G (1993) Evidence for general instability of past climate form a 250-kyr ice-core record. Nature 364:18–220

Desprat S, Sanchez-Goñi MF, Loutre MF (2003) Revealing climatic variability of the last three millennia in northwestern Iberia using pollen influx data. Earth Planet Sci Lett 213:63–78

Dorado Valiño M, Valdeolmillos Rodríguez A, Ruiz Zapata MB, Gil García MJ, Bustamante Gutiérrez I (1999) Evolución climática durante el Holoceno en la Cuenca Alta del Guadiana (Submeseta Sur Ibérica). Cuat Geomorf 13(1–2):19–32

Dorado Valiño M, Valdeolmillos A, Ruiz Zapata MB, Gil García MJ (2001) El paisaje vegetal de la Llanura Manchega durante el Tardiglaciar. In: Meléndez G, Herrera Z, Delvene G, Azanza B (eds) Los fósiles y la Paleogeografía. Univ. Zaragoza, Spain

Dorado Valiño M, Valdeomillos Rodríguez A, Ruiz Zapata MB, Gil García MJ, Bustamante I (2002) Climatic changes since the Lateglacial/Holoceno transition in the Mancha plain (South Central Iberian Peninsula, Spain) and their evidence in Las Tablas Daimiel marshlands. Quat Int 93–94:73–84

Dorado Valiño M, Valdeolmillos Rodríguez A, Alonso Zarza AM, Ruiz Zapata MB (2004) Registro paleoambiental de los estadios isotópicos 8 y 9 (Pleistoceno Medio) en el Parque Nacional de las tablas de Daimiel (Ciudad Real). Act. XX Jor. Soc. Sp. Paleont. 43–50, Alcalá de Henares

Duplessy JC, Cortijo E, Kallel N (2005) Marine records of Holocene climatic variations. C R Geosci 337:87–95

Dupré M (1988) Palinología y paleoambiente. Nuevos datos españoles, vol 84. Diputación Provincial de Valencia, Valencia
Faegri K, Kaland PE, Krzywinski K (1989) Textbook of pollen analysis, 4th edn. Wiley, Chichester
Frenzel B, Pécsi M, Velichko AA (eds) (1992) Atlas of paleoclimates and paleoenvironments of the Northern Hemisphere. Late Pleistocene–Holocene. Verlag, Budapest-Stuttgart
Galop D (1998) La forêt, l'homme et le troupeau dans les Pyrénées. 6000 ans d'histoire de l'environnement entre Garonne et Méditerranée. Geode, Toulouse
García Antón M, Morla Juaristi C, Ruiz Zapata MB, Saínz Ollero H (1986) Contribución al conocimiento del paisaje vegetal Holoceno en la Submeseta Sur Ibérica: Análisis polínico de sedimentos higroturbosos en el Campo de Calatrava (Ciudad Real, España). In: López Vera F (ed) Quaternary climate in Western Mediterranean. Universidad Autonoma, Madrid
García Antón M, Maldonado Ruiz J, Morla Juaristi C, Sainz Ollero H (2002) Fitogeografía histórica de la península ibérica. In: Pineda FD, De Miguel JM, Casado MA, Montalvo J (eds) La Diversidad Biológica de España. Prentice Hall, Madrid
García-Hidalgo JF, Temiño J, Segura M (2007) Holocene aeolian development in Central Spain; chronology, regional correlations and causal processes. Quat Sci Rev 26:2661–2673
Gil García MJ, Dorado Valiño M, Valdeolmillos Rodríguez A, Ruiz Zapata MB (2002) Lateglacial and Holocene palaeoclimatic record from Sierra de Cebollera (northern Iberian Range, Spain). Quat Int 93–94:13–18
Gil García MJ, Ruiz Zapata MB, Santiesteban JI, Mediavilla R, Lopez-Pamo E, Dabrio CJ (2007) Late Holocene environmentes in Las tablas de Daimiel (South Central Iberian Peninsula, Spain). Veg Hist Archaeobot 16:241–250
Girard M (1973) Pollens et Paléoethnologie. In: Sauter M (ed) L'homme, hier et aujourd'hui. Recueil d'études en hommage à André Leroi-Gourhan. Cujas, Paris
Girard M, Renault-Miskovsky J (1969) Nouvelles techniques de préparation en palynologie appliquées à trois sédiments du Quaternaire final de l'Abri Cornille (Istres, Bouches du Rhône). Bull Soc Fr Etude Quat 6(4):275–284
Gómez-Orellana L, Ramil-Rego P, Muñoz-Sobrino C (2007) The Würm in NW Iberia, a pollen record from Area Longa (Galicia). Quat Res 67:438–452
González-Sampériz P (2001) Análisis palinológico aplicado a la reconstrucción paleoclimática en medios mediterráneos y eurosiberianos. Ph.D. Univ. Zaragoza
González-Sampériz P (2004) Evolución paleoambiental del sector central de la Cuenca del Ebro durante el Pleistoceno Superior y Holoceno. Inst. Piren. Ecol. CSIC. Univ Zaragoza, Zaragoza
González-Sampériz P, Valero-Garcés BL, Moreno A, Morellón M, Navas A, Machín J, Delgado-Huertas A (2008) Vegetation changes and hydrologucal fluctuations in the Central Ebro Basin (NE Spain) since the Late Glacial period: Saline Lake records. Palaeogeogr Palaeoclimatol Palaeoecol 259:157–181
Grimm EC (1987) CONISS: a FORTRAN 77 program for stratigraphically constrained cluster analysis by the method of incremental sum of squares. Comput Geosci 13(1):13–35
Gutierrez-Elorza M, Peña Monné J (1998) Geomorphology and Late Holocene climatic change in Northeaster Spain. Geomorphology 23:205–217
Heinrich H (1988) Origin and consequences of cyclic rafting in the northeast Atlantic Ocean during the past 130, 000 years. Quat Res 29:142–152
Huntley BA, Allen JRM, Pollard D, Tzedakis PC, de Beaulieu JJ, Grürger E, Watts B (2003) European vegetation during Marine Oxygen Isotope Stage-3. Quat Res 59:195–212
Johnsen S, Clausen HB, Dansgaard W, Fuhrer K, Gundestrup N, Hammer CU, Iversen P, Jouzel J, Stauffer B, Steffensen JP (1992) Irregular interstadials recorded in a new Greenland ice core. Nature 359:311–313
Julià R, Negendank JFW, Seret G, Brauer A, Burjachs F, Endress C, Giralt S, Parés JM, Roca JR (1994) Holocene climatic changes and desertification in the Western Mediterranean region. Terra Nostra 1:81–84
Julià R, Burjachs F, Dassí MJ, Mezquita F, Miracle MR, Roca JR, Seret G, Vicente E (1998) Meromixis origin and recent trophic evolution in the Spanish mountain lake La Cruz. Aquat Sci 60:279–299

López Sáez JA, van Geel B, Martín Sánchez M (2000) Aplicación de los microfósiles no polínicos en Palinología Arqueológica. In: Oliveira Jorge V (ed) Contributos das Ciências e das Technologias para a Arqueologia da Península Ibérica. Act. 3rd Congresso de Arqueologia Peninsular. Adecap, Porto

López Sáez JA, López García P, Burjachs F (2003) Arqueopalinología: Síntesis crítica. Polen 12:5–35

Luzón A, Pérez A, Mayayo MJ, Soria AR, Sánchez Goñi MF, Roc AC (2007) Holocene environmental changes in the Gallocanta lacustrine basin, Iberian Range, NE Spain. Holocene 17(5):649–663

Macklin MG, Benito G, Gregory KJ, Johnstone E, Lewin J, Michczynska DJ, Soja R, Starkel L, Thorndycraft VR (2006) Past hydrological events reflected in the Holocene fluvial record of Europe. Catena 66:145–154

Martín Arroyo T, Ruiz Zapata MB, Pérez-González A, Valdeolmillos A, Dorado Valiño M, Benito G, Gil García MJ (1999) Paleoclima y paleoambiente durante el Pleistoceno Superior y el Tardiglaciar en la región central peninsular. In: Pallí L, Roqué C (eds) Avances en el estudio del Cuaternario español (Secuencias, Indicadores Paleoambientales y Evolución de Procesos). University of Gerona, Catalonia

Menéndez Amor J, Florchütz A (1968) Estudio palinológico de la turbera de Daimiel. La Prehistoire. Problemes et tendences. CNRS, Paris, pp 291–294

Molina F, Nájera T, Aranda G, Sánchez M, Haro M (2005) Recent fieldwork at the Bronze Age fortified site of Motilla de Azuer (Daimiel, Spain). http://antiquity.ac.uk/projgall/aranda/index.html. *Antiquity* 79, Proj. Gallery

Montserrat J (1992) Evolución glaciar y postglaciar del clima y la vegetación en la vertiente sur del Pirineo: Estudio palinológico. Inst. Pirenaico Ecol, Zaragoza

Moore PD, Webb JA, Collinson ME (1991) Pollen analysis, 2nd edn. Blackwell, London

Moreno A, Cacho I, Canals M, Prins MA, Sánchez-Goñi MF, Grimalt JO, Weltje GJ (2002) Saharan dust transport and high-latitude glacial climatic variability: the Alboran sea record. Quat Res 58:318–328

Moreno A, Cacho I, Canals M, Grimalt JO, Sánchez-Goñi MF, Shackleton N, Sierro FJ (2005) Links between marine and atmospheric processes oscillating on a millennial time-scale. A multi-proxy study of the last 50, 000 yr from the Alboran Sea (Western Mediterranean Sea). Quat Sci Rev 24:1623–1636

Peñalba MC, Maurice A, Guiot J, Duplessy JC, de Beaulieu JL (1997) Termination of the last glaciation in the Iberian Peninsula inferred from the pollen sequence of Quintanar de la Sierra. Quat Res 48:205–214

Pons A, Reille M (1988) The Holocene and Upper Pleistocene pollen record from Padul (Granada, Spain): a new study. Palaeogeogr Palaeoclimatol Palaeoecol 66:243–263

Ramos J (1988) Romanización de Castilla La Mancha, vol 4. I Congreso de Historia de Castilla La Mancha, Toledo, pp 53–69

Rendell MM, Pérez-González A, Calderón T, Benintez P (1996) Late Quaternary aeolian activity in the Manchega Plain, Central Spain. In: 8th International Conference on Luminiscence and Electron Spin Resonance Dating. Canberra, Australia 220–221

Reille M (1992) Pollen et Spores d'Europe et d' Afrique du Nord. Lab. Botan. Histor. Palynol, Marseille

Rendell HM, Calderon T, Pérez-Gonzalez A, Gallardo J, Millán A, Townsend PD (1994) Thermiluminiscence and optically stimulated luminiscence dating of Spanish dunes. Quat Sci Rev 13:429–432

Riera S, Wansard G, Julià R (2004) 2000-year environmental history of a karstic lake in the Mediterranean Pre-Pyrenees the Estanya lakes (Spain). Catena 55:293–324

Ruddiman WF, McIntyre A (1981) The north Atlantic Ocean during the last deglaciation. Palaeogeogr Palaeoclimatol Palaeoecol 35:145–214

Ruiz Zapata MB, Gil García MJ, Dorado M (1996) Climatic changes in the Spanish Central zone during the last 3000 BP based on pollinic analysis. In: Angelakis AN, Issar AS (eds) Diachronic climatic impacts on water resources with emphasis on Mediterranean Region. Springer-Verlag, Berlin

Ruiz Zapata MB, Pérez-González A, Dorado Valiño M, Valdeolmillos Rodríguez A, Bustamante Gutiérrez I, García G (2000) Caracterización climática de las etapas áridas del Pleistoceno Superior en la Región Central Peninsular. Geotemas 1(4):273–278

Ruiz Zapata MB, Gil García MJ, Dorado Valiño M, Valdeolmillos A, Pérez-González A (2002) Clima y vegetación durante el pleistoceno superior y Holoceno en la Sierra de Neila (Sistema Ibérico Noroccidental)". Rev C & G 16(1–4):9–20

Ruiz Zapata MB, Gil García MJ, Dorado Valiño M Valdeolmillos A (2005) El paisaje vegetal durante el Pleistoceno medio en el interior de la península. Zona Arqueológica 5:214–221. University of Rioja

Ruiz Zapata MB, Vegas J, García Cortes A, Gil García MJ, Torres T, Ortiz JE, Galan L, Pérez-Gonzáez A (2008) Comportamiento de la vegetación, durante el Último Máximo Glaciar, en la secuencia FU-1 (Laguna del maar de Fuentillejo, Campo de Calatrava, Ciudad Real). Polen 18:37–49

Sánchez-Goñi MF, Eynaud F, Turon JL, Shackleton NJ (1999) High-resolution palynolical record off the Iberian margin: direct land–sea correlation for the Last Interglacial complex. Earth Planet Sci Lett 171:123–137

Sánchez-Goñi MF, Turon JL, Eynaud F, Gendreau S (2000) European climatic response to millennial-scale changes in the atmosphere-ocean system during the Last Glacial Period. Quat Res 54:394–403

Sánchez-Goñi MF, Cacho I, Turon JL, Guiot J, Sierro FJ, Peypouquet JP, Grimalt JO, Shackleton NJ (2002) Synchroneity between marine and terrestrial responses to millennial scale climatic variability during the last glacial period in the Mediterranean region. Clim Dyn 19:95–105

Schulte L, Julià R, Burjachs F, Hilgers A (2008) Middle Pleistocene to Holocene geochronology of the River Aguas terrace sequence (Iberian Peninsula): fluvial response to Mediterranean environmental change. Geomorphology 98:13–33

Tzedakis PC, Andrieu V, De Beaulieu JL, Crowhurts S, Follieri NJ, Hooghiemstra H, Magri D, Reille M, Sadori L, Shackleton NJ, Wijmstra TA (1997) Comparison of terrestrial and marine records of changing climate of the last 500, 000 years. Earth Planet Sci Lett 150:171–176

Tzedakis PC, Mcmanus PC, Hooghiemstra H, Oppo DW, Wijmstra TA (2003) Comparison of changes in vegetation in northeast Greece with records of climate variability on orbital and suborbital frequencies over the last 450, 000 years. Earth Planet Sci Lett 212:197–212

Tzedakis PC, Frogley MR, Preece LIT, RC CI, De Abreu L (2004) Ecological thresholds and patterns of millennial-scale climate variability: the response of vegetation in Greece during the last glacial period. Geology 32:109–112

Tzedakis PC, Hooghiemstra H, Pälike H (2006) The last 1.35 million years at Tenaghi Philippon: revised chronostratigraphy and long-term vegetation trends. Quat Sci Rev 25:3416–3430

Valdeolmillos A (2004) Registro paleoclimático y paleoambiental de los últimos 350.000 años en el Parque Nacional de las Tablas de Daimiel (Ciudad Real). Ph.D. Univ. Alcalá. Alcalá de Henares

Valdeolmillos A, Dorado Valiño M, Ruiz Zapata MB, Gil García MJ, Bustamante I, Pérez González A (2002) Evolución de la vegetación durante las etapas áridas del Pleistoceno Superior-Holoceno, en la llanura Manchega. In: Moreno S, Elvira B, and Moreno JM (eds) Libro de textos completos del XIII Simposio APLE, Universidad Politécnica de Cartagena, Cartagena

Valdeolmillos A, Dorado Valiño M, Ruiz Zapata MB, Bardají T, Bustamante I (2003) Paleoclimatic record of the Last Glacial Cycle at Las Tablas de Daimiel National Park (Southern Iberian Meseta, Spain). In: Ruiz Zapata MB, Dorado Valiño M, Valdeolmillos A, Gil García MJ, Bardají T, de Bustamante I, Martínez Mendizábal I (eds) Quaternary climatic changes and environmental crises in the Mediterranean Region. Ser. Pub. Univ. Alcalá, Alcalá de Henares

Valdés B, Diez MJ, Fernández I (eds) (1987) Atlas polínico de Andalucía Occidental. Instituto de Desarrollo Regional, nº 43. Univ. Sevilla, Spain

Valero-Garcés BL, González Sampériz P, Delgado Huertas A, Navas A, Machín J, Kelts K (2000) Lateglacial and Late Holocene environmental and vegetational change in Salada Mediana, central Ebro Basin, Spain. Quat Int 73–74:29–46

Valero-Garcés BL, Moreno A, Navas P, Mata J, Machín A, Delgado Huertas P, Gonzalez Sampériz A, Schwalb M, Morellón H, Cheng RL (2008) The Taravilla lake and tufa deposits (Central Iberian Range, Spain) as palaeohydrological and palaeoclimatic indicators. Palaeogeogr Palaeoclimatol Palaeoecol 259:136–156

Vegas J, García-Cortés A, Galán L, Pérez-González A, Martín-Serrano A (2006) El registro sedimentario lacustre del maar de Fuentillejo (Ciudad Real). Bol Geolog Min 117:339–349

Vegas J, Ruiz ZapataMB, Ortiz JE, Galan L, Torres T, García-Cortés A, Gil García MJ, Perez González A, Gallardo-Millán JL (2009) Identification of arid phases during the last 50 kyr Cal BP from the Fuentillejo maar lacustrine record (Campo de Calatrava Volcanic Field, Spain). J Quat Sci. doi:10.1002/jqs/1262

Chapter 3
Climate and Hydrologic Trends: Climate Change Versus Hydrologic Overexploitation as Determinants of the Fluctuating Wetland Hydrology

S. Sánchez-Carrillo and M. Álvarez-Cobelas

Abstract Hydrology is the key to wetland persistence and conservation. Las Tablas de Daimiel National Park is among the many freshwater wetlands around the world that experienced strong hydrological alterations since the 1970s. Wetland hydrological functioning changed from a groundwater-dependent system to one that is maintained by artificial external water diversions. Throughout this chapter, using available data from 1904 to 2008, we analyze climate trends and hydrology of the wetland in order to quantify the changes occurred in the water budget components as well as to assess the main causes of its hydrological disturbance. Finally, through a hydrological model, developed to simulate the current wetland hydrologic functioning, the hydroperiod restoration is evaluated using multiple scenarios.

Both rainfall and water flow variability are the main hydrological patterns in the wetland area, whilst evapotranspiration appeared as the main hydrologic variable controlling the TDNP annual water storage. The historical analysis of the annual inundation patterns at TNDN has discriminating between four distinctive hydroperiods, showing the importance of groundwater discharges to sustain the pristine hydroperiod. While drought frequency – as measured by the Standard Precipitation Index (SPI) – did not increase in the region, groundwater pumping for irrigation appeared as the main cause of groundwater overexploitation and wetland hydrological degradation. Through a hydrological model developed to simulate wetland hydrology, it was possible to test the restoration of the wetland hydroperiod, using scenarios which promote cut-sedge recovery and open water areas. Hydrological restoration could be achieved by means of an external annual water supply of close to 54 Mm^3. Future stream and groundwater flows at the Upper Guadiana basin will depend on the land use management, while climate change effects will be secondary, considering the actual hydrological situation.

S. Sánchez-Carrillo (✉)
Spanish National Research Council (CSIC), Institute of Natural Resources,
Serrano 115 dpdo, E-28006, Madrid, Spain
e-mail: sanchez.carrillo@ccma.csic.es

S. Sánchez-Carrillo and D.G. Angeler (eds.), *Ecology of Threatened Semi-Arid Wetlands: Long-Term Research in Las Tablas de Daimiel*, DOI 10.1007/978-90-481-9181-9_3,

3.1 Introduction

Wetlands are systems that depend basically on hydrology. Hydrology determines its occurrence through the landscape and its persistence over seasons and years. All wetlands are the result of physiographic settings and water balances that favor the accumulation or retention of soil water and/or surface water for a period of time (Winter 1988). The hydrology of a wetland creates the unique physicochemical conditions that make such an ecosystem different from both well-drained terrestrial systems and deepwater aquatic systems (Mitsch and Gosselink 2000). Water sources differ between wetland types, as they characterize ecological properties and biogeochemical processes. Floodplain wetlands maximize a mix of water proceeding from river as well as groundwater sources. Floodplain construction produces a mosaic of wetland types characterized by variations in water flux and sedimentology (Bradley 1997). Clearly, the occurrence of floodplain wetlands depends on floodplain evolution since the last Glaciation (Brown et al. 1994). After the Holocene, the European landscape was covered by wide areas of floodplain wetlands (Brown et al. 1994) but nowadays they are on decline, particularly in the Mediterranean semi-arid region.

The existing imbalance of water resources makes water shortage a great problem in many regions of Europe under semi-arid climate (Iglesias et al. 2005). Water management is often controversial because of the increasing imbalance between its supply and demands. Excessive use of the aquifers causes overexploitation problems with the consequent negative environmental, social and economic impact (Esteller and Díaz-Delgado 2002). One of immediate effects is wetland degradation or disappearance (Llamas 1988). Because the hydrologic system is a continuum, any modification of one component will have an effect on contiguous components. Disturbances commonly affecting the hydrologic system as it relates to wetlands, include weather modification, alteration of plant communities, storage of surface water, road construction, drainage of surface water and soil water, alteration of groundwater recharge and discharge areas, and pumping of groundwater (Winter 1988). Besides, in semi-arid regions, rational hydrological wetland management may be impossible due to the seemingly chaotic behavior of the climate (Molles et al. 1992). Effective water management of wetlands requires a thorough understanding of the way in which the water budget components interact (Gilman 1994; Acreman et al. 2000).

Las Tablas de Daimiel National Park (TDNP) is probably the best example of wetland degradation through the cumulative effects of several disturbances on the hydrologic system as a result of unsustainable economic development (Álvarez-Cobelas et al 2001). From the 1970s, conflicts between wetland conservation and groundwater exploitation dominate the controversial water management of the basin (Llamas 1988). Almost 20 years later, the aquifer was declared to be overexploited, after which several management plans were implemented, such as groundwater pumping restrictions, water diversions from the neighboring basin and dam constructions inside the wetland; in spite of all these measures, TDNP is suffering nowadays from severe hydrologic damage. Climate change or water overexploitation is often cited as the main cause of water scarcity, influenced by factors from the social sector. In this chapter hydrologic functioning of TDNP as well as historical changes are studied. We determine the main causes of wetland hydrologic degradation and assess the hydrological keys required for wetland hydroperiod restoration.

3.2 Material and Methods

TDNP meteorological data were obtained from an automatic station located at the wetland (details of the recorded variables can be founded in Sánchez-Carrillo et al. 2004). Meteorological data from Ciudad Real and Daimiel were provided by the AEMET (The National Agency of Meteorology). Data on river flows were provided by the Guadiana River Basin Authority through measurements on calibrated gauging stations. Groundwater levels were taken from annual reports of the IGME (Geological Survey of Spain) as well as by measurements performed by the Guadiana River Basin Authority. Inundation and water-level were supplied by the TDNP staff.

Data on open water evaporation for the period 1993–1998 were taken from Sánchez-Carrillo et al. (2004), measured by means of the energy-budget method. TDNP staff also provided data on evaporation through a class A evaporation tank located in the wetland area. Transpiration rates were assessed by Sánchez-Carrillo et al. (2001), who measured leaf conductance and transpiration rates of the main emergent macrophytes (reed, cut-sedge and cattail) in TDNP, using a steady-state diffusion porometer and taking the leaf area index (LAI) throughout the wetland into account. Also, daily transpiration estimates were computed using the model by Sánchez-Carrillo et al. (2001) based on short-wave radiation, relative humidity and air temperature.

Distinctive wetland hydroperiods were discerned through a cluster analysis, using monthly inundation data. Standardized Precipitation Index (SPI) was calculated following McKee et al. (1993) using annual precipitation records at TDNP, Ciudad Real and Daimiel for the period 1904–2008.

Finally, the hydrological model of TDNP was performed using the monthly hydrological database for the period 1993–2004. Data for the period 1993–1998 were used for parameter estimations while those for 1999–2004 were employed for model calibration. Equations used for parameter estimations were obtained through mathematical functions adjusted by least squares using the software CurveExpert 1.3 (Hyams 2001). All statistical analyses were carried out using STATISTICA v7.0 (Stat Soft Inc).

3.3 Long-Term Climate Trends

An extensive study of TDNP climate can be found in Álvarez-Cobelas and Verdugo (1996) for the period 1904–1992. Climatology of the La Mancha Húmeda region is analyzed exhaustively by Pérez-González and Sanz-Donaire (1998) although for the short period 1961–1990. Weather data analyzed here correspond to the period 1904–2008; therefore, our results are somewhat different than those mentioned in those studies. The yearly average air temperature at TDNP is 14.2°C, with strong differences between summer and winter (up to 35°C). Maximum temperatures appear during July and August whilst minima are registered in December–January (Fig. 3.1). Monthly temperature did not show remarkable interannual oscillations (Fig. 3.1). Annual air temperature increased from 1904 (+0.013°C year^{-1}) although it was not statistically significant ($p > 0.05$). Average annual rainfall in TDNP is 402 mm, seasonally distributed between winter and spring (Fig. 3.2). Interannual differences in monthly rainfall are quite accentuated and statistically significant

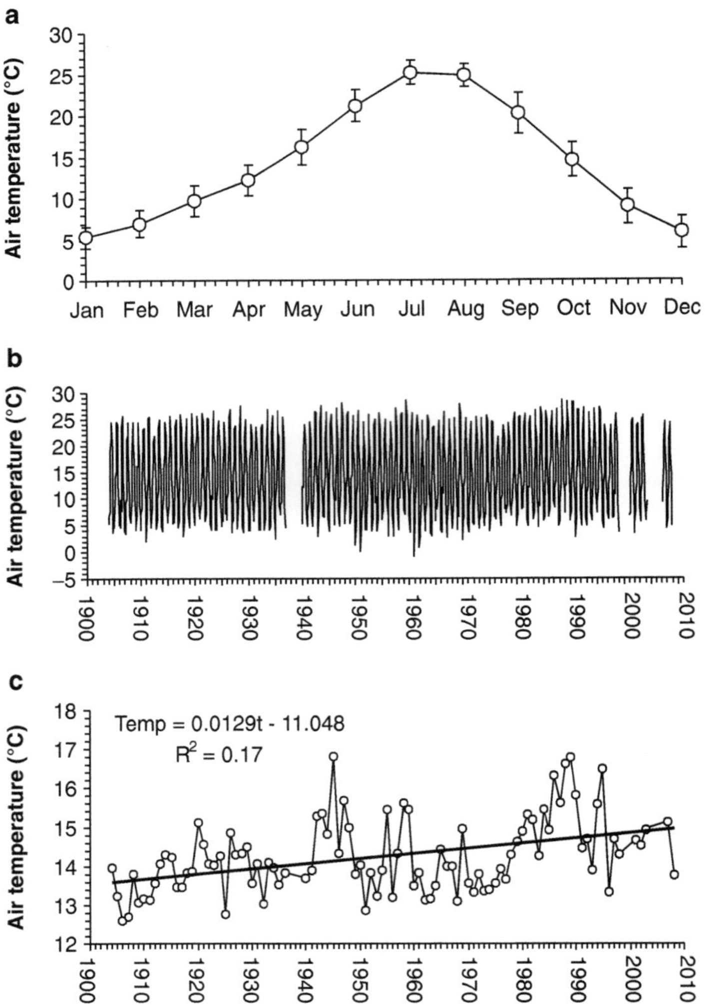

Fig. 3.1 Air temperature trends in Las Tablas de Daimiel for the period 1904–2008. (**a**) Monthly averages, (**b**) daily averages, and (**c**) yearly averages; *t* = time in years

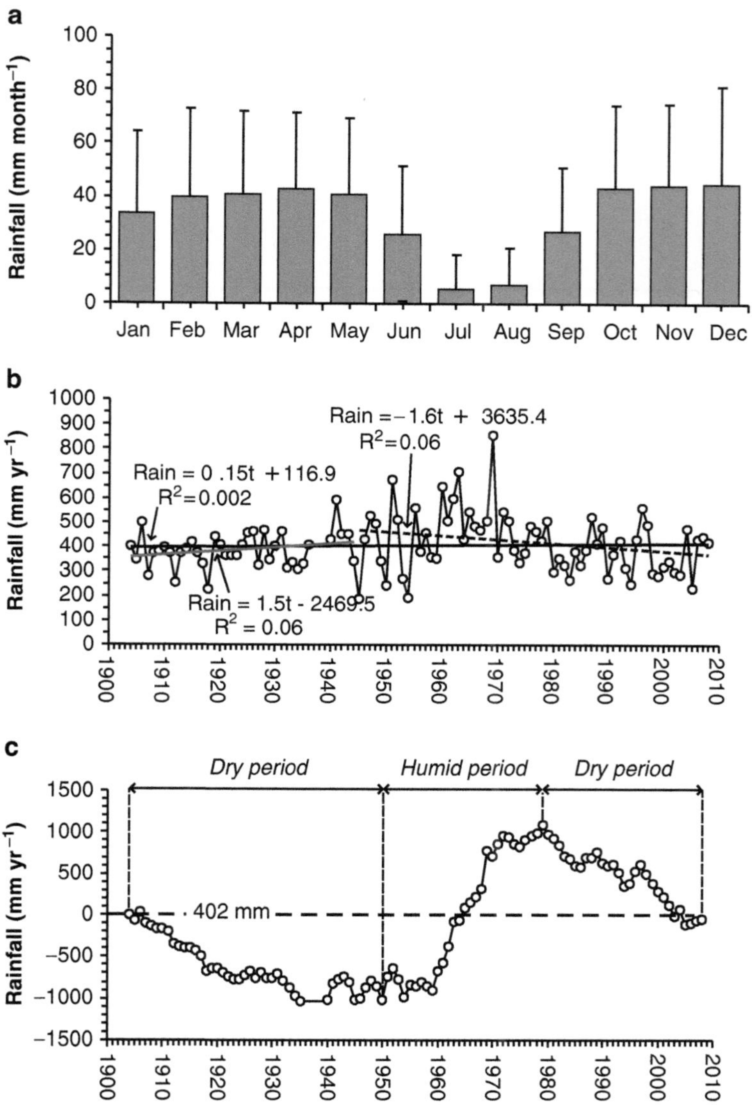

Fig. 3.2 Rainfall pattern in Las Tablas de Daimiel from 1904 to 2008. (**a**) Monthly precipitation trend, (**b**) annual rainfall and trends in the period 1904–1944 (*grey line*), 1945–2008 (*dotted line*) and 1904–2008 (*black line*), and (**c**) accumulated deviation of annual rainfall showing major distinctive humid and dry periods; t = time in years

(Rank test $p < 0.05$; Fig. 3.2). Similarly to air temperature, annual rainfall showed an increase from 1904 to nowadays; however, from 1945 onwards, a slight negative trend appeared in rainfall (Fig. 3.2). Increases or decreases, however, were not significant ($p > 0.05$; Fig. 3.2). The accumulated deviations of rainfall from the average proved the existence of three hydrological periods between 1904 and 2008

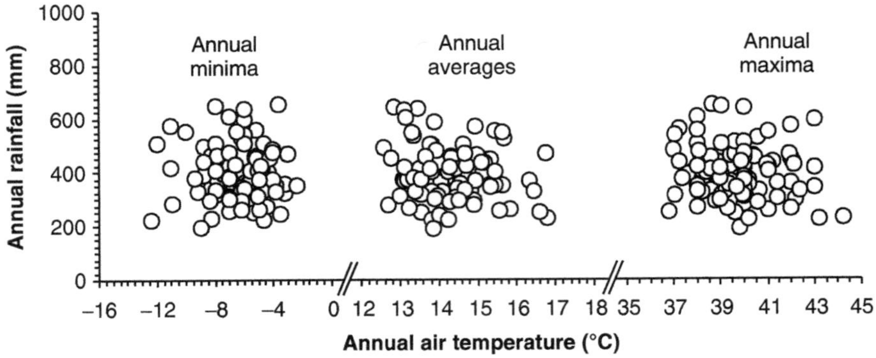

Fig. 3.3 Relationships between rainfall and air temperature in Las Tablas de Daimiel during 1904–2008

(Fig. 3.2): a dry period until 1950, a humid one from 1951 to 1979, and a dry one from 1980 until today. No significant relationships were found between annual rainfall and either mean annual, maxima or minima annual air temperatures ($p > 0.05$; Fig. 3.3).

The Thornthwaite potential evapotranspiration value of 850 mm year^{-1} reveals an annual water deficit of close to 450 mm. Climate of TDNP can be classified then as semi-arid, receiving low (<500 mm year^{-1}) and variable annual rainfall and characterized by potential evapotranspiration that is far greater than precipitation (ratio of mean annual precipitation to mean potential evapotranspiration <0.5; United Nations Environment Programme 1992).

3.4 An Historical Analysis of Wetland Hydroperiod and Inundation Patterns: The Importance of Groundwater

The hydroperiod is the seasonal pattern of either water level or inundation area of a wetland and is like a hydrological signature of each wetland type (Mitsch and Gosselink 2000). The hydroperiod integrates all inputs and outputs of water as influenced by watershed geomorphology, climate and ecohydrology, but also by local conditions of the wetland. Wetlands vary in their hydroperiods from very short (holding water for less than a few weeks in the year) to very long or permanent (lakes and ponds). Between these extremes, there are wetlands that hold water for various lengths of time during the year, including some wetlands that dry out only in years when there is very little precipitation. Inland wetlands show water levels varying considerably from year to year, depending on the variability of climate and groundwater interactions (Rosenberry and Winter 1997). Wetlands in semi-arid climate exhibit a highly variable balance between precipitation and evaporation, indicating that their ecosystem stability is more sensitive to climate change than other wetland types (Bergkamp and Orlando 1999). Some semi-arid wetlands greatly respond to seasonal as well as long-term changes in the balance between precipitation and evaporation. Many of these wetlands dry up in summer and/or autumn. Other wetlands are fairly stable in response to changing climatic conditions and dry-up only after prolonged, severe drought. The hydroperiod stability of some semi-arid wetlands is closely linked to the stable flux of groundwater contributing to water inputs (Winter and Llamas 1993; Rosenberry and Winter 1997). Furthermore, in floodplain wetlands, additional fluxes may include subsurface seepage to and from a river, infiltration of floodwater and spatially variable groundwater flows (Bradley 2002), increasing the importance of these latter flows for wetland hydroperiod as aridity rises.

Inundation at TDNP showed recurring seasonal patterns between 1945 and 1973, with roughly the same range from year to year; after this period, the inundation area became more fluctuating, increasing the frequency of low water-level situations (Fig. 3.4). The local aquifer ceased its water discharges to TDNP after 1986 when inundation changed dramatically to a more uneven pattern (Fig. 3.4). From then on, draw down periods increased and flooding depended almost exclusively on water diversions from other catchments (Fig. 3.4). In the long term, the average annual inundation at TDNP showed a decreasing trend (Fig. 3.5a); these trends of decreased inundation appeared more clearly in the maxima annual inundation (Fg. 2.5b) than in the minima (Fig. 3.5c). Seasonal patterns of the hydroperiod were dramatically changing over time: firstly, with a peak in mid-spring and a drought in late summer and early autumn (1945–1973); in the period 1974–1980 the summer dry up was longer for smaller flooded areas and, from then on, seasonal inundation changed dramatically (Fig. 3.6a). Frequent drought periods (1981–1988, 1991–1995 and 2006–2008) kept the wetland almost dry during the whole year and only water diversions or unusually humid years briefly restored the hydroperiod (Fig. 3.6a). From 1974, water deficit increased at TDNP, not only during summer but also throughout the rainy seasons – winter and spring – (Fig. 3.6b). It appeared as a consequence of 04.04 aquifer overexploitation (see Chapter 1) that disrupted

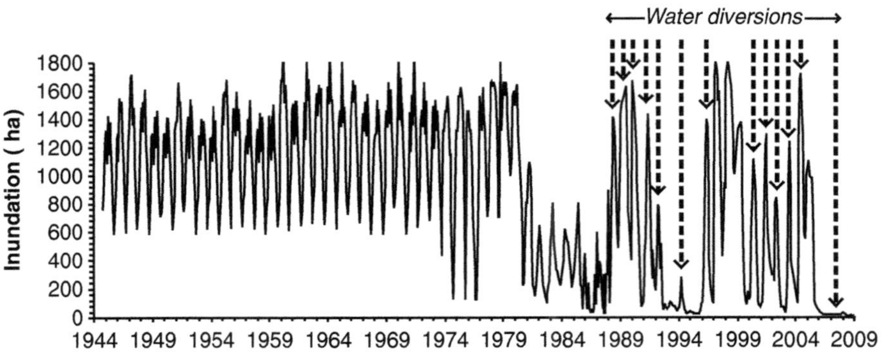

Fig. 3.4 Monthly inundation pattern in Las Tablas de Daimiel from 1944 to 2008, showing water diverted from a Northern catchment to the wetland, to sustain inundation

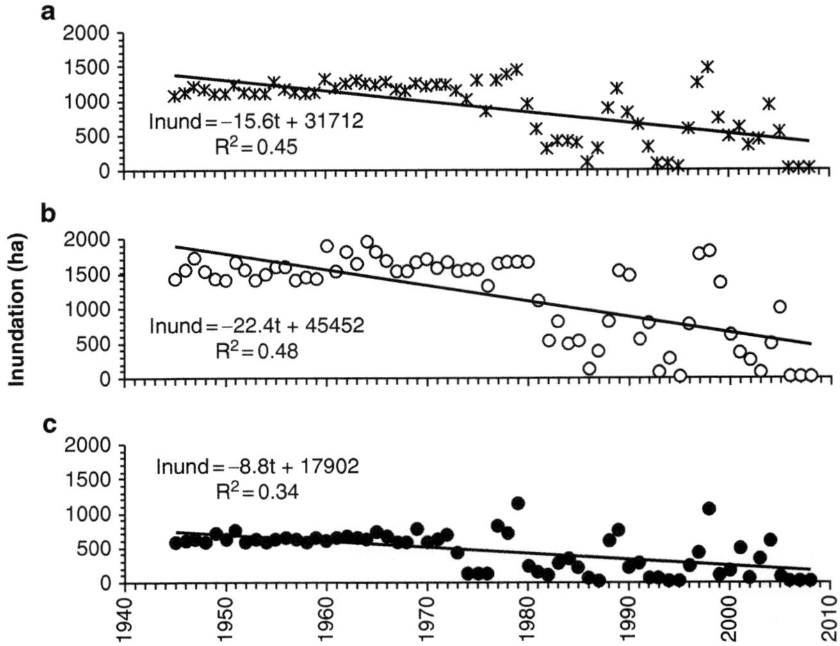

Fig. 3.5 Annual inundation trends in Las Tablas for the period 1944–2008 (**a**) mean annual inundation, (**b**) maximum annual inundation, and (**c**) minimum annual inundation; t = time in years

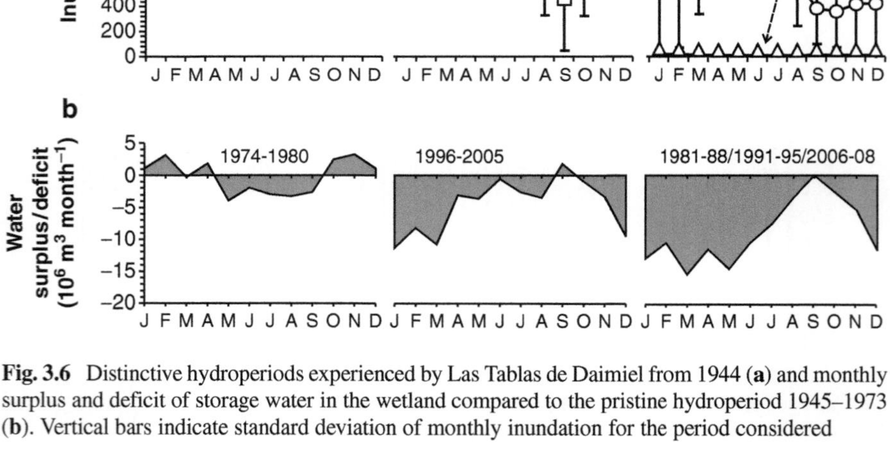

Fig. 3.6 Distinctive hydroperiods experienced by Las Tablas de Daimiel from 1944 (**a**) and monthly surplus and deficit of storage water in the wetland compared to the pristine hydroperiod 1945–1973 (**b**). Vertical bars indicate standard deviation of monthly inundation for the period considered

both groundwater discharges into the wetland and the subsurface seepage to the rivers (Álvarez-Cobelas et al. 2001). All recognized hydroperiods were statistically distinctive (Friedman ANOVA test; $p = 0.004$).

The pristine seasonal flooding pattern at TDNP was a mixture of the seasonality of the Gigüela and Guadiana river inflows, because in winter and spring the Gigüela inflows were more important, whereas in summer and early autumn the groundwater discharge via Guadiana course dominated flooding (Álvarez-Cobelas et al. 2001). Figure 3.7a shows the importance of the groundwater spring from Ojos del Guadiana in the TDNP hydroperiod until 1973. Water inflows from the Gigüela were irregular, both seasonally and annually, disappearing during the dry summer months. Groundwater inputs from Ojos del Guadiana exerted a strong control on the seasonal hydroperiod at TDNP throughout the pristine period ($R^2 = 0.94$ $p < 0.05$; Fig. 3.7b) while Gigüela river discharges only explained 68% of seasonal inundation; Fig. 3.7c). After 1973 the relationship between groundwater and hydroperiod was significant, until 1980, where groundwater discharges were discontinued (Fig. 3.7b). Unfortunately, groundwater level records before 1974 were not available for the 04.04 Aquifer. Nevertheless, data on groundwater table depths registered close to Ojos del Guadiana (Zuacorta site) clearly show how overexploitation decreased aquifer water levels until 1995 (Fig. 3.8). In connection to the TDNP area, groundwater level trends did not evidence any noticeable changes from 1974, except for one strong water levels decline which took place during the severe drought between 1992 and 1996 (Fig. 3.8). Furthermore, during pristine wetland conditions, most aquifer discharges might have taken place through specific wetland zones instead of throughout the entire wetland, as observed in other wetlands and aquifers (Rosenberry and Winter 1997; Swanson et al. 2006). For example, changes in wetland substrate composition affect subsurface flow through the wetland since hydraulic properties of peat vary with vegetation composition, and this complicates any attempt to quantify horizontal and vertical water fluxes (Bradley 1997). Finally, although a slight aquifer recovery could be found after the aforementioned drought period, the groundwater level rise was not enough to reinstate groundwater discharges into the wetland; it was more evident, however, that there was an increase of seasonal fluctuations of groundwater level from then on (Fig. 3.8).

3.5 Long-Term Changes in the Wetland Water Budget: Hydrologic Degradation

Variations in the water flux provide the background for evaluating and classifying individual floodplain wetlands (Bradley 1997). Major components of the wetland water budget are precipitation, surface-water flow, ground-water flow, and evapotranspiration. Wetlands and uplands continually receive or lose water through exchange with the

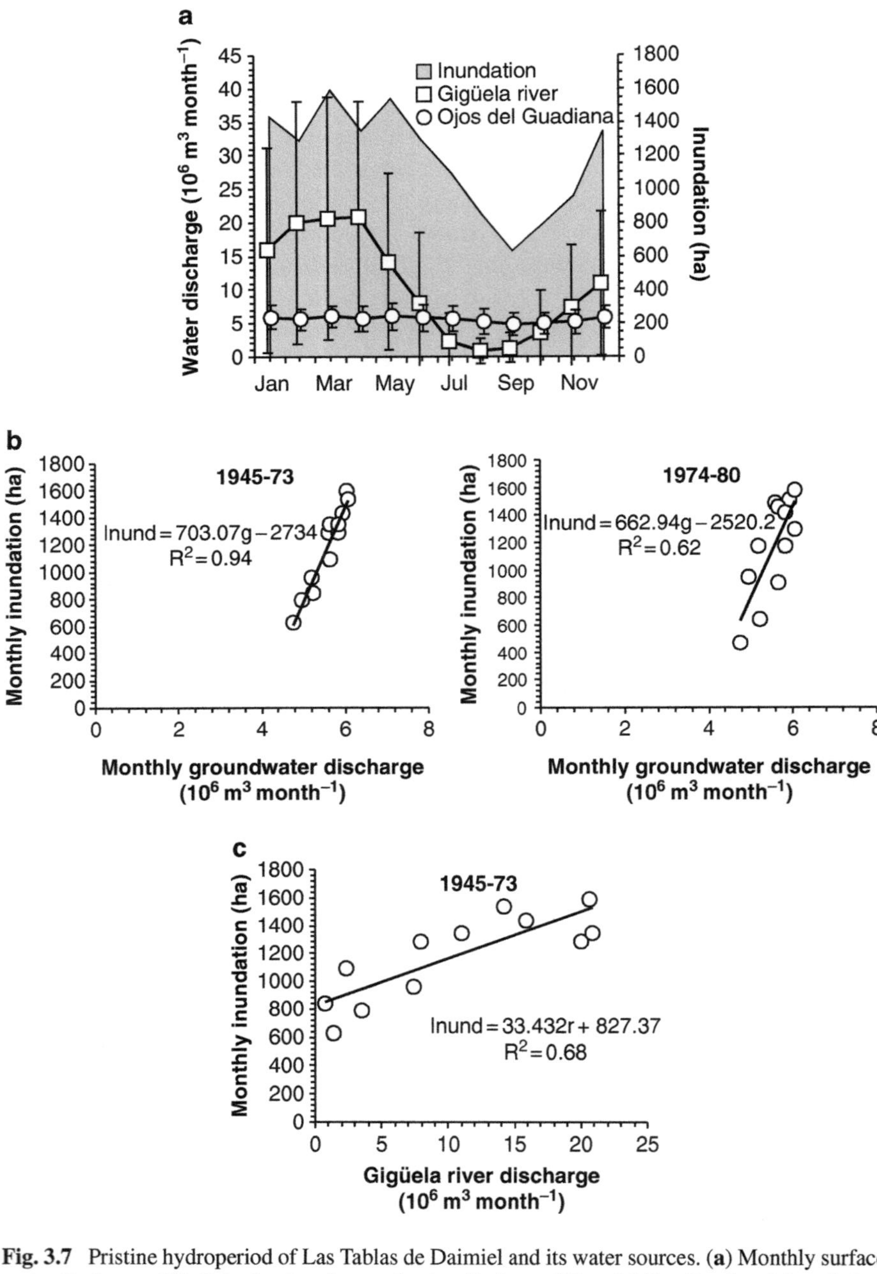

Fig. 3.7 Pristine hydroperiod of Las Tablas de Daimiel and its water sources. (**a**) Monthly surface water discharges (Gigüela river) and groundwater discharge (Ojos del Guadiana) and inundation, (**b**) relationships between groundwater discharge and inundation for the periods 1945–1973 and 1974–1980, and (**c**) Relationships between Gigüela river discharge and inundation during 1945–1973; g is monthly groundwater discharge and *r* is Gigüela river discharge both in 10^6 m^3 month^{-1}

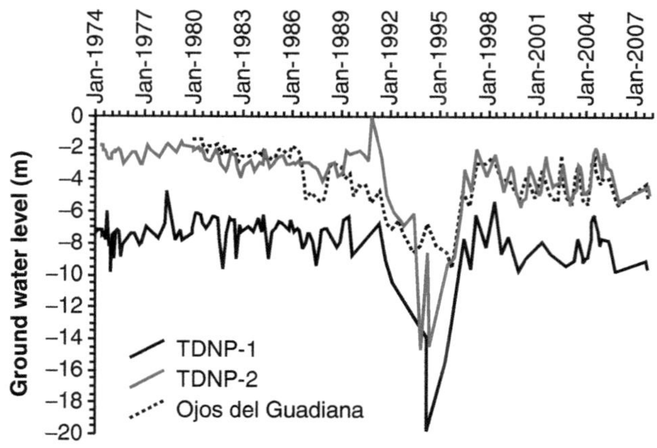

Fig. 3.8 Depth of groundwater table in different locations of 04.04 Aquifer during 1974–2007: two sites are located within Las Tablas de Daimiel (TDNP-1 and TDNP-2) and the other is close to the Ojos del Guadiana spring site

atmosphere, streams, and groundwater. Both a favorable geologic setting and an adequate and persistent supply of water are considered necessary for the existence of wetlands (Winter 1988). However, the relative importance of each component of the water budget differs from wetland to wetland (Mitsch and Gosselink 2000). Although the water balance of floodplain wetlands is dominated by river flooding (Winter and Woo 1990), an adequate reflection of the runoff generation and water balance processes is necessary for a successful evaluation of the water balance (Krause and Bronstert 2005). Furthermore, short-term wetland water balance is recognized to be strongly affected by groundwater processes as well as by the interactions between groundwater and surface water (Sophocleous 2002) at temporally variable spatial extents (Krause and Bronstert 2004). Results of these short-term processes characterize the whole wetland hydroperiod as response to water flux seasonality.

Long-term hydrological data are not available usually in wetlands worldwide, due to their acknowledgment as valuable ecosystems in only recent years. However, long-term overall patterns of wetland water fluxes provide the only approach to identify the nature (natural or anthropogenic) of wetland hydrological fluctuations as related to watershed hydrological alterations.

Long-term data on water flows in TDNP reveal a pattern with strong variability between years, including those that represent hydrological pristine conditions (Fig. 3.9). Annual rainfall did not show as irregular patterns as those of flows. Since the 1970s, all surface and groundwater flows experienced an important reduction linked to an increase of irregularity. Mean annual water storage at the wetland changed dramatically from then, with prolonged drought periods appearing (Fig. 3.9). The Gigüela River contributed 81–89% of TDNP inflows, averaging 135×10^6 m^3 year^{-1} in the most pristine scenario. From the period 1974–1980,

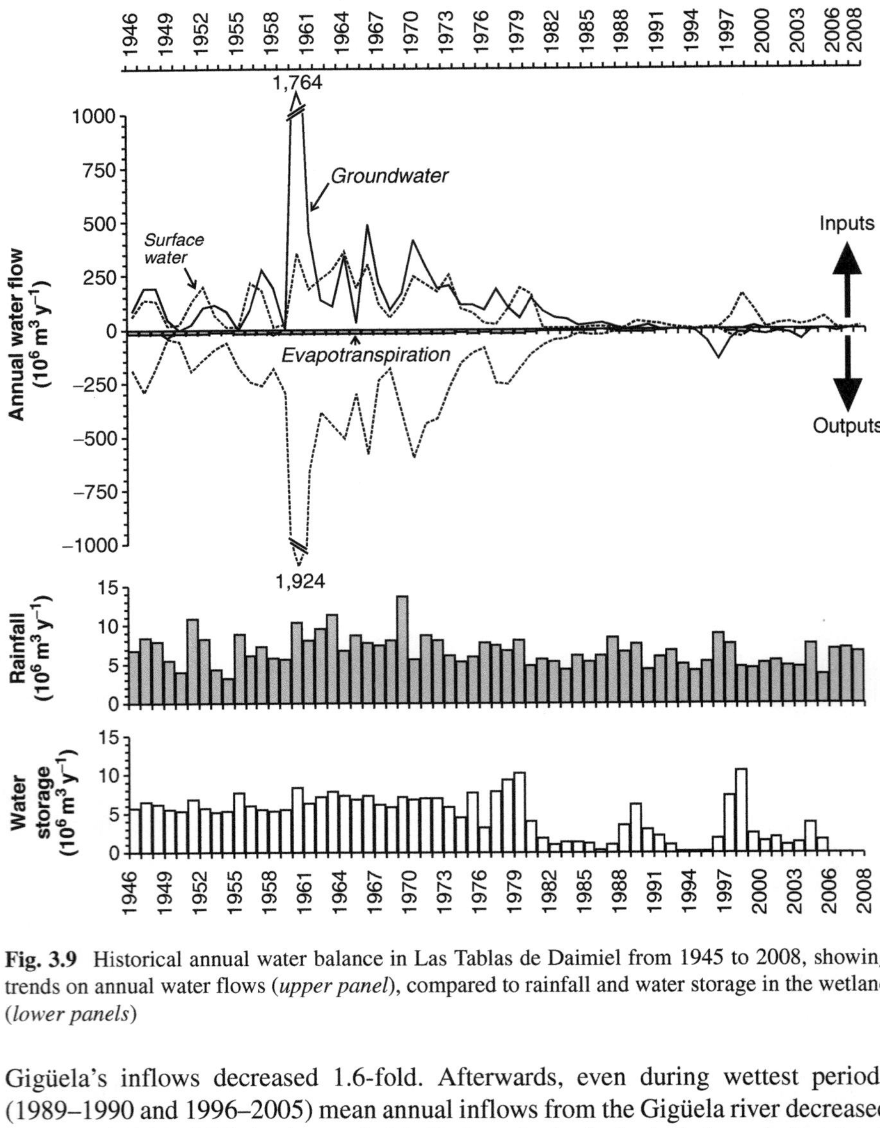

Fig. 3.9 Historical annual water balance in Las Tablas de Daimiel from 1945 to 2008, showing trends on annual water flows (*upper panel*), compared to rainfall and water storage in the wetland (*lower panels*)

Gigüela's inflows decreased 1.6-fold. Afterwards, even during wettest periods (1989–1990 and 1996–2005) mean annual inflows from the Gigüela river decreased by around 75% (97% in dry periods). Before the hydrological degradation (until 1973) water crossing TDNP annually ranged 44–1924 × 10^6 m^3 $year^{-1}$ (Table 3.1). Most water supplied by subterranean sources amounted to 60% of annual water inputs (until 92%; Table 3.1, Fig. 3.10). Mean groundwater discharge at TDNP achieved 225 × 10^6 m^3 $year^{-1}$ in that period (Table 3.1). After 1973, groundwater discharge was reduced to 50% for the period 1974–1980, disappearing completely from 1986 (Table 3.1 and Fig. 3.9). At this time, surface inflow became the main water source to TDNP (95% of water inputs) while infiltration appeared as the main water output, being between 60% and 89% of the total wetland water loss when stagnant water was present (Table 3.1, Figs. 3.9 and 3.10). Average wetland water

Table 3.1 Summary of the water budget in TDNP during different hydroperiods: *A* – 1945–1973; *B* – 1974–1980, *C* – 1981–1988/1991–1995/2006–2008, *D* – 1989–1990/1996–2005. Hydroperiods were depicted through changes in wetland inundation patterns. Data in parenthesis indicate the range. All data are in 10^6 m^3 $year^{-1}$

		Inflows			Outflows		
	Storage	Rainfall	Surface	Groundwater	EVT	Surface	Groundwater
A:	6.4	7.6	151.8	224.9	32.8	350.1	2.7
	(5.2–8.3)	(3.1–13.7)	(7.0–361.0)	(0.0–1763.9)	(30.0–36.4)	(43.7–1924.5)	(0.0–37.8)
		Total inputs: 384.2 (10.1–2138.6)			Total outputs: 385.7 (73.7–1998.8)		
B:	6.7	6.6	94.0	112.7	32.3	167.8	0.0
	(3.1–10.2)	(4.8–8.1)	(18.6–188.5)	(54.1–185.7)	(22.5–40.0)	(86.5–259.8)	(0.0–0.0)
		Total inputs: 213.3 (77.5–382.4)			Total outputs: 200.1 (109.0–299.8)		
C:	1.0	5.9	5.4	15.8	6.9	17.2	6.9
	(0.0–3.4)	(4.2–8.3)	(0.0–19.0)	(0.0–85.0)	(0.7–21.7)	(0.0–73.6)	(0.0–58.7)
		Total inputs: 27.1 (4.2–112.4)			Total outputs: 31.0 (0.7–154.1)		
D:	3.6	5.7	42.3	2.0	13.5	8.4	27.3
	(1.1–10.6)	(3.7–8.9)	(2.7–160.2)	(0.0–13.5)	(4.0–32.6)	(0.0–42.1)	(0.0–145.3)
		Total inputs: 50.0 (6.4–182.6)			Total outputs: 49.2 (4.0–219.9)		

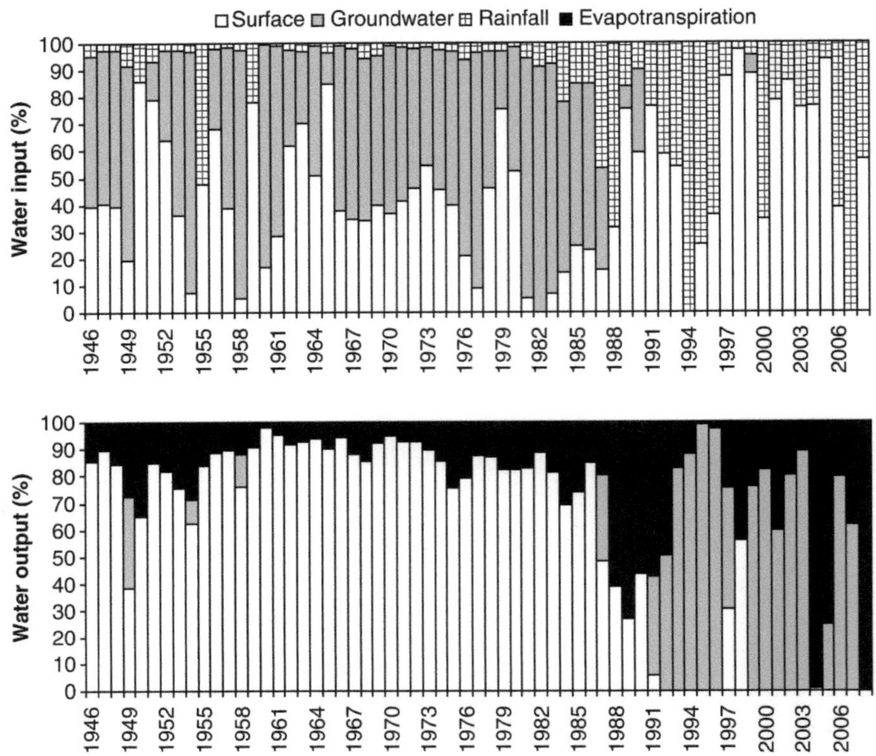

Fig. 3.10 Weight (percentage) of each hydrological component in Las Tablas de Daimiel for long-term water budget 1945–2005

resources accounted for 213 × 10^6 m^3 $year^{-1}$ during 1974–1980, later decreasing dramatically by 77–87% (Table 3.1, Fig. 3.10). Since the ratio of macrophyte cover to open water areas did not to change significantly until 1980, as well as that of dominant plant species (cut-sedge; see Chapter 7), evapotranspiration losses did not change, representing 32–33 × 10^6 m^3 $year^{-1}$ and oscillating 5–35% of the total wetland water losses (Table 3.1 and Fig. 3.9). From 1980, losses by evapotranspiration reached 30–99% of total water losses (Fig. 3.10), varying between years as macrophyte cover did (see Chapter 7). In spite of evapotranspiration increased its weight on the wetland water budget from 1980s supposed only the half of that occurring previously (Table 3.1, Fig. 3.10). The weight of rainfall on the wetland water budget increased while inflows and groundwater discharges were decreasing (3–5% to 22–43% of total water inputs during 1945–1980 and 1981–2008, respectively; Fig. 3.10). Water supplied by rainfall alone was insufficient to achieve wetland inundation (Table 3.1).

The annual water budgets presented here are quite different from those of previous estimations. López-Camacho et al. (1996), using data provided by the Geological Survey of the Hydraulic Public Works Management (*Servicio*

Geológico de la Dirección General de Obras Hidráulicas), estimated that groundwater seepage into TDNP before 1975 averaged around 45 × 10^6 m^3 $year^{-1}$, a value noticeably lower that our estimates (mean 225 × 10^6 m^3 $year^{-1}$). The difference responds mostly to an underestimation of evapotranspiration losses (9 versus 33 × 10^6 m^3 $year^{-1}$, respectively) and outflows (253–342 × 10^6 m^3 $year^{-1}$, respectively). The water budget for the period 1988–1989 offered by that study did not differ from our estimates for the hydroperiod 1989–1990/1996–2005. Our data on evapotranspiration accounted for water losses of 13 × 10^6 m^3 $year^{-1}$, while López-Camacho et al. (1996) estimated 10 × 10^6 m^3 $year^{-1}$. Likewise, infiltration losses are quite similar in both water balances (33 and 32 × 10^6 m^3 $year^{-1}$, respectively). The main difference between both studies appears on surface water flow measurements, where our data show lower values (inflows: 42–124 × 10^6 m^3 $year^{-1}$; outflows: 3–88 × 10^6 m^3 $year^{-1}$; respectively). Data disagreements did not correspond to the period considered in this study, as neither annual recorded outflow data from 1981 to 1982 exceed 40 × 10^6 m^3 $year^{-1}$. Even registered annual inflow data in TDNP from 1980 to 1997 did not surpass 30 × 10^6 m^3 $year^{-1}$, only occurring in 1997–1998 and 2004 from then on. A recent study (Castaño-Castaño et al. 2008) estimated mean daily infiltration rate in TDNP to be 10 mm day^{-1} (3,650 mm $year^{-1}$) through a simple daily water balance model-inundation area using compiled data from the 1996 and 1997 summers. Applying that infiltration rate for the period 1988–2008, the averaged annual infiltration estimates by these authors appeared lower than our own (18.9 ± 15.5 and 24.5 ± 31.4 × 10^6 m^3 $year^{-1}$, respectively). Again the main differences correspond to the evapotranspiration term used in the water budget, which Castaño-Castaño and coworkers assumed to be equal to water evaporated by a Class-A tank, which is noticeably different, since the vegetation cover changes (Sánchez-Carrillo et al. 2004). Another possible inconsistency emerged because these authors did not consider the outflow occurring through the Molemocho area (31–32 × 10^6 m^3 $year^{-1}$; see Fig. 1.4 for location) where the ground subsidence resulting from groundwater overexploitation changed the river flow direction upstream towards the Ojos del Guadiana spring (Sánchez-Carrillo 2000). Considering the spatially heterogeneous hydraulic properties of wetland soils (Winter 1999, Bradley 2002), and particularly in TDNP (García-Rodríguez 1996), global estimates on infiltration rates can be undertaken by the resulting component of the water budget when the remaining hydrologic components are measured accurately.

In the pristine condition, all wetland water cycle components, except groundwater inflows, are significantly correlated with annual water storage (1945–1973; Table 3.2). Evapotranspiration appeared as the main hydrologic variable controlling TDNP annual water storage (Table 3.2). Outflows also became the other main hydrological variable, after wetland hydrological degradation, controlling annual inundation. At present, despite the importance of infiltration in wetland water loss, surface water losses control the TDNP water budget (Table 3.2). Since 1974, in fact, a remarkable correlation exists between both variables. The lack of statistical significance of groundwater inflow controlling TDNP water reserves can be

Table 3.2 Correlations (R values of Spearman rank order correlations) between water sources/sinks and the mean annual water storage in distinctive hydroperiods: *A* – 1945–73; *B* – 1974–80, *C* – 1981–88/1991–95/2006–08, *D* – 1989–90/1996–2005 (*correlations significant at $p < 0.05$)

	A	B	C	D
Rainfall	0.596*	0.184	0.044	0.071
Inflows	0.381*	0.063	0.054	0.282
Groundwater inputs	0.129	0.046	0.164	0.081
EVT	0.922*	0.990*	0.768*	0.689*
Outflows	0.533*	0.617*	0.416*	0.618*
Infiltration	0.212*	0.000	0.172	0.330

Table 3.3 Summary of principal component analysis (PCA) with varimax factor rotation of overall annual hydrological data for TDNP water cycle components during the period 1945–2008

	Factor loading	Eigenvalue	% Total variance	Cumulative Eigenvalue	Cumulative %
Factor 1		4.72	67.45	4.72	67.45
Gigüela inflows	0.716				
Puente Navarro outflows	0.930				
Molemocho outflows	0.883				
Evapotranspiration	0.944				
Factor 2		1.59	22.75	6.31	90.20
Azuer inflows	0.870				
Rainfall	0.792				
Groundwater in/outflows	–0.828				

explained by the water budget method applied (tending to accumulate errors in the unknown variable).

Using the entire database, a principal component analysis revealed that water budget components can be grouped into two main factors (Table 3.3). Major inflows and outflows and evapotranspiration explained 67% of the hydrological variance, while rainfall, Azuer inflows, groundwater inflows and Molemocho outflows explained almost 23%. Evapotranspiration and outflows represented a larger loading than the Gigüela inflows, the main water source to TDNP nowadays (Table 3.3).

Finally, multiple stepwise regressions revealed again the importance of evapotranspiration to control wetland water storage (Table 3.4). Results contrast with previous findings. Here, outflows and groundwater inputs are statistically significant but only during the hydroperiod 1945–1973; on the contrary, for the hydroperiod 1989–1990/1996–2005, which represents the wetland hydrologic degradation in humid periods, inflows and rainfall were significantly linked to evapotranspiration, thus explaining wetland water reserves (Table 3.4). Despite the high R^2 values in multiple regressions, error on wetland water storage estimates oscillate from 2% to 21%, being the largest during drought cycles (Table 3.4).

Table 3.4 Multiple stepwise regressions (forward selection) explaining mean annual water storage at TDNP in each distinctive hydroperiod. N = the number of years

	Step	Multiple R-square	F to enter	p-level
(A) 1945–1973:				
EVT	1	0.943	433.5	0.000
Outflows	2	0.954	5.5	0.027
Groundwater inputs	3	0.961	4.3	0.049
Storage = –6.10 + 0.81 * *EVT* + 0.45 * outflows + 0.33 * groundwater inflows				
N: 28; Std. Error = 0.879				
(B) 1974–1980:				
EVT	1	0.988	397.0	0.000
Storage = –7.53 + 0.93 * EVT				
N: 7; Std. Error = 0.178				
(C) 1981–1988/1991–1995/2006–2008:				
EVT	1	0.916	152.7	0.000
Storage = –0.07 + 1.01 * EVT				
N: 16; Std. Error = 0.261				
(D) 1989–1990/1996–2005:				
EVT	1	0.868	66.1	0.000
Inflows	2	0.940	10.8	0.009
Rainfall	3	0.976	12.0	0.008
Storage = –1.43 + 0.526 * EVT + 0.364 * Inflows + 0.180 * Rainfall				
N: 12; Std. Error = 0.354				

3.6 Causes of Wetland Hydrologic Degradation: Climate Change Versus Hydrologic Overexploitation

Water-level fluctuations and their ecological and socio-economic consequences have been investigated in several aquatic ecosystems, including small shallow lakes and wetlands (e.g. Woo and Rowsell 1993; Rosenberry and Winter 1997; Naselli-Flores and Barone 2005; Van del Valk 2005). The reasons and causes of water-level fluctuations can be due to several climatic changes in the constellation of large atmospheric pressure systems (e.g. North Atlantic and El Niño Southern Oscillation), or, most frequently, to seasonal variations in meteorological conditions. But they can also be the result of anthropogenic use of water resources, as demonstrated in the case of the Aral Sea (Usmanova 2003).

Hydrologically induced water level fluctuations are the result of a change in the water budget and, therefore, these fluctuations depend on the amounts of precipitation and evaporation, catchment size and on discharge conditions (inflow versus outflow) of the basin. The time scales of the hydrologically induced water fluctuations range from days to centuries (and even up to geological time scales) (Hofmann et al. 2008). One of the most common factor inducing hydrological fluctuation in semi-arid regions is drought (Le Houérou 1996). Although drought can be defined differently, we restrict it here to the meteorological and hence hydrological. Drought is a normal feature in semi-arid regions, having three major characteristics – intensity, duration and spatial extent – which differ from a seasonal drought (Wilhite 2000). Nevertheless, supraseasonal droughts are considered as aberrant and unpredictable phenomenons (Lake 2003). Assuming that instrumental weather records began approximately 150 years ago, there are no indications of any long-term trends in rainfall from the global view point; there are, however, local middle-term trends in rainfall reduction accompanied by an increase in drought frequency in most semi-arid regions, as a possible consequence of climate change (Le Houérou 1996; Hanson and Weltzin 2000; Bond et al. 2008). For example, there was a decrease in rainfall in the South-Eastern portion of the Iberian Peninsula, together with an increase in variability, particularly between 1890 and 1940 (Conte 1993; Oñate 1993; Oñate and Pou 1996; Trigo and Palutikov 2001). Several studies have identified dry periods that have affected Spain in different centuries (Vicente-Serrano and Cuadrat 2007). The recent drought period between 1991 and 1995, which was the most intense of the twentieth century in Spain, is an excellent example (Almarza et al. 1999).

Several drought indices have been used to test drought effects on water resource availability, most of them based on normal rainfall deviations (Alley 1984) and focusing on spatial distribution (Hanson and Weltzin 2000). Drought, therefore, and its frequency can indicate the extent of climate changes causing the hydrological degradation of TDNP, as opposed to water resource overexploitation. To quantify and analyze droughts with precipitation data, we used the simplified Standardized Precipitation Index (SPI) for the period 1904–2008. Because Spanish droughts have a strong regional component (Vicente-Serrano and Cuadrat 2007), local meteorological conditions close to TDNP were used as reference of the regional climate.

55% of the annual SPI values for TDNP area were below 0 (from near normal to slight drought) and 20% were lower than –0.2 (considered as drought; Fig. 3.11a). Certainly, after 1971 droughts increased their occurrence, with SPI registering 11 cases below –0.2 (29%); comparatively, until 1970 (1904–1970) only 29% of SPI values (9 cases) appeared below –0.2 (Fig. 3.11a). Nevertheless, analyzing each time period separately, the percentage of SPI values below 0 always stayed constant at around 55%. During 1971–2008, although moderate droughts were more frequent (13% versus 11% < –0.3, respectively), strong droughts (<0.4) were higher during 1904–1970 (Fig. 3.11a). From 1980, it is possible to observe that extended duration of drought appeared more frequently: 1980–1983, 1990–1993–1994, 1998–1999, 2000–2002–2003–2005 (Fig. 3.11a). The atmospheric circulation pattern known as the North Atlantic Oscillation (NAO) affects precipitation over large areas of central and south-western Spain (Trigo et al. 2004). Similarly, it has been concluded that the temporal variability in the NAO played a major role in explaining droughts with positive values tending to cause dry conditions (Vicente-Serrano and Cuadrat 2007). Over recent decades, the NAO has undergone major low frequency variations, superimposed on an upward trend in NAO index and resulting in more frequent droughts (Karl and Trenberth 2003; Rind et al. 2005). Under this scenario, it would be possible to expect a strong effect on TDNP hydrology through water scarcity when approaching the twenty-first century, as observed in other semi-arid wetlands (e.g. Conly and Van der Kamp 2001). Until 1974, when wetland hydrological degradation is thought to have started, there is a moderately strong

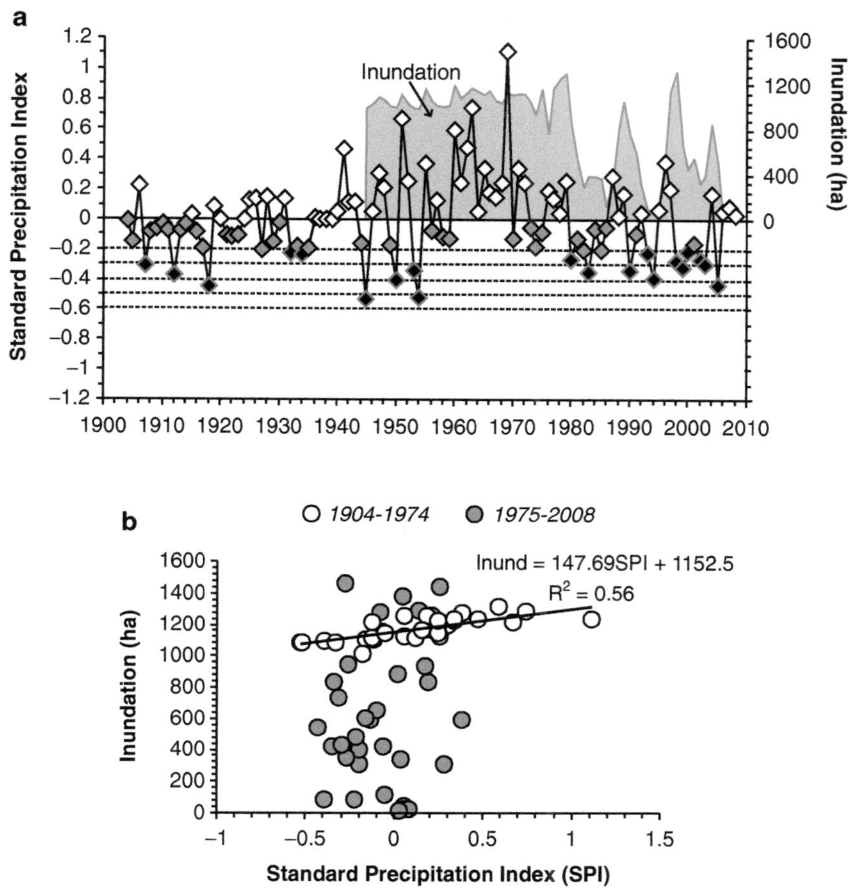

Fig. 3.11 Relationship between the Standardized Precipitation Index (SPI) calculated on annual time scales and the inundation in Las Tablas de Daimiel. (**a**) SPI and inundation trends for the period 1904–2008 (white diamond: SPI > 0, grey diamond: SPI > –0.2, black diamond: SPI < –0.2), and (**b**) correlation between SPI and inundation by the distinctive periods 1904–1974 (before hydrological degradation) and 1975–2008 (degraded wetland)

statistically significant ($p < 0.05$) correlation between inundation and SPI (Fig. 3.11b). However, hydrological degradation must be viewed from another approach, since such a correlation did not seem to be significant for the period 1975–2008 ($p < 0.05$; Fig. 3.11b). Effects of drought on river inflows or groundwater were also unclear (Fig. 3.12). While the Gigüela and Azuer inflows showed a slight drought influence, both the Guadiana inflows, which depend strongly on aquifer discharges, and the groundwater table inside TDNP did not reflect it (Fig. 3.12). Temporal and spatial distributions of rainfall were very variable along the catchment (Conan et al. 2003) and hence drought influence on Gigüela discharges requires a broader and detailed analysis. Besides, since 1954, the increase of water storage in the basin in order to satisfy the large water consumption by

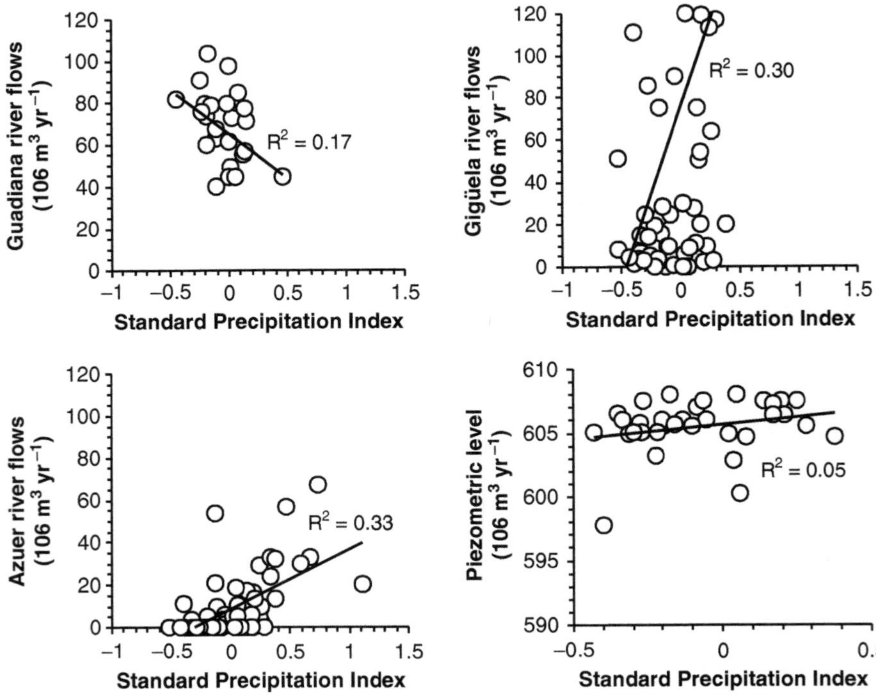

Fig. 3.12 Relationships between Standard Precipitation Index (SPI) river discharge and aquifer piezometric level measured in Las Tablas de Daimiel

irrigation, has significantly modified the natural regime of the river (Kilsby et al. 2007). Nevertheless, direct runoff is usually very low in the upper Guadiana basin and stream-flows are strongly linked to groundwater release to the streams (Conan et al. 2003). Assuming that groundwater recharge depends on climate variations (i.e. droughts; Bouraoui et al. 1998), the SWAT model representing the upper Guadiana basin, developed by Conan et al. (2003), demonstrated that aquifer depletion reduced or eliminated the stream-flow in the catchment. Although the Gigüela river discharges into TDNP showed a slightly significant inverse relationship with 04.04 aquifer groundwater abstractions ($R^2 = 0.13$, $p < 0.05$), flow effects can only be evaluated by means of watershed hydrological modeling, which is not the aim of this chapter.

Groundwater consumption of the 04.04 Aquifer for irrigation has often been cited as the main cause of the hydrological degradation of TDNP and La Mancha Húmeda wetlands (see Chapter 1; Amezaga and Santamaría 2000; Fornés et al. 2000). Figure 3.13 shows a simple exercise of aquifer water balance with groundwater recharge estimates based on the SPI values, adjusting values by trial and error to match with IGME (2004, 2008) observations on annual groundwater reserve variations and irrigation pumping. Average annual recharge was estimated to be 215×10^6 m^3 year^{-1}, while maximum recharge achieved 365×10^6 m^3 year^{-1} and a

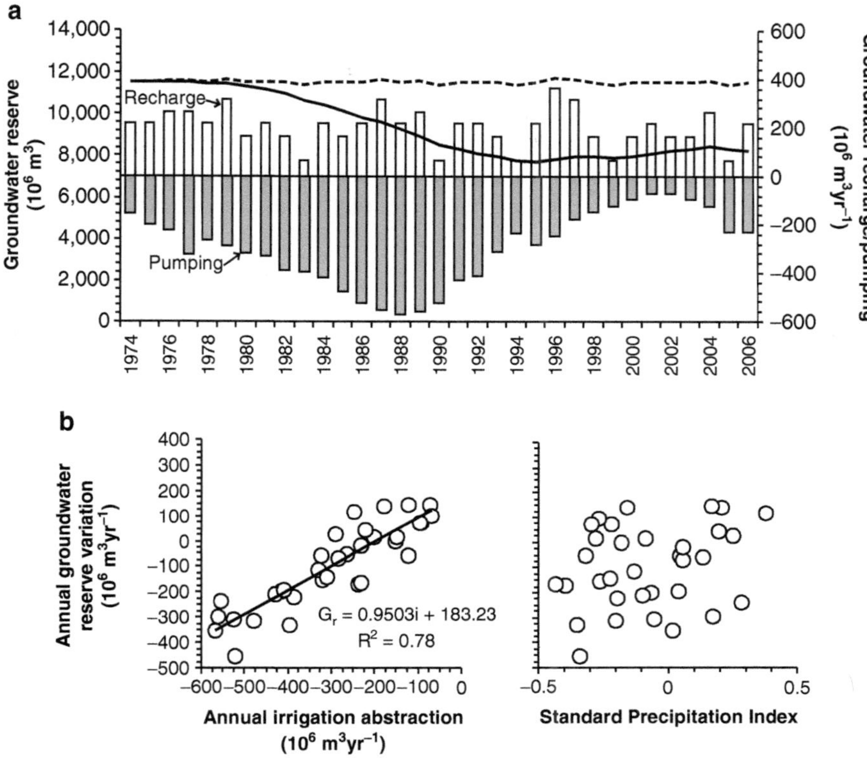

Fig. 3.13 Aquifer 04.04 overexploitation. (**a**) An attempt of water balance from 1974 to 2006, showing the current evolution of groundwater reserves (*continuous line*) and the aquifer reserves trend if pumping levels occurring in 1974 would have been maintained (*discontinuous line*) and (**b**) relationships between annual variation of groundwater reserves (G_r) and both annual irrigation abstraction (i) in and Standard Precipitation Index

minimum of 65 × 10^6 m^3 $year^{-1}$. It has been proven that recharge variability did not compensate the excess of groundwater abstraction, decreasing total groundwater reserves until 66% in 1995 or to 71% nowadays. The fact that irrigation pumping is above the aquifer sustainability (or compensated by recharge) can be demonstrated if groundwater pumping would be maintained at the levels existing in 1974 (175 × 10^6 m^3 $year^{-1}$) where groundwater reserves did not change over time (Fig. 3.13). From 1974, 78% of the variance of aquifer depletion is explained by irrigation pumping (Fig. 3.13b), indicating that climate variability could only be responsible for less than 22%. In fact, SPI did not show a significant relationship with the annual groundwater reserve variation (Fig. 3.13b). Unfortunately, statistical analyses of immediate effects of groundwater overexploitation on TDNP hydrology cannot be easily assessed because, before 1974, data of the 04.04 aquifer water reserves were not available.

Cruces et al. (1997) estimated that wetland disappearance annually saves *ca.* 150×10^6 m^3 year^{-1} to the groundwater reserves, because the reduction of evapotranspiration in wetland areas. This quantity appears then close to groundwater pumping existing in 1974 (175×10^6 m^3 year^{-1}) which could be considered as close to groundwater sustainability. In fact, in the hydrological context of La Mancha Húmeda, those 150×10^6 m^3 year^{-1} can be viewed as the water subtracted to the wetlands for irrigation practices and the water that should be returned to them for their hydrological restoration.

3.7 Evapotranspiration Controls of Wetland Hydrology: A Conceptual Model

Evapotranspiration is probably the most studied hydrological variable in TDNP, given that it is the main factor controlling ecosystem dynamics. Evapotranspiration is cited as a major component of the water balance in semi-arid wetlands (Winter 1988); however, few studies have addressed its dynamics in regions with this kind of climate. As evapotranspiration is typically the largest consumer of the incoming energy in wetlands (Pribán and Ondok 1985; Sánchez-Carrillo et al. 2004), it also has a great influence on water temperature, water level and salinity, dictating the functioning of wetland ecosystems. Habitat quality is influenced by the ratio of emergent macrophyte cover to open water. For example, a ratio near unity reflects a high quality habitat for waterfowl, since areas of open water are present for feeding and brood rearing, while emergent cover offers protection and nesting materials (Poiani and Johnson 1993). Hence, it is usual to manipulate cover-to-water ratios in wetlands by managing vegetation to improve conditions for nesting waterfowl and wildlife (Weller 1989); however, managers may not recognize the potential impact of such practices on the natural hydrological functioning of wetlands.

Independent estimates of evaporation and transpiration are uncommon, due to the technical difficulties entailed. Nevertheless, independent measurements were performed in the TDNP wetland to assess the relative importance of evaporation and transpiration in wetland water budgets, taking into consideration the composition of the vegetation, landscape features and climatic conditions.

Daily cycles of transpiration rates for each emergent macrophyte species in summer (August) and autumn (October) in 1997 and 1998 are shown in Fig. 3.14. In summer, reed shows a curve with maximum transpiration around midday. In contrast, cut-sedge presented maximum values in the morning and evening hours, with a midday depression. In the cattail stands, transpiration rates were lowest during hours of highest radiation and temperature. This phenomenon, which can be frequently observed in plants, is a physiological response associated with the regulation of stomatal conductance during periods of maximum thermal stress (Collatz et al. 1991). In autumn, total daily transpiration was significantly reduced, owing basically to the reduction of the light period and lower solar incidence. Cut-sedge presented a transpiration response which exhibited maxima during midday (Fig. 3.14)

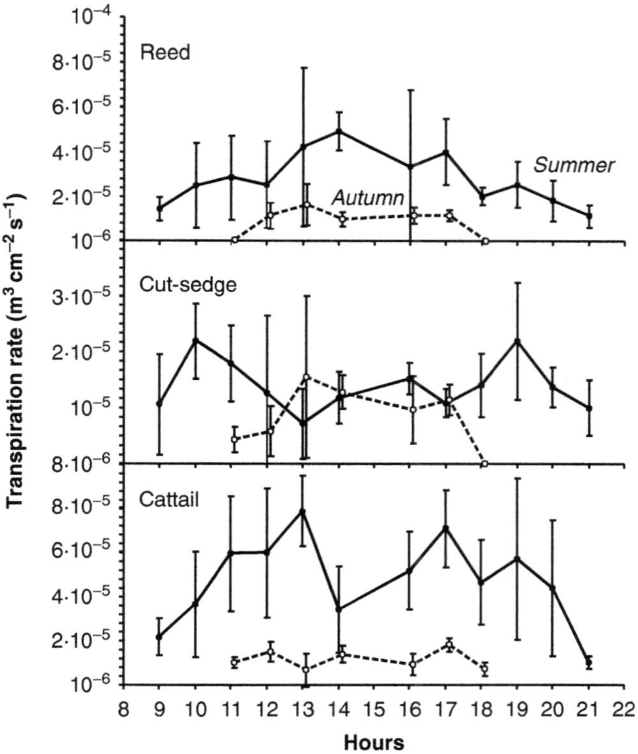

Fig. 3.14 Hourly transpiration rates for the main emergent macrophytes at TDNP measured in summer (August) and autumn (October) of 1997 and 1998

Table 3.5 Average ± SD of leaf area index (LAI) and annual transpiration rates (T) in vegetation stands in TDNP during 1997 and 1998

	Reed		Cut-sedge		Cattail	
	LAI ($m^2\ m^{-2}$)	T (mm day^{-1})	LAI ($m^2\ m^{-2}$)	T (mm day^{-1})	LAI ($m^2\ m^{-2}$)	T (mm day^{-1})
1997	8.28 ± 5.06	7.1 ± 1.1	10.59 ± 2.71	11.7 ± 5.0	5.26 ± 1.96	6.4 ± 0.2
1998	9.13 ± 5.35	7.8 ± 1.2	15.13 ± 1.95	17.8 ± 7.1	3.68 ± 0.89	4.3 ± 0.7

with values coinciding with those registered in summer. Reed and cattail experienced lower transpiration rates in autumn and, furthermore, showed a daily plot with a central plateau, indicating that transpiration rates remained practically constant during the day (Fig. 3.14). Transpiration rates in vegetation stands exhibited an important variability among plant species depending on LAI (Table 3.5). Cut-sedge presented the highest mean rates, while cattail showed the lowest. Reed transpiration rates measured in TDNP are similar to those reported by Burba et al. (1999) and Tuschl (1970). Transpiration data of the European cut-sedge reported in the study of Sánchez-Carrillo et al. (2001) are unique, and, therefore, cannot be compared with data from other ecosystems. Data reported by Koch and Rawlik (1993) for the sawgrass *Cladium jamaicense* – an American species morphologically and anatomically similar to *Cladium mariscus* – showed lower transpiration rates (5.57 ± 0.45 mm day^{-1}) which is probably related to micrometeorological differences between TDNP and The Everglades. It may also seem appropriate to consider the differences on LAI values of macrophyte stands between both ecosystems for observed differences (Idso and Anderson 1988). However, most published papers did not present this kind of information, while foliar density controlled the value of global wetland transpiration. In regard to the similarity of data, the method described by Sánchez-Carrillo et al. (2001) provides results with a lower error, considering the in situ origin of the data – i.e. an oasis effects or clothes-line effect (Pribán and Ondok 1985). Estimates of transpiration rates are similar to those calculated with the mass transfer and energy budget methods.

Open water evaporation measured in TDNP for the period 1993–1998 (Sánchez-Carrillo et al. 2004) obtained information on the relationship between transpiration and evaporation in the wetland for different water level scenarios, thereby determining their hydrological effects on the ecosystem. Long-term data are needed for a holistic understanding of the variability of evaporation and transpiration in aquatic ecosystems; such data would provide the necessary framework for modeling and forecasting hydrological dynamics and ecosystem responses to climate and human disturbances. However, long-term data sets for evapotranspiration in wetlands are scarce.

Mean annual open water evaporation rates in TDNP ranged from 7.6 to 8.7 mm day^{-1} (average: 8.0 ± 3.1 mm day^{-1}; Sánchez-Carrillo et al. 2004). Interestingly, the average evaporation rate observed in this study is higher than those reported in the literature for wetlands (1.5–6.9 mm day^{-1}; Souch et al. 1998; Parkhurst et al. 1998; Burba et al. 1999). The water loss by evapotranspiration depended on both spatial and temporal scales, with the relationships between evaporation and transpiration being conditioned by plant dynamics (Table 3.6). Although the maximum transpiration rates corresponded to cut-sedge, most transpiration water loss at the ecosystem level was attributable to reed, due to its greater coverage (Table 3.6). Transpiration was the main factor for water loss during periods of low flooding, reaching up to 28 times that of evaporation, while evaporation water loss exceeded twice the transpiration rate in high water level periods. On a monthly basis, the relationship between water loss by evapotranspiration and flooded area was not statistically significant ($p > 0.5$). At annual scales, however, this relationship was significant ($R^2 = 0.89$; $p < 0.05$). This indicates that evapotranspiration cannot be sufficiently explained by short-term (monthly) vegetation changes.

The distribution of species and plant communities within a wetland is primarily a function of water depth (Van del Valk et al. 1994; Álvarez-Cobelas et al. 2001) and competitive traits; flood-tolerant species outcompete less flood-tolerant early-succession species in high water level situations. Such phenomena impinge on wetland evapotranspiration. In TDNP, macrophyte cover responses to inundation patterns suggest an inverse relationship with reed cover and a direct relationship

Table 3.6 Annual evapotranspiration at TDNP ecosystem level and yearly changes in macrophyte cover and open water area during 1993–1998. 'Open water area' refers to the inundated area without emergent plants; 'inundation' refers to the total wetland area in which water covers space containing both helophyte and non-halophyte species. Areas of macrophyte cover did not always occur in the inundated areas (data from Sánchez-Carrillo et al. 2004)

	1993	1994	1995	1996	1997	1998
Cover (km^2 $year^{-1}$)						
Reed	5.55	5.74	11.10	5.88	3.50	3.40
Cut-sedge	0.35	0.44	0.25	1.40	1.10	1.15
Cattail	0.02	0.03	0.02	0.02	0.04	0.05
Open water	0.36	0.22	0.18	1.65	7.44	7.74
Inundation	0.74	0.84	0.39	6.74	13.19	14.57
Plant/open water	16.4	28.2	63.2	4.4	0.6	0.6
Water (10^6 m^3 $year^{-1}$)						
Reed	7.96	8.41	15.73	8.35	4.55	4.89
Cut-sedge	1.01	1.37	0.60	4.09	2.36	3.76
Cattail	0.02	0.02	0.02	0.02	0.05	0.04
Transpiration (T)	8.99	9.81	16.35	12.46	6.96	8.69
Evaporation(E)	1.03	0.68	0.57	4.66	21.09	21.49
T/E	8.7	14.5	28.6	2.7	0.3	0.4
Total	10.03	10.49	16.92	17.12	28.05	30.18

for cut-sedge and cattail (Álvarez-Cobelas et al. 2001; Sánchez-Carrillo et al. 2004; see Chapter 8).

Linacre's theory (1976) suggests that there is a reduction in water loss by evapotranspiration when emergent macrophyte cover increases. Evapotranspiration decreases because of the specific micrometeorology of the macrophyte-belt (high air humidity, limited air exchange due to low wind velocities, lower air temperatures and weaker solar irradiance) in aquatic ecosystems. The micrometeorological conditions of open water (free air flow, lower humidity, optimum solar irradiance and greater temperature fluctuations) are diametrically opposite to those occurring in macrophyte stands and considerably increase water consumption. Such relationships were observed for a wide range of temperate wetlands (Price 1994) where changes in macrophyte cover and species composition are mostly due to factors other than inundation area. In semi-arid wetlands, such as TDNP, although larger areas of open water also increase water consumption by evapotranspiration, it was not possible to confirm that increases in macrophyte cover generate reductions in evapotranspiration (see Fig. 5 in Sánchez-Carrillo et al. 2004) – as suggested by Linacre's theory. The relationship between transpiration to open water evaporation ratio (T/E) and evapotranspiration in TDNP is inverse. However, T/E values of up to 4 marginally reduce total evapotranspiration loss (Sánchez-Carrillo et al. 2004). In drought years, when reeds dominates, the macrophyte cover/open ratio water area increases, leading to reduced water loss by evapotranspiration, but this does not occur in wet years, when cut-sedge dominates. The study of Sánchez-Carrillo et al. (2004) suggested that variations in the T/E ratio might depend on the cover dynamics of a single macrophyte species (related to inundation patterns). This expands our understanding of Linacre's theory and its applicability in aquatic ecosystems, while being inadequate for explaining evapotranspiration at ecosystem scale in highly fluctuating, semi-arid wetlands.

At the whole-wetland scale, evapotranspiration depends on water level fluctuations which establish the open water/macrophyte cover ratio, but it also controls seasonal inundation patterns through open water evaporation and the transpiration rates of macrophyte species (according to their flood tolerance). Evapotranspiration is not limited when inundation increases, but with low standing water levels it decreases because open water evaporation is limited and macrophyte species with lower transpiration capacities colonizes the wetland (Sánchez-Carrillo et al. 2004). Such vegetation changes occur annually, thus, an interannual comparison is needed to reveal the evapotranspiration patterns for the whole wetland. In summary, in very shallow water ecosystems where emergent macrophytes have a tendency to increase their coverage, the physiological strategies used by plants to tolerate inundation play a key role in the hydrological dynamics of the ecosystem.

3.8 A Wetland Hydrological Model for Hydroperiod Restoration

Wetland hydrology is perhaps the key element and often the main focus in wetland restoration projects (Acreman et al. 2000). The reestablishment of flood pulsing in riverine and tidal wetlands is recognized as an essential step in the restoration of wetlands worldwide. Hydrological restoration is usually difficult, because there are no previous long-term records of wetland hydrological data and restoration experts often do not accurately know hydrological dynamics of the wetland that they are trying to restore (Middleton 2002).

It is unclear yet to scientists exactly how much of the natural hydrological regime has to be restored; it is necessary to know if partial improvements can restore the biota and biological functions of wetlands (Zedler 2000). It is well known that small changes in wetland hydrology can significantly affect the chemical and physical properties of ecosystems, such as nutrient availability, degree of substrate anoxia, soil salinity, sediment properties, and pH (Mitsch and Gosselink 2000). When hydrologic conditions of wetlands change, plant species composition can shift (Álvarez-Cobelas et al. 2007) and ecosystem productivity can decline (Mitsch and Gosselink 2000). Because hydrology plays a vital role in the structure of a wetland's ecosystem, particularly by acting as the main pathway in which nutrients are transported in and out of the system, the vegetation and species composition are significantly affected when natural or man-made hydrologic alterations occur. Mitsch and Gosselink (2000) list several principles underscoring the importance of hydrology in wetlands: (i) hydrology leads to a unique vegetation composition but can limit or enhance species richness, (ii) primary productivity and other ecosystem functions in wetlands are often enhanced by flowing conditions and a pulsing hydroperiod and are often depressed by stagnant conditions, (iii) accumulation of organic material in wetlands is controlled by hydrology through its influence on primary productivity and decomposition, and (iv) export of particulate organic matter, and nutrient cycling and nutrient availability are both significantly influenced by hydrologic conditions. Despite recognition of wetland hydrology, we still know very little about how to manipulate or simulate certain hydrological conditions (Zedler 2000).

Wetland restoration is often centered on establishing hydrological compensations, because of the important role it plays in wetland function (Middleton 2002). In most instances, researchers are still collecting the essential life history data that will aid in building a case for the need to recreate flood-pulsed hydrology in wetland restoration projects (Middleton 2002). The flood pulse concept, firstly developed to describe seasonal changes in water levels on Amazonian floodplains and their relationships to functional dynamics and the maintenance of species diversity (Junk 1982, 1997; Junk and Howard-Williams 1984; Junk et al. 1989; Bayley 1995), highlights the critical importance of floodplain inundation supporting production, decomposition, and consumption (Grubaugh and Anderson 1988; Sparks et al. 1990) but also driving plant succession (Van del Valk 1981; Middleton 1999). However, the importance of re-establishing wetland water regimes synchronous with seasonal climate fluctuation and water flow in riverine wetlands has not been fully appreciated in wetland restoration (Middleton 2002).

From the 1980s, hydrological remediation plans applied in TDNP attempted to maintain wetland inundation during drought years through annual water diversions from the neighbour Tagus basin (see Chapter 10). Estimates of annual wetland water requirements did not take into account TDNP hydrological and ecological functioning and ecosystem structure. Furthermore, they did not consider that water could not always be available in the other basin for diversion into the wetland. Therefore, the success of the wetland hydrological remediation has so far been very limited.

Habitually, one of the first challenges of a wetland hydrological model is to achieve high accuracy on the water cycle components explaining inundation patterns. As monthly water budgets computed at TDNP showed variable disarrangements between the inputs and outputs, water balance then has to close through residuals, which represent 19–76% of annual water inputs and 3–43% of annual water outputs (Fig. 3.15). Most residuals were supplied by unaccounted creek discharges and groundwater pumping into TDNP, as well as from the inaccurate temporal scale at which both Gigüela river inflows and TDNP outflows are measured (weekly, biweekly or even monthly or bimonthly). These premises must be taken into account in the results provided by the wetland hydrological model for any restoration purpose.

The hydrological model of TDNP, showing monthly wetland water requirements (i.e. river inflows) to support any intended inundation, was computed from a rational function (Fig. 3.16a). Considering TDNP as an impoundment system, outflows start when water-level or inundation reach a determined value. Thus, TDNP outflows are estimated from an exponential model of wetland inundation (Fig. 3.16b). Open water evaporation was computed from class A evaporation measurements

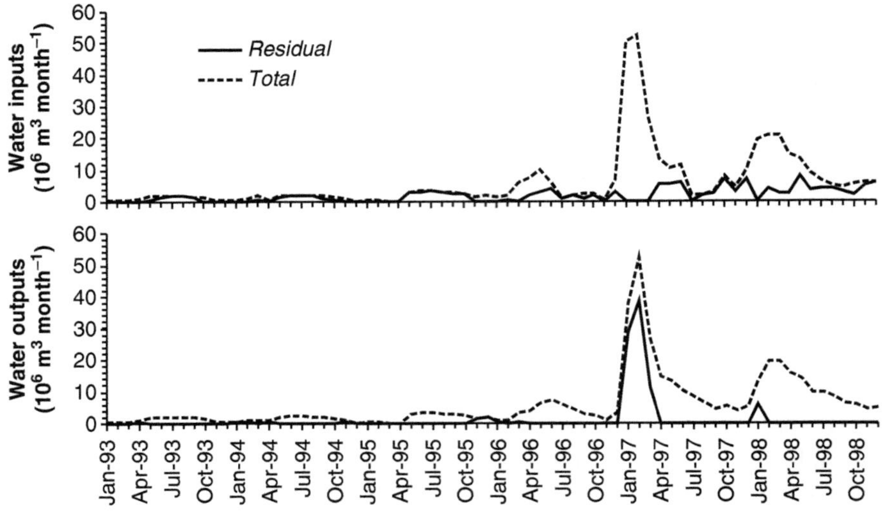

Fig. 3.15 Hydrological accuracy of monthly water fluxes measured in Las Tablas de Daimiel throughout 1993–1998

recorded by TDNP staff (Fig. 3.16c); class A evaporation was significantly related to mean monthly air temperatures (Fig. 3.16d). Macrophyte transpiration rates calculated by Sánchez-Carrillo et al. (2001) were simplified following the relationship between plant cover and total wetland water loss by transpiration (Fig. 3.16e and f). Mean infiltration was assumed to be of 10 mm month^{-1} as reported by Castaño-Castaño et al. (2008). Finally, mean wetland inundation might be estimated from river inflows through a rational function (Fig. 3.16g).

Figure 3.17 shows the model calibration using data for the period 1999–2004. Taking into account the above mentioned limitations, provided by the accuracy of the measured data, the model represented the wetland hydrological dynamics reasonably well. Inflow estimates fluctuated according to seasonal discharges, including those proceeding from water diversions (Fig. 3.17). Although the model computed some outflows during this period, it did not agree with observed data. Again, this uncoupling probably responds more to the uncertainty provided by the manual recording of data than to model weakness. The goodness-of-fit of inundation estimates using inflow data were robust (R^2 = 0.93, slope = 1.13). Test results showed that differences between calculated and observed inundation were within 150 ha 75% of the simulation time.

Hydroperiod restoration must firstly promote equilibrium between macrophyte cover species and open water areas, to support wetland function and structure (Krause and Bronstert 2005). Cut-sedge and reed covers are significantly related to the average inundation in the preceding year, albeit in an opposite way (see Chapter 7 and Fig. 7.10). Therefore, cover types could be modelled as a function of inundation (Fig. 3.18). This chart provides criteria for assessing wetland water requirements in

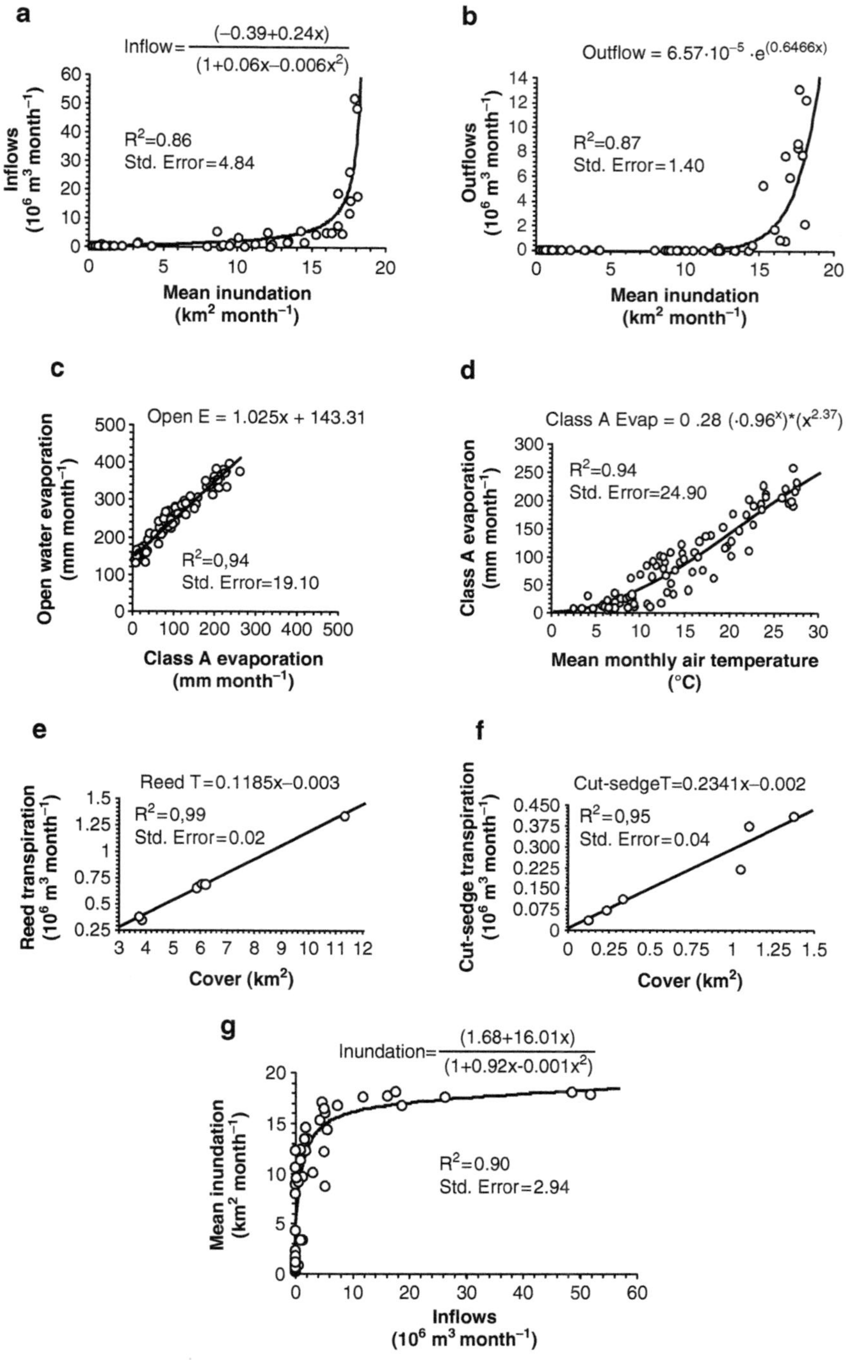

Fig. 3.16 Mathematical relationships used for parameter estimations of the TDNP hydrological model: (**a**) river inflows and inundation, (**b**) wetland outflows and inundation, (**c**) open water evaporation and Class A evaporation, (**d**) evaporation obtained from a Class A evaporation tank and mean monthly air temperature, (**e**) reed transpiration and cover, (**f**) Cut-sedge transpiration and cover, and (**g**) mean inundation and inflows

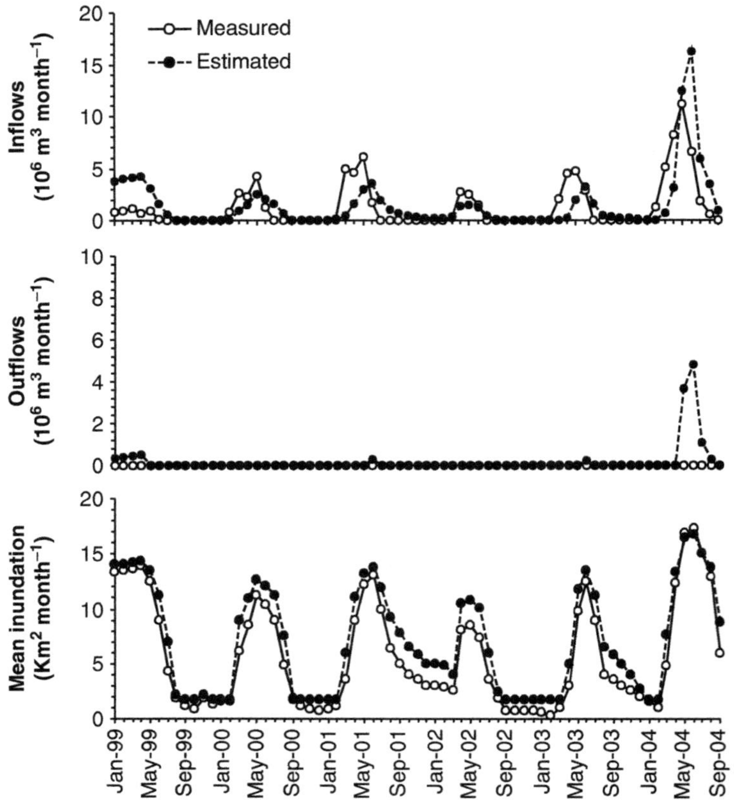

Fig. 3.17 Model calibration of the TDNP hydrological model using data for the period 1999–2004: inflow estimates (*upper panel*), outflow estimates (*middle panel*), and inundation estimates (*lower panel*)

order to sustain any cover mixture. For example, to maintain a cut-sedge cover of up to 200 ha, a minimum annual inundation of 600 ha would be required. When mean annual inundation is over 1,400 ha, then cut-sedge cover grows up to 300 ha. To control reed cover of fewer than 200 ha, the mean annual inundation must be of up to 800 ha; when mean annual inundation exceeds 1,000 ha, then reed cover has an extension of less than 100 ha. More than 300 ha of open water areas can be supported in the wetland when mean inundation is up to 800 ha. Accordingly, a mean annual inundation close to 10 km^2 appears then as the ideal hydroperiod for TDNP restoration. Obviously, these hydrological water-level based criteria must be combined with water quality criteria, which also determine the macrophyte cover in the wetland, as well as with waterfowl dynamics (see Chapter 8).

The TDNP hydrological model was run then considering several scenarios which take these water requirements into account, to restore macrophyte cover throughout a hydroperiod most similar to that existing in the 1950s (Table 3.7). At present, to achieve hydroperiods similar to those in 1944–1973 or 1974–1980, an

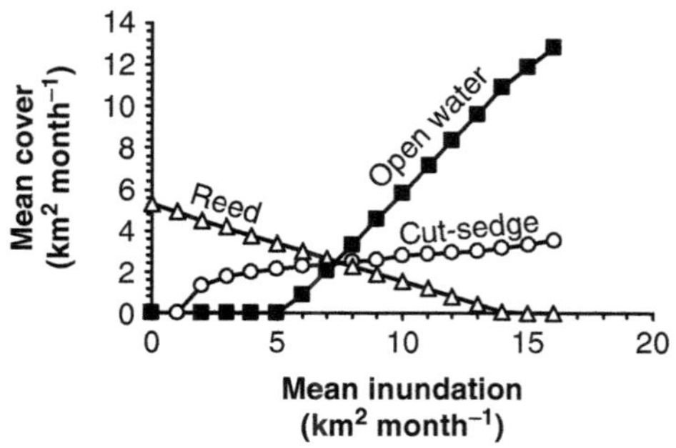

Fig. 3.18 Mathematical equations used to predict changes in the main cover types (reed, cut-sedge and open water areas) in TDNP throughout variations in the annual inundation area

$$\text{Reed cover} = 5.15 - 0.35 \cdot \text{Inundation}$$
$$R^2 = 0.99, \text{ Std. Error: } 0.17$$

$$\text{Cut-sedge cover} = \frac{(0.06 + 35.06 \cdot \text{Inundation}^{0.43})}{(33.75 + \text{Inundation}^{0.43})}$$
$$R^2 = 0.99, \text{ Std.Error: } 0.007$$

$$\text{Open water} = 6.78 + 7.26 \cdot \cos(0.19 \cdot \text{Inundation} + 2.70)$$
$$R^2 = 0.99, \text{ Std. Error: } 0.24$$

external water inflow of up to 80 × 10^6 m^3 $year^{-1}$ is needed (scenarios 1 and 2). Cut-sedge restoration to cover around 430 ha and reed declining under 100 ha following these hydroperiod demonstrated enormous water requirements, most of them devoted to maintaining evapotranspiration losses (scenarios 8 and 9). Open water areas occupying above 6 km^2 of wetland cover can only be achieved with mean annual inundation of up to 11 km^2 (scenarios 1, 2, 7, 8 and 9). The lower the mean annual inundation, the smaller the open water area in the wetland (scenarios 12, 13), pointing out the importance of plant cover management to promote free-vegetation areas where submerged macrophytes (mostly charophytes) can develop. Annual surface water discharges of 35 × 10^6 m^3 $year^{-1}$ into TDNP could maintain 220–230 ha of cut-sedge, but reed cover must be kept high (300–400 ha, scenarios 4, 5 and 6). Lower than 40 × 10^6 m^3 $year^{-1}$ of water inflows imply wetland dry-out during summer months. Under this scenario, the lower the average annual inundation, the longer the summer desiccation. Although summer dry-out shortens wetland water requirement through decreasing evapotranspiration, other biogeochemical and plant dynamic implications must be considered in wetland management. The hydroperiod modeled following these scenarios revealed an increase of wetland water requirements during summer months when macrophyte cover restoration is applied, particularly if cut-sedge and reed reach covers like those existing before wetland degradation (Fig. 3.19).

Considering the water scarcity in this region, the need for cut-sedge restoration, as well as the promotion of open water areas, scenario 3 may appear as the most suitable to be used in a hydrological remediation plan. However, since TDNP is an

Table 3.7 Summary of TDNP water requirements and plant cover for the hydroperiod restoration under different management scenarios: (1) hydroperiod of 1944–1973, (2) hydroperiod of 1974–1980, (3) hydroperiod of 1996–2005, (4) external aportation of 35×10^6 m^3 $year^{-1}$ and mean inundation of 6.8 km^2, (5) external aportation of 35×10^6 m^3 $year^{-1}$ and mean inundation of 7 km^2, (6) external supply of 35×10^6 m^3 $year^{-1}$ and mean inundation of 7.2 km^2, (7) 2.5 km^2 of cut sedge, 4 km^2 of reed and the hydroperiod of 1944–1973, (8) 4.3 km^2 of cut sedge, 1 km^2 of reed and the hydroperiod 1944–1973, (9) 4.3 km^2 of cut sedge, 1 km^2 of reed and the hydroperiod 1974–1980, (10) mean inundation of 6.5 km^2 maintaining 2 km^2 flooded in summer, (11) mean inundation of 6.8 km^2 maintaining 4 km^2 flooded in summer, (12) mean inundation of 4 km^2 drying the wetland during summer, (13) mean inundation of 3.1 km^2 with starting inundation close to 0 and 1 km^2 of summer inundation, and (14) ideal seasonally modulated hydroperiod. Annual rainfall was considered constant in all scenarios attaining 4.85×10^6 m^3 $year^{-1}$. ET is evapotranspiration

Scenarios	Average annual inundation (range) km^2	Months dry out	Open water	Reed	Cut-sedge	Inflows 10^6 m^3 $month^{-1}$	ET	Infiltration	Outflows	Total inputs	Total outputs
1	11.8 (15.9–6.3)	0	8.17	1.33	2.77	85.28	42.18	39.64	4.13	90.13	90.13
2	11.7 (15.8–4.7)	0	8.05	1.36	2.77	83.44	40.55	39.07	4.27	88.28	88.28
3	7.8 (13.1–3.6)	0	2.91	2.74	2.35	54.13	30.03	27.50	0.11	58.98	58.98
4	6.8 (15.3–0)	5	1.87	3.07	2.23	35.00	12.27	21.39	2.31	39.85	39.85
5	7.0 (15.3–0)	5	2.11	2.99	2.26	35.00	10.74	26.14	2.18	39.85	39.85
6	7.2 (15.2–0)	5	2.28	2.94	2.28	35.00	11.03	26.72	2.10	39.85	39.85
7	11.8 (15.9–6.3)	0	8.17	4.00	2.50	81.40	38.30	43.82	4.13	86.25	86.25
8	11.8 (15.9–6.3)	0	6.48	1.00	4.30	246.30	203.19	43.82	4.13	251.15	251.15
9	11.7 (15.8–4.7)	0	6.44	1.00	4.30	244.70	201.81	43.47	4.27	249.54	249.54
10	6.5 (10–2)	0	1.58	3.18	2.19	42.70	23.37	24.18	0.00	47.55	47.55
11	6.8 (10–4)	0	1.91	3.06	2.23	45.22	24.64	25.42	0.00	50.07	50.07
12	4.0 (8–0)	3	0.00	4.04	1.80	26.98	16.79	15.04	0.00	31.83	31.83
13	3.1 (8–0.1)	0	0.00	4.37	1.61	22.65	15.99	11.50	0.00	27.49	27.49
14	8.1 (15–2)	0	3.30	2.62	2.39	54.57	28.52	30.07	0.82	59.41	59.41

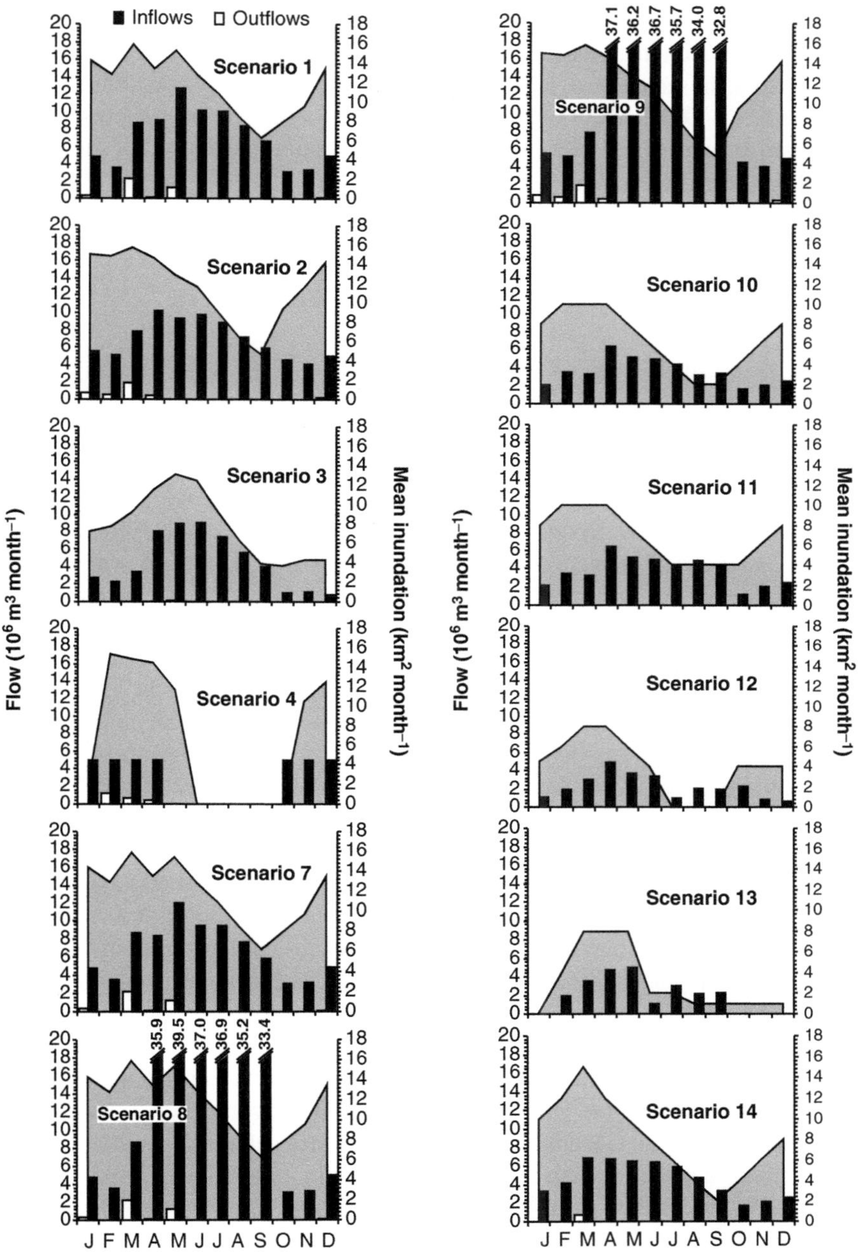

Fig. 3.19 Hydroperiods and monthly water inflow requirements and resulting outflows computed from each different simulated scenario for the hydroperiod restoration. See Table 3.7 for simulation criteria and plant/open water covers

impounded system, annual outflows must be allowed in order to reduce sediment and nutrient accumulations (see Chapter 5). Therefore, summer dry-out must be considered as in scenarios 4, 5 and 6 (Fig. 3.19). In emergency situations similar to those in 1994–1996 and 2006–2009, which TDNP is suffering at the present, most conservative scenarios (12 and 13) at least must ensure inundation for the greatest part of the year (Fig. 3.19).

3.9 Future Scenarios of TDNP Hydrology

Independently of how climate change will affect regional water resources in the Upper Guadiana basin, considering that water resources are already strongly depleted, any additional water stress derived from climate change, such as a variability increase, will only intensify the competition for water resources.

Since groundwater inflows were important for TDNP hydrology, any sustained hydroperiod restoration depends on aquifer recovery. Cruces et al. (1997) estimated that groundwater table recovery to levels before 1970s requires a water surplus (recharge higher than abstraction) around 250–350 × 10^6 m^3 $year^{-1}$ during the next 20 years, in order to compensate for the 6,000 × 10^6 m^3 $year^{-1}$ of water that were withdrawn from the aquifer. Modelling land use and climate change effects, Mulligan and Burke (1999) reported that groundwater recharge is strongly controlled by land use and that there was no clear impact from climate changes on the aquifer recharge. The model performed by these authors also demonstrated the important effect of temperature, which reduces groundwater recharge due to enhanced evapotranspiration. There is clearly a complex land use-dependent pattern of climate change impacts on recharge. Therefore, independently of climate change scenarios, the short-term recovery of the 04.04 Aquifer would need a change from crops with high water requirements, such as sugar beet, alfalfa or corn, to lower-consumption crops such as grapes or sunflower, which could increase surface waters in the basin by around 8% (Conan et al. 2003). Also, improving irrigation operations by harmonizing water resource exploitation could be another way to reduce groundwater overexploitation (Martín de Santa Olalla Mañas et al. 1999).

Scenarios of climate change predictions applied to the upper Guadiana River basin hydrology are, so far, very highly uncertain. While some models based on the interpolation from the General Circulation Model grid cell output predict no major precipitation changes except a slight increase in autumn/winter precipitation (Wigley 1992; Cubasch et al. 1997), other approach, based on the effects of the future increases on atmospheric CO_2, predict a significant rainfall raise in winter and in early autumn, when concentration doubles the contemporaneous values (Bouraoui et al. 1998). More recently, Kilsby et al. (2007), by means of a regional climate model with bias correction and a circulation-pattern model based on rainfall stations over the Guadiana basin, predicted important reductions in the mean annual rainfall (30.5% and 15.1%, respectively) for the period 2070–2100 as compared to the present values. Likewise, evapotranspiration would increase annually across the region by 40% or more, although the large increases in summer will have little effect on actual evapotranspiration, as available soil moisture is low at that time of year (Kilsby et al. 2007).

Under these contradictory scenarios, surface flow predictions also disagree between studies. While the first model predicted no significant changes on groundwater recharge, and hence on river surface flows (Mulligan and Burke 1999), the second approach predicted 200% stream-flow increases Conan et al. 2003). The latest models, however, estimated significant reductions in stream-flows through the year (26–21% for the Guadiana basin; Kilsby et al. 2007).

Some authors (e.g. Mulligan and Burke 1999) sustain that, between 40 and 400 years ago, climate variability in Central Spain exerted a stronger control on water resource availability in the basin than those derived from climate change or irrigation practices. Contrarily, all the literature on the Upper Guadiana basin hydrology discerned that future stream flows and groundwater flows will depend on the land use management implemented by policy makers. Surface flows are almost negligible nowadays in the Upper Guadiana basin, and any future rainfall reduction will result in their further reduction, as well as that of groundwater recharge. If current groundwater extractions exceed the groundwater recharge, and no stream flow occurs except in infrequent humid years, future predictions indicate that available water at the basin must be rationed in order to ensure, not only water discharges to TDNP, but also to provide enough water for human consumption. Recovery and long-term sustainability of those important ecological hotspots represented by La Mancha Húmeda wetlands and, particularly by TDNP, depend on a significant improvement of the water management strategies to be implemented. Because there is still a great deal of uncertainty on the hydrological dynamics at both watershed and wetland scales, scientific research must support the accuracy of future projected water policies. Besides, any water policy must be undertaken at the whole basin. Water sustainability can only be achieved if water users, ecological requirements, and water availability are integrated in the management.

References

Acreman M, Almagro J, Alvarez J, Bouraoui F, Bradford R, Bromley J, Croke B, Crooks S, Cruces J, Dolz J, Dunbar M, Estrela T, Fernandez-Carrasco P, Fornes J, Gustard G, Haverkamp R, Hernández-Mora N, Llamas R, Martinez CL, Papamasorakis J, Ragab R, Sánchez M, Vardavas I, Webb T (2000) Groundwater and River Resources Programme on a European Scale (GRAPES), Technical Report to the European Union ENV4 – CT95-0186. Institute of Hydrology, Wallingford

Alley WM (1984) The Palmer drought severity index: limitations and assumptions. J Clim Appl Meteorol 23:1100–1109

Almarza C, Chazarra A, Pedraza B (1999) Adaptación del SPI para el análisis de la variabilidad intra-anual de periodos secos. In: Raso JM, Martín-Vide J (eds) La climatología española en los albores del siglo XXI. Oikos-Tau, Barcelona

Álvarez-Cobelas M, Verdugo M (1996) Clima. In: Álvarez-Cobelas M, Cirujano S (eds) Las Tablas de Daimiel: Ecología Acuática y sociedad. Ministerio de Medio Ambiente, Madrid

Álvarez-Cobelas M, Cirujano S, Sánchez-Carrillo S (2001) Hydrological and botanical man-made changes in the Spanish wetland of Las Tablas de Daimiel. Biol Conserv 97:89–97
Álvarez-Cobelas M, Sánchez-Carrillo S, Cirujano S, Angeler DG (2007) Long-term changes in spatial patterns of emergent vegetation in a Mediterranean floodplain: natural versus anthropogenic constraints. Plant Ecol 194:257–271
Amezaga JM, Santamaría L (2000) Wetland connectedness and policy fragmentation: steps towards a sustainable European wetland policy. Phys Chem Earth Pt B 25:635–640
Bayley PB (1995) Understanding large river-floodplain ecosystems. Bioscience 45:153–158
Bergkamp G, Orlando B (1999) Wetlands and climate change. The Ramsar convention on wetlands (http://www.ramsar.org/key_unfccc_bkgd.htm#1)
Bond NR, Lake PS, Arthington AH (2008) The impacts of drought on freshwater ecosystems: an Australian perspective. Hydrobiologia 600:3–16
Bouraoui F, Vachaud G, Chen T (1998) Prediction of the effect of climatic changes and land use management on water resources. Phys Chem Earth Pt B 23:379–384
Bradley C (1997) The hydrological basis for conservation of floodplain wetlands: implications of work at Narborouth, UK. Aquat Conserv Mar Freshw Ecosyst 7:41–62
Bradley C (2002) Simulation of the annual water table dynamics of a floodplain wetland, Narborough Bog, UK. J Hydrol 261:150–172
Brown AG, Keough MK, Rice RJ (1994) Floodplain evolution in the East Midlands, United Kingdom: the lateglacial and Flandrian alluvial record from the Soar and Nene valleys. Philos Trans R Soc A 348:261–293
Burba GG, Verma SB, Kim J (1999) A comparative study of surface energy fluxes of three communities (Phragmites australis, Scirpus acutus, and open water) in a prairie wetland ecosystem. Wetlands 19:451–457
Castaño-Castaño S, Martínez-Santos P, Martínez-Alfaro PE (2008) Evaluating infiltration losses in a Mediterraneal wetland: Las Tablas de Daimiel National Park, Spain. Hydrol Process. doi:10.1002/hyp. 7214
Collatz GJ, Ball JT, Grivet C, Berry JA (1991) Physiological and environmental regulation of stomatal conductance, photosynthesis and transpiration: a model that includes a laminar boundary layer. Agric Forest Metrol 54:107–136
Conan C, de Marsily G, Bouraoui F, Bidoglio G (2003) A long-term hydrological modelling of the Upper Guadiana river basin (Spain). Phys Chem Earth Pt B 28:193–200
Conly FM, Van der Kamp G (2001) Monitoring the hydrology of Canadian prairie wetlands to detect the effects of climate change and land use changes. Environ Monit Asess 67:195–215
Conte M (1993) Regional scale short term climatology. In: Medalus I, Programme on Climate and Natural Hazards – Final Report. EEC, London
Cruces J, Casado ME, Llamas MR, de la Hera A, Martínez-Cortina L (1997) El desarrollo sostenible de la Cuenca Alta del Río Guadiana: aspectos hidrológicos. Revista Obras Públicas 3362:7–18
Cubasch U, Caneill J-Y, Filiberti MA, Hegerl G, Johns TC, Keen A, Parey S, Thual O, Ulbrich U, Voss R, Waskewitz J, Wild M, van Ypersele JP (1997) Anthropogenic climate change, Project No. EV5V-CT9200123 DG. Final report. Office for official publication of the European Communities, Luxembourg
de Santa M, Olalla Mañas F, Ramos A, Cortés C, González DC, Córcoles H (1999) Improvement of irrigation management towards the sustainable use of groundwater in Castilla-La Mancha, Spain. Agric Water Manage 40:195–205
Esteller MV, Díaz-Delgado C (2002) Environmental effects of aquifer overexploitation: a case of study in the highlands of Mexico. Environ Manag 29:266–278
Fornés J, Rodríguez JA, Hernández N, Llamas MR (2000) Possible solutions to avoid conflicts between water resources development and wetland conservation in the La Mancha Húmeda biosphere reserve (Spain). Phys Chem Earth Pt B 25:623–627
García-Rodríguez M (1996) Hidrogeologia de Las Tablas de Daimiel y de los Ojos del Guadiana. Bases hidrogeológicas para una clasificación funcional de los humedales rivereños (Hydrogeology of Las Tablas de Daimiel and Ojos del Guadiana spring-water. Hydrogeological

basis for a functional classification of riparian wetlands). Ph.D. thesis, Geological Sciences, University Complutense of Madrid
Gilman K (1994) Hydrology and wetland conservation. Wiley, Chichester
Grubaugh JW, Anderson RV (1988) Spatial and temporal availability of floodplain habitat: long-term changes at Pool 19, Mississippi River. Am Midl Nat 119:402–411
Hanson PJ, Weltzin JF (2000) Drought disturbance from climate change: response of United States forests. Sci Total Environ 262:205–220
Hofmann H, Lorke A, Peeters F (2008) Temporal scales of water-level fluctuations in lakes and their ecological implications. Hydrobiologia 613:85–96
Hyams D (2001) Curve expert version 1.37, a curve fitting system for windows, Hixson
Idso SB, Anderson MG (1988) A comparison of two recent studies of transpirational water loss from emergent aquatic macrophytes. Aquat Bot 31:191–195
Iglesias A, Estrela T, Gallart F (2005) Impactos sobre los recursos hídricos (Impact on water resources). In: Moreno JM (ed) Evaluación preliminar de los impactos en España por efecto del cambio climático (Preliminary assessment of impacts related to climate change in Spain). Ministerio de Medio Ambiente-Universidad de Castilla-La Mancha, Madrid
IGME (2004) Informe sobre evolución piezométrica de la Unidad hidrogeólogica 04.04, Mancha occidental, Año 2004. Instituto Geológico y Minero de España, Ministerio de Ciencia e Innovación, Madrid
IGME (2008) Informe sobre evolución piezométrica de la Unidad hidrogeólogica 04.04, Mancha occidental, Año 2008. Instituto Geológico y Minero de España, Ministerio de Ciencia e Innovación, Madrid
Junk WJ (1982) Amazonian floodplains: Their ecology, present and potential use. In: Proceedings of the international scientific workshop on ecosystem dynamics in freshwater wetlands and shallow water bodies, Scientific Committee on Problems of the Environment (SCOPE), United Nations Environment Program (UNEP), New York
Junk WJ (1997) Structure and function of the large central Amazonian River floodplains: synthesis and discussion. In: Junk WJ (ed) The Central Amazonian floodplain. Springer-Verlag, Berlin
Junk WJ, Howard-Williams C (1984) Ecology of aquatic macrophytes in Amazonia. In: Sioli H (ed) The Amazon: limnology and landscape ecology of a mighty tropical river and its basin. Dr W Junk Publishers, Dordrecht
Junk WJ, Bayley PB, Sparks RE (1989) The flood pulse concept in river floodplain systems. In: Dodge DP (ed) Proceedings of the international large river symposium. Canadian Special Publication of Fisheries and Aquatic Sciences, Otawa
Karl TR, Trenberth KE (2003) Modern global climate change. Science 302:1719–1723
Kilsby CG, Tellier SS, Fowler HJ, Howels TR (2007) Hydrological impacts of climate change on the Tejo and Guadiana Rivers. Hydrol Earth Syst Sci 11:1175–1189
Koch MS, Rawlik PS (1993) Transpiration and stomatal conductance of two wetland macrophytes (Cladium jamaicense and Typha domingensis) in the subtropical Everglades. Am J Bot 80:1146–1154
Krause S, Bronstert A (2004) Approximation of groundwater – surface water – interactions in a mesoscale lowland river catchment. Hydrology: science and practice for the 21st century, vol 2. British Hydrological Society, pp 408–415
Krause S, Bronstert A (2005) An advanced approach for catchment delineation and water balance modelling within wetlands and floodplains. Adv Geosci 5:1–5
Lake PS (2003) Ecological effects of perturbation by drought in flowing waters. Freshwater Biol 48:1161–1172
Le Houérou HN (1996) Climate change, drought and desertification. J Arid Environ 34:133–185
Linacre ET (1976) Swamp. In: Monteith JL (ed) Vegetation and the atmosphere vol II. Academic, London
Llamas MR (1988) Conflicts between wetland conservation and groundwater exploitation: two case histories in Spain. Environ Geol Water Sci 11:241–51

López-Camacho B, de Bustamante I, Valdeolmillos A (1996) Hidrologia. In: Álvarez-Cobelas M, Cirujano S (eds) Las Tablas de Daimiel: Ecología Acuática y sociedad. Ministerio de Medio Ambiente, Madrid

MCJr M, Dahm CN, Crocker MT (1992) Climatic variability and streams and rivers in semi-arid regions. In: Robarts RD, Bothwell ML (eds) Aquatic ecosystems in semi-arid regions: implications for resource management. NHRI Symposium series. Environment Canada, Saskatoon

McKee TBN, Doesken J, Kleist J (1993) The relationship of drought frequency and duration to times scales. Eight Conference on Applied Climatology, American Meteorological Society, Anaheim

Middleton BA (1999) Succcession and herbivory in monsoonal wetlands. Wetl Ecol Manag 6:189–202

Middleton BA (2002) The flood pulse concept in wetland restoration. In: Middleton BA (ed) Flood pulsing in wetland: restoring the natural hydrological balance. Wiley, New York

Mitsch WJ, Gosselink JG (eds) (2000) Wetlands, 3rd edn. Wiley, New York

Mulligan M, Burke S (1999) Modelling the future of groundwater resources in central Spain. Manag Environ Qual 10:41–51

Naselli-Flores L, Barone R (2005) Water-level fluctuations in Mediterranean reservoirs: setting a dewatering threshold as a management tool to improve water quality. Hydrobiologia 548:85–99

Oñate JJ (1993) Caracterización de los cambios climáticos en la Península Ibérica y Canarias desde principios de siglo. Universidad Autónoma de Madrid, Madrid

Oñate JJ, Pou A (1996) Temperature variations in Spain since 1901: a preliminary analysis. Int J Climatol 16:805–816

Parkhurst RS, Winter TC, Rosenberry DO, Sturrock AM (1998) Evaporation from a small prairie wetland in the Cottonwood Lake area, North Dakota – an energy budget study. Wetlands 18:272–287

Pérez-González ME, Sanz-Donaire JJ (1998) Clima y microclima de La Mancha Húmeda. Anales Geografía Universidad Complutense 18:239–256

Poiani KA, Johnson WC (1993) A spatial simulation model of hydrology and vegetation dynamics in semi-permanent prairie wetlands. Ecol Appl 3:279–293

Pribán K, Ondok JP (1985) Heat balance components and evapotranspiration from a sedge-grass marsh. Folia Geobot Phytotaxon 20:41–56

Price JS (1994) Evapotranspiration from a lakeshore Typha marsh on Lake Ontario. Aquat Bot 48:261–272

Rind D, Perlwitz J, Lonergan P (2005) AO/NAO response to climate change: 1. Respective influences of stratospheric and tropospheric climate changes. J Geophys Res 110:D12107. doi:10.1029/2004JD005103

Rosenberry DO, Winter TC (1997) Dynamics of water-table fluctuations in an upland between two praire-pothole wetlands in North Dakota. J Hydrol 191:266–289

Sánchez-Carrillo S (2000) Hidrología y sedimentación actual en Las Tablas de Daimiel (Hydrology and current sedimentation in Las Tablas de Daimiel). Ph.D. Thesis, Geological Sciences, Universidad Autónoma de Madrid, Madrid

Sánchez-Carrillo S, Álvarez-Cobelas M, Benítez M, Angeler DG (2001) A simple method for estimating water loss by transpiration in wetlands. Hydrol Sci J 46:537–552

Sánchez-Carrillo S, Angeler DG, Sánchez-Andrés R, Álvarez-Cobelas M, Garatuza-Payán J (2004) Evapotranspiration in semi-arid wetlands: relationships between inundation and the macrophyte-cover:open-water ratio. Adv Water Resour 27:643–655

Sophocleous M (2002) Interactions between groundwater and surface water: the state of the science. Hydrogeol J 10:52–67

Souch C, Grimmond CSB, Wolfe CP (1998) Evapotranspiration rates from wetlands with different disturbance histories: Indiana Dunes National Lakeshore. Wetlands 18:216–229

Sparks RE, Bayley PB, Kohler SL, Osborne LL (1990) Disturbance and recovery of large floodplain rivers. Environ Manage 14:699–709

Swanson SK, Bahr JM, Bradbury KR, Anderson KM (2006) Evidence for preferential flow through sandstone aquifers in Southern Wisconsin. Sediment Geol 184:331–342
Trigo RM, Palutikov JP (2001) Precipitation scenarios over Iberia: a comparison between direct GCM output and different downscaling techniques. J Climate 14:4422–4446
Trigo RM, Pozo-Vázquez D, Osborn TJ, Castro-Díez Y, Gámiz-Fortis S, Esteban-Parra MJ (2004) North Atlantic oscillation influence on precipitation, river flow and water resources in the Iberian peninsula. Int J Climatol 24:925–944
Tuschl P (1970) Die Transpiration von Phragmites communis Trin. im geschlossenen Bestand des Neusiedler Sees (Transpiration of P. communis in a closed stand of Lake Neusiedlersee, in German). Wiss Arb Burgenld 44:126–186
United Nations Environment Programme (1992) World atlas of desertification. Edward Arnold, London
Usmanova RM (2003) Aral Sea and sustainable development. Water Sci Technol 47:41–47
Van del Valk AG (1981) Succession in wetlands: a Gleasonian approach. Ecology 62:688–696
Van del Valk AG (2005) Water-level fluctuations in North American prairie wetlands. Hydrobiologia 539:171–188
Van del Valk AG, Squires L, Welling CH (1994) Assessing the impacts of an increase in water level on wetland vegetation. Ecol Appl 4:525–34
Vicente-Serrano SM, Cuadrat JM (2007) North Atlantic oscillation control of droughts in north-east Spain: evaluation since 1600 A.D. Climatic Change 85:357–379
Weller MW (1989) Waterfowl management techniques for wetland enhancement, restoration and creation useful in mitigation procedures. In: Kusler JA, Kentula ME (eds) Wetland creation and restoration, the status of the science. Island Press, Washington, DC
Wigley TML (1992) Future climate of the Mediterranean Basin with particular emphasis on changes in precipitation. In: Jeftic L, Milliman JD, Sestini G (eds) Climate change and the Mediterranean. UNEP-Routledge, New York
Wilhite DA (2000) Drought as a natural hazard: concepts and definitions. In: Whilhite DA (ed) Drought: a global assessment, vol 1. London, Routledge
Winter TC (1988) A conceptual framework for assessing cumulative impacts on the hydrology of nontidal wetlands. Environ Manage 12:605–620
Winter TC (1999) Relation of streams, lakes and wetlands to groundwater flow systems. Hydrogeol J 7:28–45
Winter TC, Llamas MR (1993) Introduction to the 28th International Geological Congress Symposium on the Hydrogeology of Wetlands. J Hydrol 141:1–3
Winter TC, Woo M-K (1990) Hydrology of lakes and wetlands. In: Wolman MG, Riggs HC (eds) Surface water hydrology, the geology of North America, vol O-1. Geological Society of America, Boulder, CO
Woo M-K, Rowsell RD (1993) Hydrology of a praire slough. J Hydrol 146:175–207
Zedler JB (2000) Progress in wetland restoration ecology. Tree 15:402–407

Chapter 4
The Effects of Anthropogenic Stressors on Wetland Loss and Habitat Quality Deterioration in the Upper Guadiana River Basin: A Long-Term Assessment (1970–2000)

R. Sánchez-Andrés, M.O. Viedma, and S. Sánchez-Carrillo

Abstract During the last four decades, around 1,500 km^2 of dry croplands have been transformed, and are now irrigated in the Upper Guadiana river basin, causing hydrologic overexploitation and wetland desiccation. However, there are no estimations on how anthropogenic stressors have been changing the wetland landscape in the recent past. This chapter focuses on the understanding of how the changes on land-use land-cover (LULC), economic activities and population have driven wetland losses and habitat degradation in the basin from the 1970s. Our results show that 40.5% (2,041.6 ha) of the 5,321 ha of wetlands existing in the early 1970s had disappeared in the last 30 years (1970–2000). Most wetland losses occurred through the period 1978–1990, which registered a rate 127 ha of wetland lost per year. Most affected were floodplain wetlands (47% of total loss) and rain-fed temporary ponds (24%). During the entire period 1978–1999, the loss of wetlands could be significantly related to the loss of natural vegetation, as well as to the reduction of agricultural employment. Habitat quality of wetlands showed a clear pattern of nutrient over-enrichment, as well as a trend towards salinization, the later related to the greater disappearance of most freshwater wetlands (0–2,500 $\mu S\ cm^{-1}$). LULC, economic activities and demography explained around 50% of wetland loss and habitat quality deterioration. Until 1990, the pressure of population growth, combined with the agricultural sector, explained the disappearance of most wetland area. From then on, habitat quality has been more impacted in areas where industry and building sectors had more weight in the socioeconomic development (also densely populated watersheds).

R. Sánchez-Andrés (✉)
Spanish National Research Council (CSIC), Royal Botanic Garden, Pza. Murillo 2, E-28014 Madrid, Spain
e-mail: rsanchez@rjb.csic.es

S. Sánchez-Carrillo and D.G. Angeler (eds.), *Ecology of Threatened Semi-Arid Wetlands: Long-Term Research in Las Tablas de Daimiel*, DOI 10.1007/978-90-481-9181-9_4,

4.1 Introduction

Human transformation of the Earth's land surface has far-reaching and results in serious consequences for the functioning of hydrological and biogeochemical processes in watersheds (Calder 1998). Changes in vegetation cover and land uses at watershed-scale have been cited to significantly alter downstream hydrological fluxes (DeFries and Eshleman 2004) and water quality (Berka et al. 2001; Houlahan and Findlay 2004), modifying aquatic habitats (Wang et al. 2008; Allan et al. 1997). In fact, land use–land cover changes (LULC hereafter) have been considered a major cause of the global environmental change (Hollis 1992; Riebsame et al. 1994).

It is recognized that historical land use and land management by humans is the first cause of land-cover change around the world (Dale et al. 2000), with agricultural expansion as the principal agent of change (Lambin et al. 2001). Human alteration of the land affects aquatic ecosystems through multiple processes that likely operate from regional to local spatial scales (Allan et al. 1997). Land cover and landscape configuration and evolution respond to multiple human and natural factors (e.g. climate oscillations and economic markets) operating from global to local scales but, also, since LULC patterns are indeed consequences of socioeconomic systems, they are highly dynamic and rarely in a stable equilibrium (Niehoff et al. 2002). Wetland habitats within a watershed respond then, to a greater or lesser degree, to each of these socioeconomic changes, and it is difficult to discriminate between the main drivers of degradation. A clear implication of this perspective is that local conditions are under some degree of regional influence and hard to estimate (Hildrew and Giller 1994). Patterns of LULC and socioeconomic variables are complex, often complicating the understanding of the relationship between those anthropogenic stressors and wetland habitat degradation (Houlahan and Findlay 2004). It includes the scale effects of anthropogenic stressors on wetland degradation. For example, we know that the effects of adjacent land-use on water quality can extend over comparatively large distances (Houlahan and Findlay 2004; Angeler et al. 2008) but, in the practice, it is very difficult to establish. Geographic Information Systems combined with multivariate statistical methods have been shown as useful tools for the assessment of what anthropogenic stressors, such as LULC, population or economy, bear an influence on wetland loss and degradation (Houlahan et al. 2006).

The Upper Guadiana basin is an excellent natural laboratory to evaluate the effects of human-related activities on wetland degradation patterns. The large-scale decline in groundwater levels has been related to the strong increase of the irrigation surface at the basin, as well as to changes in cultivated crop types (high water requirements; Llamas 1988). During the nineteenth century, it has led to the disappearance of half of the existing wetlands in the area (see Chapter 1). These major changes in LULC experienced by the basin were driven by changing social and political circumstances, fluctuating crop prices and various regional, national and international incentive programs (Mulligan and Burke 1999, see Chapter 10). In addition to the increase in irrigation for the production of high-value crops, set-aside policies and rural–urban migration have also resulted in large scale abandonment of previously cultivated land (Molinillo et al. 1997) which slowly reverts to a semi-natural shrub community, held in check by grazing and fire. All accompanied with an increase in the use of agrochemicals, a raise of the second and third economic sectors, large discharges of poorly treated sewages, etc. All these impacted on wetlands through a dynamic and complex mosaic of LULCs, acting at different spatial and temporal scales. However, we do not know how it occurred and how these anthropogenic stressors have performed on the current wetland landscape.

These issues motivated this study, whose main goal focuses on understanding what main anthropogenic stressors (LULC, economic activities and population growth) have driven wetland losses and habitat degradation in the Upper Guadiana basin from 1970s to 2000, spanning 30 years. For this purpose, we first investigated the loss of wetlands in the basin, as well as habitat quality deterioration for this period, grouping by hydrodynamic types (i.e. water source) and using a multi-scale approach (watershed and local) and multivariate statistical methods.

4.2 Material and Methods

Besides the Upper Guadiana basin, the adjacent Calatrava catchment (Fig. 4.1a) was also included for the analyses in order to incorporate wetlands with contrasting physical and biological functioning (wetlands developed over volcanic rocks) but with similar anthropogenic stressors. This allows us to evaluate the effects of stressors under different physical and biological settings. Since the entire Upper Guadiana basin extends for ca. 2 million hectares, occupying 185 watersheds, and many of them are of no interest in the absence of wetlands of a certain entity (>0.5 ha), the analyses were therefore applied to those watersheds that met the following criteria: (1) those containing large wetlands of high ecological value (e.g. Las Tablas de Daimiel National Park, TDNP hereafter) and (2) those containing temporary ponds of small-medium size (e.g. Ruidera, Calatrava, Riansares and Las Animas). Therefore, five watersheds, subdivided into 44 sub-watersheds (≈40% of total area), containing 94 wetlands (up to 95% of total wetland area), were extracted to be analyzed in this study (Table 4.1).

Because there is no reliable estimate of the wetland extent in the Upper Guadiana basin before the 1990s, we first tried to obtain the wetland distribution before this date, as a starting point for the analyses of wetland changes. However, because wetland losses by desiccation were strongest for the period 1950–1970, the starting point of the wetland extension in the basin had to be restricted to the early 1970s, since bibliographic sources could only corroborate the existence of the ancient wetland during this date. Therefore, wetland extent in the early 1970s was obtained

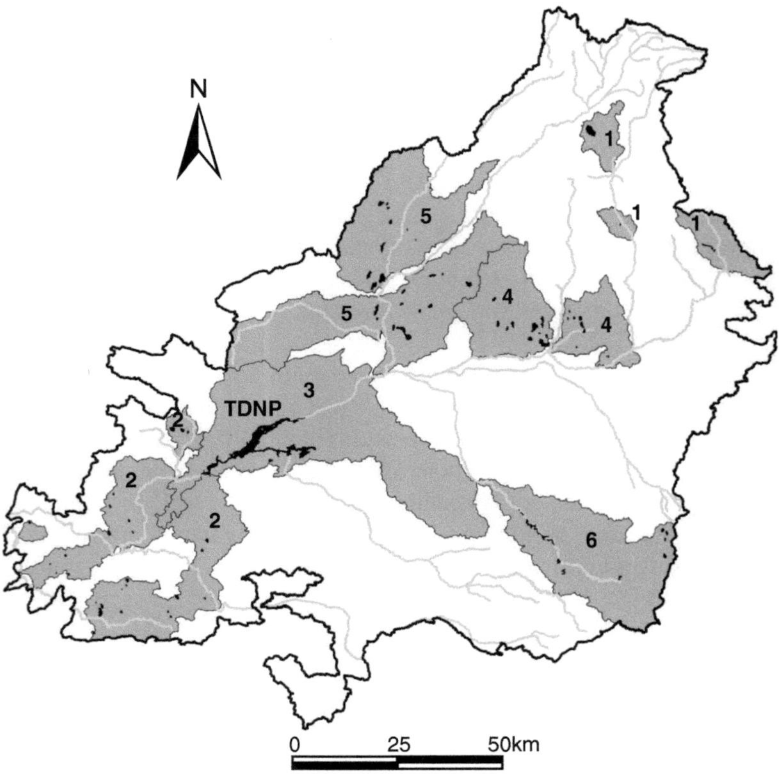

Fig. 4.1 Distribution of wetlands in the Upper Guadiana basin in the early 1970s. Grey areas represent the studied watersheds (1: Blanco, 2: Calatrava, 3: Daimiel, 4: Las Animas, 5: Riansares, and 6: Ruidera)

Table 4.1 Watersheds of the Upper Guadiana basin and associated wetlands considered in this study during early 1970s

Selected watersheds	Watershed area (ha)	Number of water bodies	Total wetland area (ha)	Mean wetland area (ha)	Major and minor wetlands (ha)
Blanco	34,781	4	400.1	100.0 ± 166.1	Capellanes (348) Hito (4)
Calatrava	127,034	27	910.9	33.7 ± 26.6	Nava Grande (106) Dehesa (4)
Daimiel	65,402	3	1,532.0	495.0 ± 582.2	TDNP (1,600) Escoplillo (47)
Las Animas	77,370	19	815.0	42.9 ± 39.1	Manjavacas (139) Hoya Honda (3)
Riansares	176,116	20	1,254.5	62.7 ± 56.0	Taray (231) Salobral (5)
Ruidera	73,439	21	404.5	19.3 ± 22.0	Colgada (103) Redondilla (4)
Total	554,142	94	5,320.8	56.6 ± 157.1	

using a combination of the land use cover, from the 1978 National Land Use/Land Cover map – Spanish Ministry of Agriculture – through identifying the existing unproductive areas, the digital elevation model (DEM, identifying lowland areas), the lithological map (identifying peat deposits or riverbed sediments), the river network (identifying those floodplain wetlands) as well as the topographic maps elaborated during the 1950s (http://www.ign.es/ign/es/IGN/ cartoteca_MapTopo.jsp). Subsequent estimates of wetland area in the study area were computed from the CORINE database for 1990 and 2000 (EEA 2005), corrected by the DEM, lithologic information, bibliographic sources and recent aerial photograph surveys provided by the Castilla-La Mancha regional government in the SIGPAC application (http://pagina.jccm.es/agricul/sigpac.htm). DEM was provided by the Spanish National Centre of Geographical Information (CNIG) at 25 m of resolution, whereas lithological data were taken from digital geologic maps (Magna series 1:50,000) from the Spanish Geological Survey (IGME). LULC were also obtained using the 1978 National Land Use/Land Cover map, and the revised CORINE for 1990 and 2000. Four broad LULC classes were defined, according to the level of naturalness: natural vegetation (comprised by conifers, deciduous forests, holm oaks, scrubland and grasslands), dry croplands (mainly wheat), irrigated croplands (sunflower, corn, and other vegetables, as well as olives and vineyards), and dry-irrigated croplands (a mixture of dry and irrigated croplands). LULC maps at sub-watershed scale were created through establishing synthetic classes defined by dominant land covers.

To test whether the loss or degradation is linked to a particular wetland type, they were hydro-dynamically grouped according their main water sources: rainfall, surface water (river flow) and groundwater, following the hydrogeomorphic classification criteria from Brinson (1993). Lithological information, river network, terrain elevation and groundwater aquifers were combined into a GIS to obtain the following criteria: wetlands fed exclusively by rainfall – rain fed – (relied on impervious substrates – clay, gypsum, metamorphic and volcanic rocks – and away from rivers), those fed mainly by groundwater – groundwater-fed – (relying on permeable substrates – conglomerate, limestone, gravel and sand – leaning on aquifers and located in lowland areas), those fed jointly by rain and groundwater – rain and groundwater fed – (relying on semi-permeable rocks – marl, pyroclastic rock, phreatomagmatic rock and silt) and, finally, those riverside wetlands – floodplains – (connected to the river network), subdividing those floodplain wetlands fed exclusively by river waters –floodplain (surface water) – (relying on impervious rocks) of those receiving contributions from surface and groundwater sources – floodplain (surface and groundwater) – (relying on aquifer substrates).

The degree of wetland habitat degradation by anthropogenic pressure was estimated through water conductivity and phosphorus load. Conductivity was used as an indicator of changes on wetland water sources (e.g. reduction of any water source by prolonged droughts or aquifer overexploitation), while phosphorus was used as a measure of wetland eutrophication, by both point and non-point pollution sources. The data used proceeded from own sources and from Armengol et al. (1975),

Cirujano (1980a, b, c), Velayos et al. (1984 and 1989), Cirujano (1990), Cirujano et al. (1990), MOPU (1990), Cirujano et al. (1992, 1996), Vicente and Miracle (1998), Florin and Montes (1999), Álvarez-Cobelas et al. (2001), and Cirujano and Medina (2002) as well as those provided by the Guadiana Water Authority. Average values measured during the spring season were used for analyses, taking the year with more available data as the most representative for each decade (1970s, 1990s and 2000s). Anthropogenic stressors also included some socioeconomic indicators, grouped by sub-watersheds and municipalities (where the wetland is included), such as population changes and main economic activity (workforce engaged in agriculture, building, industry and service sectors). All information was obtained from the INE base of the National Statistical Institute (INE; http://www.ine.es/intercensal). All spatial information was processed using the ArcGIS 9.2 software package (ESRI 2006).

Pearson product-moment correlation was used at sub-watershed scale to evaluate the effects of changes in LULC, population and economic activity (independent variables) on the wetland loss rate per periods (1978–1990, 1990–2000 and 1978–2000). Changes were computed from differences between periods, taking positive values when the variable increased and negative when it decreased. The mechanistic patterns underlying wetland's habitat quality (water quality: phosphorus loading and conductivity) and wetland losses from 1970 to 2000 were determined using redundancy analysis (RDA). This method allows to determine the relative importance of local and regional conditions on the spatial-temporal variability of our dependent variables (habitat quality and wetland area). Partial RDA analyses allowed decomposing the inertia (variance) into various components attributed to different sets of explanatory variables (Ter Braak 1988). Local conditions were those of existing LULC from a buffer of 100 m around the wetlands, and socioeconomic variables were those represented by the municipalities intersected by the wetlands. Regional conditions were those of its sub-watershed by LULC and socioeconomic variables. Therefore, we decomposed variance attributable to explanatory variables as follows:

1. Local variation [L]: variation explained by local conditions that covary with regional variables.
2. Regional variation [R]: variation explained by regional conditions that covary with local variables.
3. Pure local variation [L|R]: the fraction of variation that can be explained by local conditions independently of any regional condition.
4. Pure regional variation [R|L]: the fraction of variation that can be explained by regional conditions independently of any local characteristic.
5. The variation explained by combined local and regional variables [L + R].
6. The shared/confounded variation explained simultaneously by local and regional conditions [L∩R] ([L∩R] = [L] – [L|R] = [R] – [R|L]).
7. The total percentage of unexplained variance (100 – [L + R]).
8. We run RDA models for water quality parameters and for wetland area independently. We explicitly considered all variables every year.

4.3 Wetland Cover and Extent of Hydrodynamic Types in the Early 1970s

The Upper Guadiana basin landscape consists of a spatially discontinuous complex of numerous small temporary ponds, occupying a very small area in the studied watersheds (<1%), with the exception of TDNP, which occupies ≈2% of (Fig. 4.1). Although the elevation gradient of catchments ranges 430–1,203 m, most wetlands are concentrated in lowland areas, between 600 and 700 m. Only the Blanco and Ruidera watersheds record wetland habitats at higher elevations (>800 m; Fig. 4.1).

Our estimates indicated that, in the early 1970s, the total wetland area in the studied watersheds extended over 5,321 ha, distributed into 94 water bodies (Table 4.1). On that date, the mean surface of each individual wetland reached 56.6 ha, although highly variable spatially (Table 4.1). For example, the region most prone to wetland development appeared linked to La Mancha Occidental aquifer (23 or 04.04, see Fig. 1.4), concentrating 40.3% of the total wetland area. Here, the mean surface of a water body reached 119.3 ha. By contrast, other zones such as those linked to Altomira aquifer represented only 0.8% of the total wetland area. Small wetlands appeared in the Campo de Montiel aquifer (≈20 ha). 39.7% of wetland area (37 water bodies) in the basin was not linked to any aquifer system and could be considered as temporary rain-fed wetlands.

Other previous approaches to the total wetland area in La Mancha Húmeda were performed before the contemporary degradation amounting 25,000 ha (see Chapter 1). In the early 1970s, however, the observed differences with our estimates are not contradictory, considering that our methodology allowed no return to the landscape situation prior to the generalized wetland desiccation in the region, which started from the 1950s. Most of the large floodplain wetlands associated to the Gigüela and Záncara rivers (see Fig. 1.2) were already lost in the early 1970s (e.g. Tablas de Arenas de San Juan, Tablas de Villarta de San Juan, Tablas de Villarrubia and Tablas del Záncara; Heras et al. 1971; Cirujano 1980b; Álvarez-Cobelas and Cirujano 1996). However, the extent of these wetlands before desiccation is not precisely known.

From the hydrodynamic point of view, in the early 1970s most wetlands in the Upper Guadiana basin were temporary water bodies, exclusively fed by rainwater (39.5%; Table 4.2). In contrast, 36% of wetlands developed over floodplains, most of them receiving water from both surface and groundwater sources (Table 4.2). Wetlands fed jointly by rain and ground waters accounted for 14.5% of the total wetland area in the basin (Table 4.2). In the early 1970s, finally, wetlands fed mainly by groundwater (relying on permeable aquifers) represented around 10% of the total wetland surface (Table 4.2). Spatially, most rain fed wetlands concentrated at the Riansares watershed (50%), whereas floodplain wetlands were restricted to the Daimiel and Ruidera watersheds (96%, Table 4.2). Floodplain wetlands fed exclusively by river flows only appeared in the Blanco watershed, representing only 3% of the total wetland area on its watershed (Table 4.2). Most wetlands fed by

Table 4.2 Area of wetlands (ha) according to their hydrodynamic functioning in the studied watersheds of the Upper Guadiana basin during the early 1970s

Watershed	Rainfed	Rain and groundwater fed	Floodplain (surface water)	Floodplain (surface and ground waters)	Ground water-fed	Total
Blanco	347.9	42.0	10.3	0	0	400.1
Calatrava	481.4	239.5	0	51.3	138.7	910.9
Daimiel	0	46.9	0	1,485.0	0	1,532.0
Las Animas	201.0	395.0	0	21.7	197.3	815.0
Riansares	1,054.4	4.7	0	0	195.4	1,254.5
Ruidera	14.1	42.8	0	347.5	0	404.5
Total	2,102.7	771.0	10.3	1,905.5	531.4	5,320.8

both rain and groundwater concentrated at the Calatrava and Las Animas watersheds (82%), while wetlands depending on groundwater as the main source appeared homogenously distributed along watersheds (Table 4.2).

4.4 Land Use–Land Cover Changes and Socioeconomic Indicators During 1978–2000

Dry cropland was the dominant land cover in the basin throughout the studied period (1978–2000; Table 4.3). In 1978, this LULC occupied 70% of the territory in question, reducing its extension to 55% in 1990, at the expense of dry-irrigated cropland, which grew by 15% (Fig. 4.2a). Throughout 2000, dry croplands increased to nearly their original extension, covering 63% of total land area (Fig. 4.2a). Despite the apparent human pressure on the Upper Guadiana basin, natural vegetation appeared as the second dominant LULC, occupying 21% in 1978, and even increasing slightly to 23 % during the period 1990–2000 (Fig. 4.2a). This LULC was dominant in the Ruidera watershed, covering 52–55% of its area (Table 4.3). Irrigated croplands appeared as the third LULC in the studied area, increasing from 27,352 ha in 1978 to 37,322 ha in 2000 (from 5% to 7% of total area; Fig. 4.2a). Urban areas occupied the smallest portion of watersheds (3%), increasing slightly (≈1%) during the entire studied period (Fig. 4.2a).

From 1978 to 2000, land devoted to agriculture increased in all watersheds, except in Calatrava (from 70% to 59%) and Ruidera (from 46% to 43%) where the area of natural vegetation increased. The most complex pattern appeared in irrigated lands, which behaved differently from watersheds throughout the study period. The area devoted to irrigation showed a diminishing trend in Calatrava (from 15,710 ha in 1978 to 8,749 ha in 2000) and Las Ánimas (from 2,676 ha in 1978 to 1,709 ha in 2000). The Riansares and Daimiel irrigation croplands, however, increased strongly (from 5,187 ha to 10,274 ha (twofold) and from 2,422 to 10,579 ha (fourfold), respectively, in the period 1978–2000). Although in the Blanco watershed the land devoted to irrigation only represented 2%, it increased

Table 4.3 Dominant land uses (% of total area), population (inhabitants × 1,000) and main sector of economic activities (>50% of occupancy) in the Upper Guadiana basin in each studied watershed, from 1978 to 2000. Watersheds: Bla: Blanco, Cal: Calatrava, Dai: Daimiel, Ani: Las Animas, Ria: Riansares, and Rui: Ruidera; land uses: Dry croplands, Natural: natural vegetation; main economic activities: Agri: agriculture, Serv: services

	Dominant land use			Population			Main activity		
	1978	1990	2000	1978	1990	2000	1978	1990	2000
Bla	Dry (83%)	Dry (84%)	Dry (84%)	7.98	7.10	6.02	Agri (59%)	Agri (51%)	Serv (46%)
Cal	Dry (57%)	Dry (42%)	Dry (44%)	153.17	159.21	164.81	Serv (37%)	Serv (42%)	Serv (58%)
Dai	Dry (50%)	Dry (54%)	Dry (53%)	41.77	42.02	42.59	Agri (46%)	Serv (32%)	Serv (42%)
Ani	Dry (87%)	Dry (71%)	Dry (86%)	28.39	28.19	28.08	Agri (46%)	Serv (34%)	Serv (43%)
Ria	Dry (82%)	Dry (60%)	Dry (82%)	74.35	75.34	75.65	Agri (43%)	Serv (30%)	Serv (39%)
Rui	Natural (52%)	Natural (55%)	Natural (55%)	15.29	14.55	13.14	Agri (52%)	Agri (31%)	Serv (39%)
Total	Dry (69%)	Dry (55%)	Dry (64%)	320.95	326.41	330.29	Agri (42%)	Serv (35%)	Services (47%)

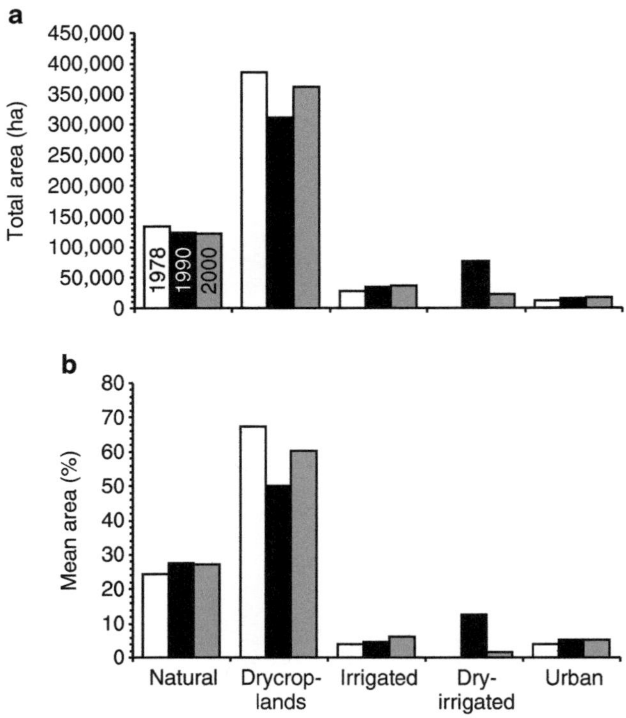

Fig. 4.2 Patterns of the main land use changes in the Upper Guadiana basin during 1978–2000. (**a**) Total area covered by the main land uses in the studied sub-catchments; (**b**) Mean percentage of occupancy of adjacent land uses around wetlands of the Upper Guadiana basin (buffer: 100 m)

twofold from 1978. The observed reduction of croplands in the Calatrava watershed could partially be explained by a tendency to a decline in agriculture, due to the application, from 1993, of the European Agro-Environmental Programs on water demands for irrigation (Programs 2078/2092 and 1257/1999). This Policy, conceived as a vehicle to recover the overexploited aquifer through farmer income compensation, was developed to reduce during 5 years the volume of water used for irrigated lands (Viladomiu and Rosell 1997). Despite compliance with the environmental policy, as has been demonstrated previously, LULC changes were not applied consistently throughout the studied watersheds. In fact, in the Daimiel watershed, which includes the core area of the overexploited Llanura Manchega aquifer, as well as the TDNP wetland, and where implementation of this measurement should have been more effective, there was no perceived reduction of the irrigated croplands.

At local wetland scale (wetland buffer of 100 m), land uses presented a similar pattern to that of its watersheds (Fig. 4.2b). Dry croplands appeared as the dominant land use in wetland surroundings, followed by natural vegetation (Fig. 4.2b). The main difference was the increased weight of the urban areas, occupying up to 5% of wetland surroundings. Anyway, overall the situation in 2000 was not very different than that in 1978: 68% vs. 71% of agricultural lands, and 27% vs. 24% of natural vegetation (Fig. 4.2). However, in some individual wetlands there was a great contrast between the patterns. In TDNP, for example, in 1978, 53% of the surrounding area was dominated by irrigation crops, whilst 42% exhibited natural vegetation; from 1990 on, 93% of the surrounding wetland was occupied by natural vegetation. In the Carrizosa wetland at the Calatrava watershed, dry-cropland occupying 100% of the wetland surroundings was changed to natural vegetation (100%) in 1990. Other wetlands experienced an increase of irrigation in the surroundings, such as the Laguna Cueva Morenilla at the Ruidera watershed (0–33%), at the expense of the natural vegetation (85–66%), or the Laguna Grande de Quero at Riansares watershed (0–37%), with dry-croplands (100–0%).

The studied watershed had a population of 320,949 in 1978, increasing by 9,339 inhabitants (2.9%) until 2000 (Table 4.3). Population density increased slightly from 57.9 to 59.6 inhabitants km^{-2} in the 1978–2000 period, below the Spanish mean (74.6–82.2 inhabitants km^{-2}, INE 2008; Fig. 4.3). Annual population growth rate was of 1.7% in the period 1978–1990, decreasing afterwards until 1.2%. Both rates were lower than those registered for these periods at national scale (3.15% and 4.19%, respectively; INE 2008). In the 1978–2000 period, the Calatrava, Daimiel and Riansares watersheds increased their population by 7.5%, 2.0% and 1.5% respectively, while Blanco, Ruidera and Animas lost population (26%, 14% and 1%, respectively; Table 4.3).

In 1978, the main economic activity in the study area was agriculture, employing 42% of the workforce, followed by the service sector (31%). This appeared as a common pattern throughout all watersheds (Fig. 4.3). The former sector suffered a serious setback, decreasing to 15% in 2000; this is compensated by employment in services (47% in 2000). This is consistent with the national trend, while the Upper Guadiana basin retained its greater importance of agriculture in the economy (the occupancy in the agriculture sector in Spain: from 22% to 7%, for the period 1978–2000; Cano-Soler et al. 2000). Employment in industry grew slightly until 18% in 1990 (14% in 1978), maintaining its occupancy at the same rate in 2000. Employment in the building sector, which was the main driver of the Spanish economy during the study period (from 6% in 1978 to 14% in 2005; INE 2008), grew here far above the national mean, from 13% to 22%. In the watersheds, agricultural employment remained as an important sector in Blanco only, whereas in the others areas it was relegated to the last place (Fig. 4.3). Highest growth in industrial employment was recorded at the Riansares watershed; Daimiel and Calatrava watersheds, however, showed a strong growth in building activity, which appeared as the second largest source of employment (Fig. 4.3).

4.5 Wetland Losses from 1970 to 2000

During the 1970–2000 period, 40.5% of the total wetland area disappeared from the studied territory. It showed a loss of 2,041.6 ha of wetlands in 30 years. Most wetland disappearances occurred throughout the period 1978–1990 (1,529.3 ha) but

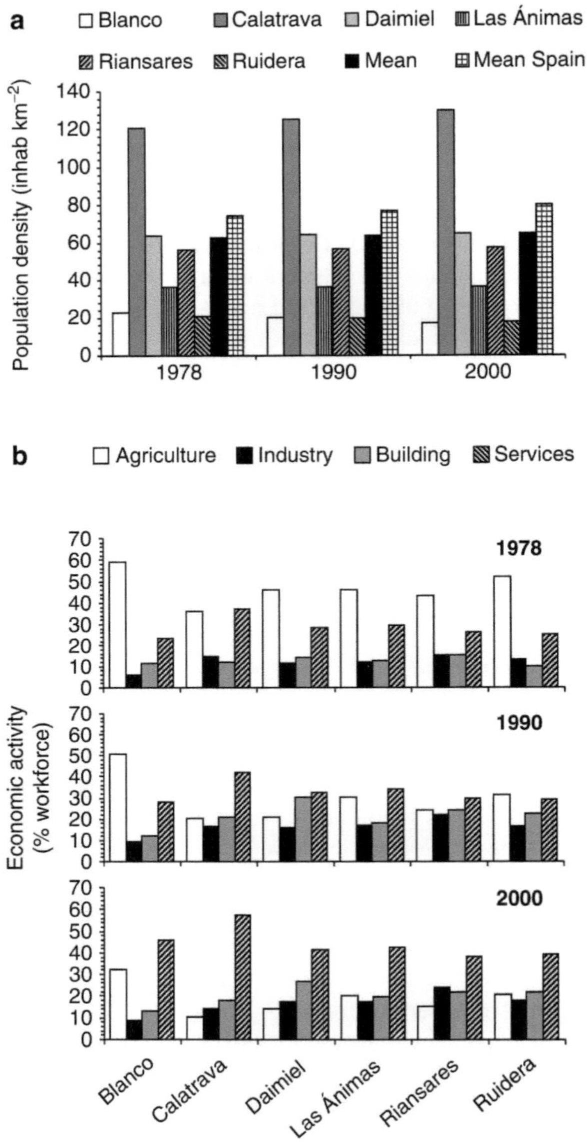

Fig. 4.3 Population growth (**a**) and economic activities (**b**) in the studied sub-catchments of the Upper Guadiana basin, from 1970 to 2000

this could not be comparatively attributed to any observed prolonged drought cycle associated to this period (see Chapter 3, Fig. 3.11a). From 1970 to 1978, 455.9 ha of wetlands were drained, and later reduced to 86.4 ha in the period 1990–2000. Annual rates of wetland losses peaked in the period 1978–1990, when 127 ha year^{-1} of wetlands were transformed, declining strongly from then on down, to 9 ha year^{-1}.

During the period 1970–1978, the wetland loss rate reached 57 ha year^{-1}. This temporal pattern coincides with other regions of North America, such as the Mississippi River delta plain (Bernier et al. 2006) rather than with northern Europe and the northern Mediterranean, where wetland losses have been ongoing for hundreds of years (Brinson and Malvárez 2002). Daimiel and Calatrava were the watersheds suffering the greatest wetland losses for the entire study period (63% and 60% of wetlands existing *ca.* 1970, respectively), most occurring in the period 1978–1990 (Fig. 4.4). On the other hand, the Blanco and Ruidera watersheds only showed wetland losses until 1978 (12% and 22% of total wetlands, respectively; Fig. 4.4). Riansares was the watershed that preserved more wetlands, losing 10% of its original area as it existed in the early 1970s (Fig. 4.4).

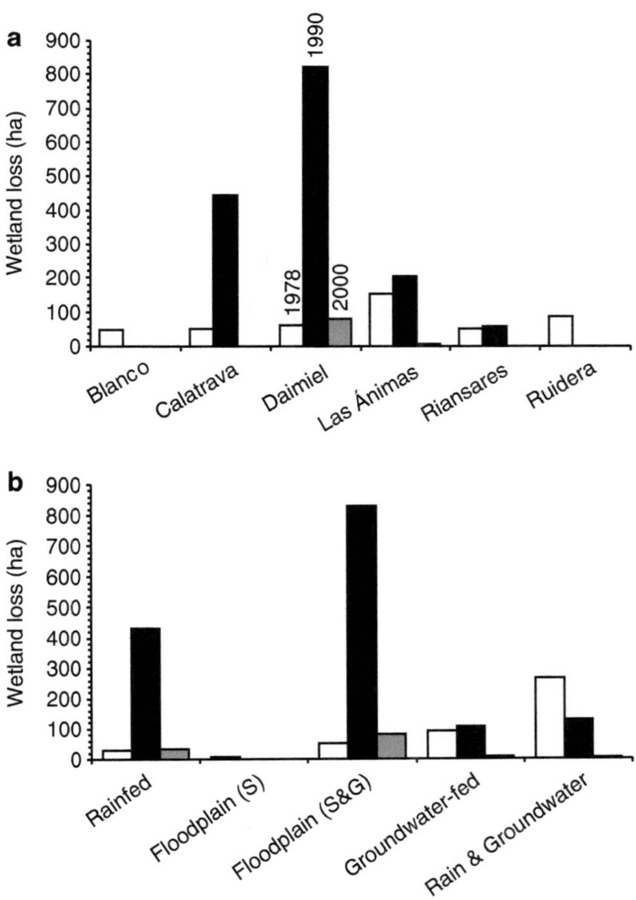

Fig. 4.4 Wetland losses in the Upper Guadiana basin for the periods 1970–1978, 1979–1990 and 1991–2000 by studied catchments (**a**) and by hydrodynamic categories (**b**)

Taking into account hydrodynamic types, wetland losses were in accordance with their relative extension in the study area. Floodplain wetlands fed jointly by surface and ground waters were the most desiccated, accounting for 47% of total wetland losses in the basin for the period 1970–2000, with its main focus on the period 1978–1990 (Fig. 4.4). Most affected by desiccation were, however, floodplain wetlands fed by surface water, whose size was reduced by 62%. Although, for the studied period, rain-fed temporary wetlands scored second in losses (24% of total losses; Fig. 4.4), in 2000, 86% of those existing in the early 1970s were maintained. The other wetland hydrodynamic type most affected by desiccation was that fed by rain and groundwater, which accounted for 19% of total wetland losses (Fig. 4.4), with 52% of its original surface disappearing until 2000. Finally, from 1970 to 2000, groundwater-fed wetlands have been reduced by 39%, but this only accounted for 10% of their total losses (Fig. 4.4). Contrarily to what has been observed in other wetland-rich regions around the world (Bedford and Preston 1988), wetland losses in the Upper Guadiana basin seem not to be influenced by ecosystem landscape positions, rather by their abundance in the landscape. While floodplain and rain-fed wetlands were subject to strong losses during the study period, it probably indicated that anthropogenic stresses affected the areas more than the distribution pattern of the wetlands in the basin, as also observed by Zhang et al. (2009) in the wetlands of the Sanjiang Plain (China).

4.6 Wetland Habitat Quality Evolution During the Period 1978–2000

The trophic state of the Upper Guadiana wetlands during the period 1978–2000 is shown in Fig. 4.5. The studied wetlands drew a clear pattern of nutrient over-enrichment. While most wetlands were mesotrophic in 1978, with a large percentage of poor nutrient (oligotrophic) wetlands, time trend has been towards an increasing wetland trophic level through P-enrichment. Therefore, the 14% of wetland area being eutrophic and hypereutrophic rose to 65% in 1990, and to 83% in 2000 (Fig. 4.5). In 1990, oligotrophic wetlands virtually disappeared (<1%; Fig. 4.5).

Water conductivity also displayed a general pattern toward an increase of salinity in the studied wetlands (Fig. 4.5). However, this should be understood by considering the conductivity of the disappearing wetlands for the entire period. From 1990 wetlands increased their conductivity, falling into the category 2,500–20,000 μS cm^{-1} (Fig. 4.5). Only two water bodies really became more saline (Laguna de Alcahozo – Las Ánimas watershed – and TDNP). This pattern appeared because majority of disappeared wetlands in the basin were those ranging from 0 to 2,500 μS cm^{-1} (84% and 70% in 1990 and 2000, respectively).

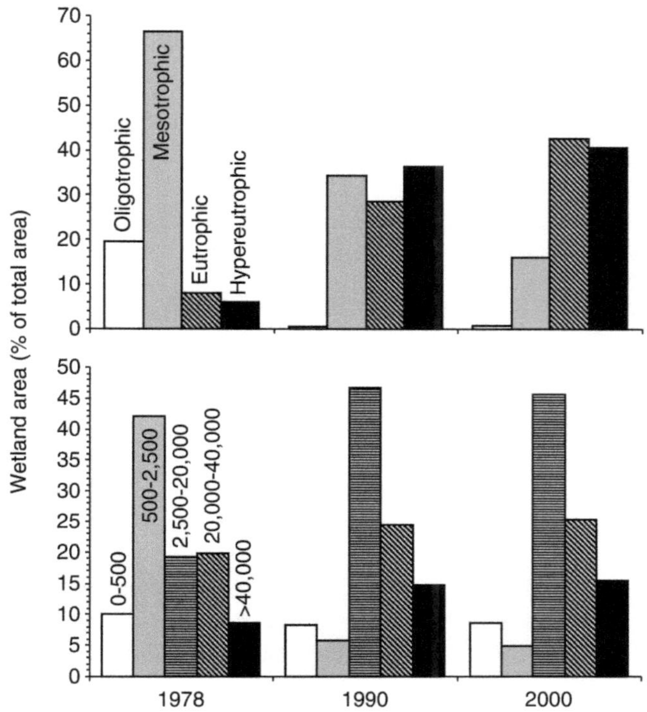

Fig. 4.5 Temporal trends on trophic status (upper) and water conductivity (lower) of wetlands in the Upper Guadiana for the period 1978–2000. Oligotrophic: <0.01 mg P l^{-1}, mesotrophic: 0.01–0.06 mg P l^{-1}; eutrophic: 0.06–0.1 mg P l^{-1}; hypereutrophic: >0.1 mg P l^{-1}; conductivity measured in µS cm^{-1}

4.7 Patterns on Wetland Loss and Habitat Quality Deterioration: the Influence of the Main Anthropogenic Stressors at Local and Regional Scales

The loss of wetlands in the Upper Guadiana basin during the period 1978–2000 could be significantly related to the loss of natural vegetation, as well as to the reduction of the employment in the agriculture sector (Table 4.4). Similarly, the increase of irrigated lands, urban areas and the growth of the industry and building activities also significantly reduced the wetland area in the basin (Table 4.4). While the period 1978–1990 did behave similarly during the entire period (including the increase of dry-croplands), for the subsequent period, the increase of irrigation, population and the strong occupancy in the service sector accounted for the wetland desiccation, along with the loss of employment in the agricultural sector (Table 4.4).

At local scale, the RDA analyses showed that wetland area variance during 1990–2000 was explained by a first axis (explaining 41% of total variance, $p < 0.001$),

Table 4.4 Pearson product–moment correlations (r values) of the effects of changes (independent variables) in land use (ha), population (inhabitants) and economic activities (inhabitants employed in each economic activity) on wetland area disappearance (ha loss per period; dependent variable) at sub-watershed scale in the Upper Guadiana from 1978 to 2000. Marked correlations are significant at $p < 0.05$

	Wetland losses		
Changes	1978–1990	1990–2000	1978–2000
Natural vegetation	0.94*	−0.58	0.93*
Dry-croplands	−0.71*	0.71	−0.20
Irrigated lands	−0.93*	−0.99*	−0.92*
Dry-irrigated	−0.44	−0.67	−0.35
Urban	−0.92*	−0.71	−0.87*
Population	−0.02	−0.96*	0.00
Agriculture	0.90*	0.97*	0.91*
Industry	−0.67*	−0.42	−0.61*
Building	−0.85*	0.66	−0.82*
Services	−0.38	−0.97*	−0.22

which included as main explanatory variables the population employed in industry and building as well as population density, all positively correlated (Fig. 4.6a and Table 4.5). Therefore, for this period, conservation of wetland's size appeared to be better in industrialized areas as compared to agricultural and services zones, where almost all wetlands were previously lost, in the period 1978–1990. The second axis, which was not statistically significant, explained 6% of the variance, and was positively related to the wetland's size in 1970–1978 (Fig. 4.6). The percentage of dry and irrigated croplands was positively correlated with this second axis, whereas the correlation was negative in the case of percentage of natural vegetation (Fig. 4.5). Hence, in watersheds dominated by agricultural uses, large wetlands were formerly present, but were lost until 1990, whereas in those watersheds with high percentage of natural vegetation there was an abundance of small ponds, which were better preserved throughout the study period.

From 1990–2000, water quality changes at local scale were at first explained by a first axis justifying 41% of total variance ($p < 0.001$; Fig. 4.6b). The main significant explanatory variables related to the axis were population dedicated to the industry and population density, positively; and population dedicated to services, negatively (Fig. 4.6b and Table 4.6). According to these, water quality conditions deteriorated in the most industrialized and populated areas for this period. Negatively related to this first axis were good quality wetlands habitats located in areas where the service sector developed strongly, and were lost in the period 1978–1990. Most of these wetlands belonged to the rain-fed hydrodynamic category. The second axis, which was not statistically significant, explained 5% of the variance, and, in 1978, was positively related to water quality conditions (Fig. 4.6b). Here, the percentage of dry and irrigated croplands was positively correlated, whereas the percentage of natural vegetation was negatively correlated. It indicated how the best wetland habitat quality was found in areas surrounded by natural vegetation.

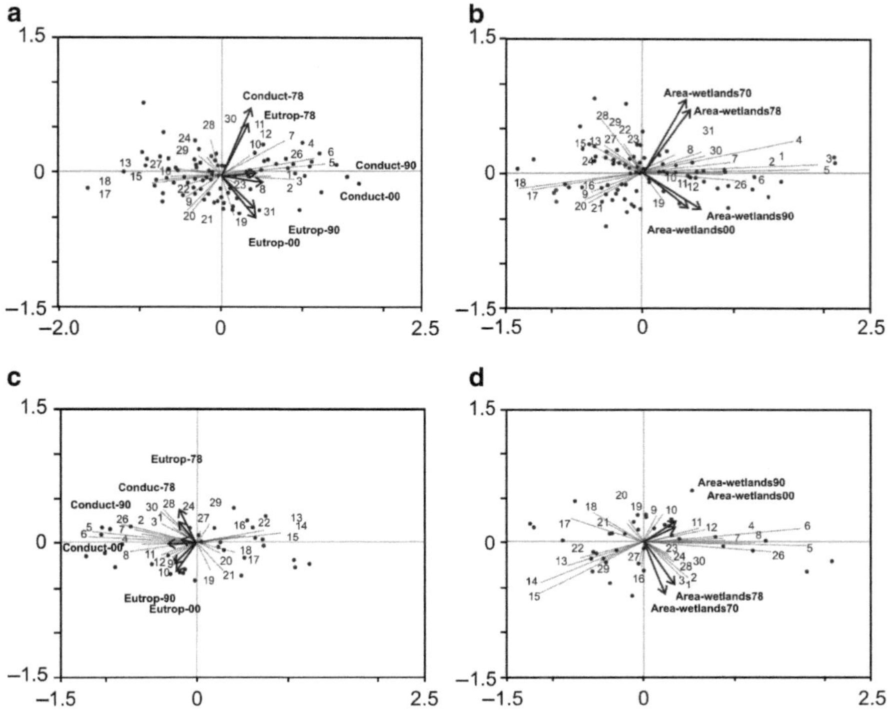

1-Pop78; 2-Pop90; 3-Pop00; 4-Ind81; 5-Ind91; 6-Ind01;7 -Build81; 8-Build91; 9-Build01; 10-Urb78; 11-Urb90; 12-Urb00; 13-Serv81; 14-Serv91; 15-Serv01; 16-Agri81; 17-Agri91; 18-Agri01; 19-Nat-78; 20-Nat-90; 21-Nat00; 22-Irri-78; 23-Irri-90; 24-Irri-00; 25-Dryir78; 26-Dryir90; 27-Dryir00; 28-Dry78; 29-Dry90; 30-Dry00; 31-Villages.

Fig. 4.6 Biplots of the partial RDAs applied on wetland area (**b** and **d**) water quality variables (**a** and **c**) using all variables (land use-land cover, economic activities and population) derived at local scale (L; panels **a** and **b**) and regional scale (R: panels **c** and **d**) for the years 1978, 1990 and 2000. Abbreviations legend: Pop: population; economical activities, Ind: industry, Serv: services, Build: building, Agri: agriculture; land-uses, Nat: natural vegetation, Dry: dry croplands, Dryir: dry-irrigation croplands, Irri: irrigation crops

Similarly to local conditions, changes in wetland area at regional scale during 1990–2000 were explained by a first axis (explaining 45% of total variance, $p < 0.001$), where population dedicated to industry and building sectors were positively related, whereas population dedicated to services was negatively (Fig. 4.6c and Table 4.5). As at local scale, industrialization at watershed scale led to a better conservation of the wetland area. On the opposite side, the growth of the services sector contributed to greater loss of wetlands (Fig. 4.6c). The second axis, explaining 5% of variance ($p < 0.001$), was negatively related to the wetland area in 1970–1978 (Fig. 4.6c). Percentage of dry croplands and population density were positively correlated with this second axis, whereas the percentage of natural vegetation was negatively correlated (Fig. 4.6c and Table 4.5). High population combined with agriculture uses contributed to the disappearance of large wetlands in 1990; contrarily, wetland conservation increased in watersheds dominated by natural vegetation.

Table 4.5 Percentage of variance explained by main explanatory variables (land use-land cover (LULC) types, economic activities and population) on temporal dynamic of wetland area during 1978–1990–2000, by partial redundancy analysis (RDA). P-level represented between parentheses

	Local (L)	Only local (L|R)	Regional (R)	Only regional (R|L)
Total R^2 (%)	32 (0.001)	NS	43 (0.001)	10 (0.038)
LULC				
Natural vegetation 1978			5 (0.004)	
Dry croplands 1990			3 (0.028)	
Economic activities				
Industry 1990	15 (0.001)		19 (0.001)	
Services 1990			6 (0.004)	2 (0.027)
Building 1978			2 (0.044)	2 (0.045)
Building 1990	5 (0.006)			6 (0.001)
Building 2000	3 (0.039)			
Population				
Number of towns	5 (0.011)			
Population 1978	4 (0.009)			
Population 2000			3 (0.030)	

At regional scale, water quality conditions during 1990–2000 were explained by a first axis (45% of total variance, $p < 0.001$; Fig. 4.6d). Here, explanatory variables were population devoted to the industry, positively, and population devoted to the services and agriculture, negatively (Fig. 4.6d and Table 4.6). As noted at local scale, industrialization significantly deteriorated water quality, while agriculture and services maintained it in better conditions. Unfortunately, most of these wetlands had already disappeared in 1990. The second axis of the RDA analysis explained 4% of variance ($p < 0.001$). In 1978–1990, it was negatively related to water quality conditions, through population (negatively) and the percentage of urban uses (positively; Fig. 4.6d and Table 4.6). Thus, the increase of population as well as the size of towns in the watershed had negative effects on wetland habitat quality.

High co-variation between local and regional factors was observed, explaining wetland area and water quality during the years of the study (Fig. 4.7). The pure effects of local conditions (L|R) in both the wetland area and water quality were not statistically significant, due to the great co-variation with the regional ones (Tables 4.5 and 4.6). However, the pure effect of regional factors (R|L) was significant in relation to the population working in the building and service sectors (Tables 4.5 and 4.6). The trend was for co-variation to descend through time, whereas the role of pure regional conditions increased. This dynamic breaks the link between local and regional conditions, reducing the total explained variance (Fig. 4.7).

Because the unexplained variance in RDA analyses appeared to be relatively high, our results proved how wetland losses and habitat quality degradation in the

Table 4.6 Percentage of variance explained by the significant explanatory variables (land use-land cover (LULC) types, economic activities and population) on the temporal dynamic of water quality (conductivity and eutrophication) during 1978–1990–2000, by partial redundancy analysis (RDA). P-level represented between parentheses

	Local (L)	Only local (L|R)	Regional (R)	Only regional (R|L)
Total R^2 (%)	25 (0.001)	NS	39 (0.001)	8 (0.006)
LULC				
Urban 2000			2 (0.060)	
Economic activities				
Industry 1990			26 (0.001)	
Industry 2000	17 (0.001)			
Services 1978	5 (0.007)			
Services 1990			3 (0.006)	2 (0.056)
Building 19901				6 (0.001)
Agriculture 1990			5 (0.006)	
Population				
Population 1978	3 (0.040)			
Population 2000			3 (0.021)	

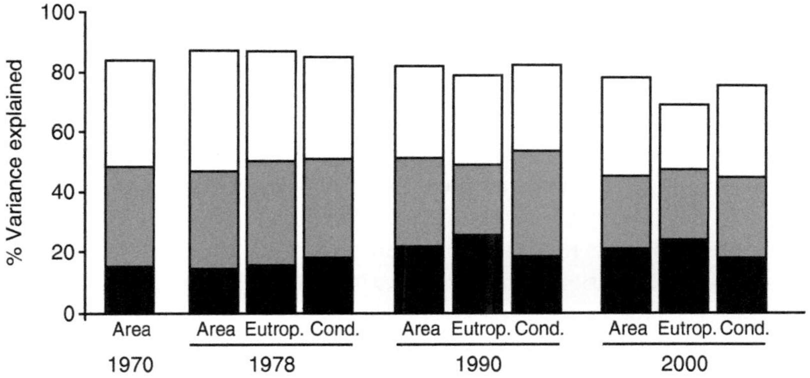

Fig. 4.7 Variance partitioning (%) of wetland variables: wetland area, eutrophication and conductivity through time by means of partial RDA. Three different components are given: pure regional variation [R|L] (*black area*), co-variation (the shared local and regional variation [L|R]; *grey area*), and the unexplained variation (100 – [L + R]; *white*). Pure local variation [L|R] was not significant and is, therefore, not shown

Upper Guadiana basin are driven, both at local and regional scales, by complex processes, difficult to estimate with the data available. Wetland disappearance and degradation have been forced by a complex matrix of anthropogenic variables, including those used here (LULC, population density and economic activities) but not only by them, which, directly or indirectly, drive landscape alteration processes. Diverse anthropogenic stressors have been shown to affect wetland integrity at watershed scale, through changes in hydrology and non-point source pollution

(Wang et al. 1997; Jones et al. 2001; Gergel et al. 2002; Tang et al. 2005; Uuemaa et al. 2007). In our case, landscape variations driven by LULC changes, intrinsically linked to economic activities and demography, were capable of explaining close to 50% of wetland changes in area and habitat quality. Pressure of population growth, combined with the agricultural sector, explained wetland area depletion until 1990. Habitat quality, which worsened from 1990 on, has had a greater impact on areas where industry and building sectors had more weight on the socioeconomic development (also densely populated). From 1990, better habitat quality conditions appeared in the areas dedicated to service and agricultural activities. Finally, throughout the study period, it was found that those areas dominated by natural vegetation appeared to strongly increase wetland conservation, improving habitat quality.

Our results agree with those obtained by others authors in similar agriculturally impacted regions (e.g. Liu et al. 2004; Zhang et al. 2009). Agricultural development, frequently the dominant human activity in wetland-rich areas, is the major cause of wetland loss (Turner 1991). Nonpoint source pollution such as that derived from agricultural practices, has been cited as the main cause for water quality degradation (Hemond and Benoit 1988; Bhaduri et al. 2000; Tuteja et al. 2003; Moss 2008). On the contrary, in the Upper Guadiana basin, wetland habitat quality has been negatively impacted, from the 1990s onwards, mainly through secondary economic activities. This could be explained by three main causes: (i) large wetlands appeared in most economically developed watersheds (with more industry, services and a dense population), (ii) semi-arid climate limits runoff at storm events and point sources from secondary and tertiary sectors, therefore, they have more weight on the total discharges entering the wetlands, and (iii) agriculture has been in economic recession from the mid-1990s. This trend concurs with the recent overview by Brinson and Malvárez (2002), who note that the more industrialized countries are likely to conserve their remaining, already impacted, wetlands, while nations with less industrialization are now experiencing accelerated losses, and may continue to do so for the next several decades.

From the 1970s, human activity in the Upper Guadiana has clearly been the main cause of wetland disappearance and degradation. High co-variation at both local and regional scales has significantly impacted on wetland degradation. However, the temporal trend shows that anthropogenic stressors operating at regional scale will have more weight on wetland conservation. Therefore, future conservation of wetland habitats depends on integrated management of natural resources at watershed scale. Although environmental policies regulating industrial and urban wastewater discharges are considered in the 7th EU Water Framework Directive, wetland sustainability in the Upper Guadiana basin requires an additional effort to adapt the availability of natural resources to economic development.

Acknowledgments The authors thank to the Junta de Castilla-La Mancha by providing spatial information on the Upper Guadiana River basin. RSA was supported by a CSIC JAE-Doc contract.

References

Allan JD, Erickson DL, Fay J (1997) The influence of catchment land use on stream integrity across multiple spatial scales. Freshw Biol 37:149–161

Álvarez-Cobelas M, Cirujano S (1996) Las Tablas de Daimiel: ecología acuática y sociedad. Ministerio de Medio ambiente, Madrid, Spain

Álvarez-Cobelas M, Cirujano S, Sánchez-Carrillo S (2001) Hydrological and botanical man-made changes in the Spanish wetland of Las Tablas de Daimiel. Biol Conserv 97:89–98

Angeler DG, Viedma O, Sánchez-Carrillo S, Alvarez-Cobelas M (2008) Conservation issues of temporary wetland Branchiopoda (Anostraca, Notostraca: Crustacea) in a semiarid agricultural landscape: what spatial scales are relevant? Biol Conserv 141:1224–1234

Armengol J, Estrada M, Guiset A, Margalef R, Planas D, Toja J, Vallespinós F (1975) Observaciones limnológicas en las lagunas de la Mancha. Boletín Estación Central de Ecología 4:11–17

Bedford BL, Preston EM (1988) Developing the scientific basis for assessing cumulative effects of wetland loss and degradation on landscape functions: status, perspectives and prospects. Environ Manage 12:751–771

Berka C, Schreier H, Hall K (2001) Linking water quality with agricultural intensification in a rural watershed. Water Air Soil Pollut 127:389–401

Bernier JC, Morton RA, Barras JA (2006) Constraining rates and trends of historical wetland loss, Mississippi River delta plain, South-Central Louisiana. In: Xu YJ, Singh VP (eds) Coastal environment and water quality. Water Resources Publications, pp 371–382. http://coastal.er.usgs.gov/gc-subsidence/historical-wetland-loss.pdf

Bhaduri B, Harbor J, Engel B, Grove M (2000) Assessing watershed-scale, long-term hydrologic impacts of land-use change using a GIS-NPS model. Environ Manage 26:643–658

Brinson MM (1993) A hydrogeomorphic classification of wetlands. Wetlands Research Program Technical Report WRP-DE-4. US Army Corps of Engineers, US. http://el.erdc.usace.army.mil/wetlands/pdfs/wrpde4.pdf. Accessed 16 Dec 2009

Brinson MM, Malvárez AI (2002) Temperate freshwater wetlands: types, status, and threats. Environ Conserv 29:115–133

Calder IR (1998) Water-resource and land-use issues. SWIM Paper 3. International Water Management Institute, Colombo, Sri Lanka

Cano-Soler D, Cendejas JL, Ruiz C, Martín D (2000) El mercado de trabajo y su medición en España. Estadística Española 42:189–204. http://www.ine.es/revistas/estaespa/146_3.pdf

Cirujano S (1980a) Estudio florístico, ecológico y sintaxonómico de la vegetación higrófila de la submeseta sur. Ph.D. thesis. Universidad Complutense Madrid

Cirujano S (1980b) Las lagunas manchegas y su vegetación I. Anales Jar Bot Madrid 37:155–192

Cirujano S (1980c) Las lagunas manchegas y su vegetación II. Anales Jar Bot Madrid 38:187–232

Cirujano S (1990) Flora y vegetación de las lagunas y humedales de la provincia de Albacete. Instituto de Estudios Albacetenses de la Excma. Diputación de Albacete. CSIC. Confederación Española de Centros de Estudios Locales. Serie I. Ensayos Históricos y científicos 52, Albacete

Cirujano S, Medina L (2002) Plantas acuáticas de las lagunas y humedales de Castilla-La Mancha. Real Jardín Botánico CSIC y Junta de Comunidades de Castilla- La Mancha, Madrid

Cirujano S, Velayos M, Carrasco MA (1990) Notas sobre higrófitos peninsulares III. Anales Jar Bot Madrid 47:519–520

Cirujano S, Velayos M, Carrasco MA (1992) Aspectos dinámicos de la flora acuática y cambios fisico-químicos del agua en dos lagunas continentales españolas: laguna de la Albardiosa (Toledo) y las Tablas de Daimiel (Ciudad Real). Historia Nat 91:249–256

Cirujano S, Casado C, Bernués M, Camargo JA (1996) Ecological study of the National Park of Las Tablas de Daimiel (Ciudad Real, Spain): changes in the physico-chemical characteristics of the waters and the vegetation between 1974–1989. Biol Conserv 75:211–215

Dale VH, Brown S, Haeuber RA, Hobbs NT, Huntly N, Naiman NJ, Riebsame WE, Turner MG, Valone TJ (2000) Ecological principles and guidelines for managing the use of land1. Ecol Appl 10(3):639–670

DeFries R, Eshleman NK (2004) Land-use change and hydrologic processes: a major focus for the future. Hydrol Process 18(11):2183–2186

EEA European Environmental Agency (2005) CORINE Land Cover 1990. CORINE Land Cover 2000. http://dataservice.eea.eu.int

ESRI (2006) ArcGIS Version 9.2 and the Spatial Analyst Extension, Environmental Systems Research Institute, Redlands, California

Florin M, Montes C (1999) Functional analysis and restoration of Mediterranean lagunas in the Mancha Húmeda Biosphere Reserve (Central Spain). Aquat Conser Mar Freshw Ecosyst 9:97–109

Gergel SE, Turner MG, Miller JR, Melack JM, Stanley EH (2002) Landscape indicators of human impacts to riverine systems. Aquat Sci 64:18–128

Hemond H, Benoit J (1988) Cumulative impacts on water quality functions of wetlands. Environ Manage 12:639–653

Heras G, Prieto E, Guerrero T, Oñate JM, Ontalba A, Cortés M, Lorenzo A, Marcilla JJ, Palomo JL, Tudanca M, Carrión J (1971) Recursos y aprovechamientos hidráulicos. Consejo Económico Sindical Interprovincial de La Mancha, Ciudad Real

Hildrew AG, Giller PS (1994) Patchiness, species interactions and disturbance in the stream benthos. In: Giller PS, Hildrew AG, Raffaelli DG (eds) Aquatic ecology: scale, patterns and process. Blackwell, Cambridge, MA

Hollis GE (1992) The causes of wetland loss and degradation in the Mediterranean. In: Finlayson CM, Hollis GE, Davis TD (eds) Managing Mediterranean wetlands and their birds. IWRB Special Publication No. 20, Slimbridge

Houlahan JE, Findlay CS (2004) Estimating the 'critical' distance at which adjacent land-use degrades wetland water and sediment quality. Landscape Ecol 19:677–690

Houlahan JE, Keddy PA, Makkay K, Findlay CS (2006) The effects of adjacent land use on wetland species richness and community composition. Wetlands 26:79–96

INE (2008) Anuario estadístico de España. Historical web-supported database (1858-nowadays). http://www.ine.es/prodyser/pubweb/anuarios_mnu.htm. Accessed 16 Dec 2009

Jones KB, Neale AC, Nash MS, Van Remortel RD, Wickham JD, Riitters KH, O'Neill RV (2001) Predicting nutrient and sediment loadings to streams from landscape metrics: a multiple watershed study from the United States Mid-Atlantic Region. Landscape Ecol 16(4):301–312

Lambin EF, Turner BL, Geist HJ, Agbola SB, Angelsen A, Bruce JW, Coomes OT, Dirzo R, Fischer G, Folke C, George PS, Homewood K, Imbernon J, Leemans R, Li X, Moran EF, Mortimore M, Ramakrishnan PS, Richards JF, Skånes H (2001) The causes of land-use and land-cover change: moving beyond the myths. Global Environ Chang 11:261–269

Liu H, Zhang S, Li Z, Lu X, Yang Q (2004) Impacts on wetlands of large-scale land-use changes by agricultural development: the small Sanjiang Plain, China. Ambio 33:306–310

Llamas MR (1988) Conflicts between wetland conservation and groundwater exploitation: two case histories in Spain. Environ Geol Water Sci 11:241–51

Molinillo M, Lasanta T, García-Ruiz JM (1997) Managing mountainous degraded landscapes after farmland abandonment in the Central Spanish Pyrenees. Environ Manage 21:587–598

MOPU (1990) Estudio de las zonas húmedas en la España peninsular, inventario y tipificación, relación con el régimen hídrico general y medidas de protección. Technical Report. Dirección General de Obras Hidráulicas, MOPU, Madrid

Moss B (2008) Water pollution by agriculture. Philos Trans R Soc B 363:659–666

Mulligan M, Burke SM (1999) Modelling the future of groundwater resources in central Spain. Environ Manage Health 10:41–51

Niehoff D, Fritsch U, Bronstert A (2002) Land-use impacts on storm-runoff generation: scenarios of land-use change and simulation of hydrological response in a meso-scale catchment in SW-Germany. J Hydrol 267:80–93

Riebsame WE, Meyer WE, Turner BL (1994) Modeling land use and cover as part of global environmental change. Clim Change 28:45–64

Tang Z, Engel BA, Pijanowski BC, Lim KJ (2005) Forecasting land use change and its environmental impact at a watershed scale. J Environ Manage 76:35–45

Ter Braak CJF (1988) Partial canonical correspondence analysis. In: Bock HH (ed) Classification and related methods of data analysis. North-Holland, Amsterdam

Turner K (1991) Economics and wetland management. Ambio 20:59–63

Tuteja NK, Beale G, Dawes W, Vaze J, Murphy B, Barnett P, Rancic A, Evans R, Geeves G, Rassam DW, Miller M (2003) Predicting the effects of landuse change on water and salt balance-a case study of a catchment affected by dryland salinity in NSW, Australia. J Hydrol 283:67–90

Uuemaa E, Roosaare J, Mander U (2007) Landscape metrics as indicators of river water quality at catchment scale. Nord Hydrol 38:125–138

Velayos M, Cirujano S, Marquina A (1984) Aspectos de la vegetación acuática de la provincia de Guadalajara. Anales Jard Bot Madrid 41:175–184

Velayos M, Carrasco MA, Cirujano S (1989) Las lagunas del Campo de Calatrava (Ciudad Real). Botánica Complutensis 14:9–50

Vicente E, Rosa Miracle MR (1998) Estudio limnológico de 28 humedales de Castilla La Mancha como base para la elaboración del Plan de Ordenación de recursos Naturales. Universidad de Valencia y Junta de Comunidades de Castilla La Mancha, Toledo

Viladomiu L, Rosell J (1997) Gestión del agua y política agroambiental: el Programa de Compensación de Rentas por reducción de regadíos en Mancha Occidental y Campo de Montiel. Revista Española de Economía Agraria 179:331–350

Wang L, Lyons J, Kanehl P, Gatti R (1997) Influences of watershed land use on habitat quality and biotic integrity in Wisconsin streams. Fisheries 22:6–12

Wang Y, Hong W, Wu C, He D, Lin S, Fan H (2008) Application of landscape ecology to the research on wetlands. J For Res 19:164–170

Zhang J, Ma K, Fu B (2009) Wetland loss under the impacto of agricultural development in the Sanjiang Plain, NE China. Environ Monitor Assess. doi:10.1007/s10661-009-0990-x

Chapter 5
A Story of the Wetland Water Quality Deterioration: Salinization, Pollution, Eutrophication and Siltation

M. Álvarez-Cobelas, S. Sánchez-Carrillo, S. Cirujano, and D.G. Angeler

Abstract Water quality has historically relied upon wetland openness and water availability. For many years until 1986, Las Tablas de Daimiel wetland was enabled to export water and materials downstream, thus diminishing the impairing effects of pollutant storage. Since 1987, when the wetland was dammed, good-quality inputs of low-salinity groundwater were discontinued, streams directed urban and agricultural wastewater towards the wetland, discharge inputs were rather low, due to groundwater exhaustion in the catchment and, hence, its internal loading had greatly increased. As a result, wetland salinization, pollution and eutrophication increased both in water and sediments, and siltation appeared as a severe threat. This chapter will review long-term patterns (1945–2007) of these processes, paying special attention to wetland connectivity and water export downstream.

Keywords Catchment impacts • internal loading • connectivity • groundwater exhaustion

5.1 Introduction

Climate has been a major driving variable affecting wetland ecosystems throughout geological time (Öquist and Svensson 1996), but changes are relatively slow. Man-made change is a more recent issue, however, with effects occurring over much shorter periods (Maltby et al. 1994). These impacts are of a very broad nature, and include direct loss of, and effects on, biodiversity, water shortage, water pollution, etc. (Mitsch and Gosselink 2000). However, these effects need to be assessed over long periods to distinguish signal from noise.

M. Álvarez-Cobelas (✉)
Spanish National Research Council (CSIC), Institute of Natural Resources, Serrano 115 dpdo, E-28006 Madrid, Spain
e-mail: malvarez@ccma.csic.es

S. Sánchez-Carrillo and D.G. Angeler (eds.), *Ecology of Threatened Semi-Arid Wetlands: Long-Term Research in Las Tablas de Daimiel*, DOI 10.1007/978-90-481-9181-9_5,

From the 1950s, many wetlands have undergone gradual eutrophication, mainly through agricultural runoff and waste water discharges. Although wetlands are known to function as filters in hydrologic systems (Phillips 1996) as well as to retain and transfer nitrogen and phosphorus from water systems (Gunatilaka 1991), increases of nutrient inputs modify or change biological communities (Mitsch and Gosselink 2000) and ecosystem function (Sánchez-Carrillo et al. 2001).

In semi-arid and arid wetlands the situation is quite different, as they are strongly dependent on hydrology for ecosystem functioning, but water levels fluctuate both seasonally and interannually. These fluctuations induce changes in nutrient inputs and hence in nutrient dynamics through complex interactions with biotic components (Howard-Williams 1985; Sánchez-Carrillo and Álvarez-Cobelas 2001). The decline of aquatic species richness is well known as an effect of diminishing water availability coupled with increasing water pollution, by means of multiple factors operating simultaneously (Breitburg et al. 1998; Álvarez-Cobelas et al. 2001). Furthermore, flooding variability changes ecosystem performance, impinging on nutrient fluxes, community metabolism and decomposition and plant interactions, and driving interactions between nutrient fluxes and plant responses in particular (Wetzel 1990).

Otherwise, wetlands show a high efficiency, retaining particulate matter and nutrient through sedimentation, which is the main sink process at the ecosystem level (Phillips 1989; Moustafa et al. 1998; Sánchez-Carrillo et al. 2001). Even though sediment deposition improves wetland water quality by reducing turbidity and retaining nutrients and pollutants (Johnston 1991), sedimentary nutrient enrichment resulting from eutrophication enhances macrophyte production and, hence, detritus accumulation, thus making the eutrophication process feed back through increased internal loading (Sánchez-Carrillo and Álvarez-Cobelas 2001). When sporadic and heavy external loads enter this closed nutrient cycle, wetland siltation appears in the short-term as a severe environmental problem (Sánchez-Carrillo et al. 2001).

Las Tablas de Daimiel National Park has a long history of water quality changes that have mostly arisen from anthropogenic changes both in the watershed and the wetland. Since the 1980s, salinity increased as a result of discontinued groundwater discharges (Álvarez-Cobelas and Cirujano 1996; Álvarez-Cobelas et al. 2001), and the wetland became hypereutrophic, due to heavy external nutrient loadings. High internal loading arising from high productivity of emergent vegetation exacerbate the eutrophication problem. This arises mainly from the limited export of materials downstream prevented by a terminal dam in the wetland (Sánchez-Carrillo and Álvarez-Cobelas 2001). Furthermore, the wetland is polluted by some organic compounds such as cyanide, which cause massive fish mortality (Álvarez-Cobelas and Cirujano 2003). This chapter describes all processes and mechanisms that caused deterioration of wetland water quality since the 1980s, assessing nutrient dynamics and sedimentation in the wetland and the spatial patterns of nutrient controls. For these purposes, unpublished data as well as those of previous studies by Sáez-Royuela (1977), INYPSA (1990), Álvarez-Cobelas and Cirujano (1996, 2003), Álvarez-Cobelas et al. (2001, 2007), Sánchez-Carrillo and Álvarez-Cobelas (2001) and Sánchez-Carrillo et al. (2001, 2000) have been reviewed.

5.2 Materials and Methods

Since 1992, sampling was conducted monthly in at least three sites of the wetland (NE inlet – Pata Gallina area, Central inlet/outlet – Molemocho area and Terminal outlet – Puente Navarro dam; see Chapter 1). Samples were collected with a Niskin bottle. Preliminary thermal profiles ensured that complete vertical mixing was taking place at each sampling site. Simultaneously, electrical conductivity was recorded *in situ* using a CRISON probe to trace wetland connectivity. Samples were preserved below 4°C during field trips and were immediately analyzed in the laboratory for N-NO_3^-, N-NH_4^+, total phosphorus (TP) and soluble reactive phosphorus (SRP ≈ PO_4^{3-}) according to standard methods (APHA 1989). Total nitrogen (TN) was measured following the method of Bachmann and Canfield (1996). Total organic carbon (TOC) and dissolved organic carbon (DOC) samples were taken using BOD borosilicate glass vials and measured through high temperature catalytic combustion in a Shimadzu Model TOC-500A. DOC was measured after water filtration using Whatman GF/C fiber glass filters (0.45 µm). Phytoplankton chlorophyll-*a* was measured spectrophotometrically after extraction with methanol (Marker et al. 1980).

Gross sedimentation was measured using cylindrical sediment traps according to Bloesch (1996) (details in Sánchez-Carrillo et al. 2001). Leaching of emergent vegetation was taken from the study of Sánchez-Carrillo and Álvarez-Cobelas (2001). Emergent macrophyte cover and biomass data were taken from Álvarez-Cobelas et al. (2001, 2007). Internal loading was calculated following Knuuttila et al. (1994). Data on discharge into the wetland and flooding area were provided by the Guadiana Water Authority and the National Park staff, respectively.

5.3 Salinization and Major Ionic Composition

As pointed out in Chapters 1 and 3, the wetland water supply was the outcome of surface and groundwater discharge, but the latter was discontinued in the mid-1980s as a result of aquifer overexploitation. Previously, there were two types of environments in the wetland, namely, a oligohaline in the NE area and a freshwater one in the SW area. This is indicated by salinity data, using electrical conductivity as a proxy (Fig. 5.1a). While groundwater entering the wetland from the Guadiana stream showed an annual average conductivity of 860 ± 125 µS cm^{-1}, conductivity

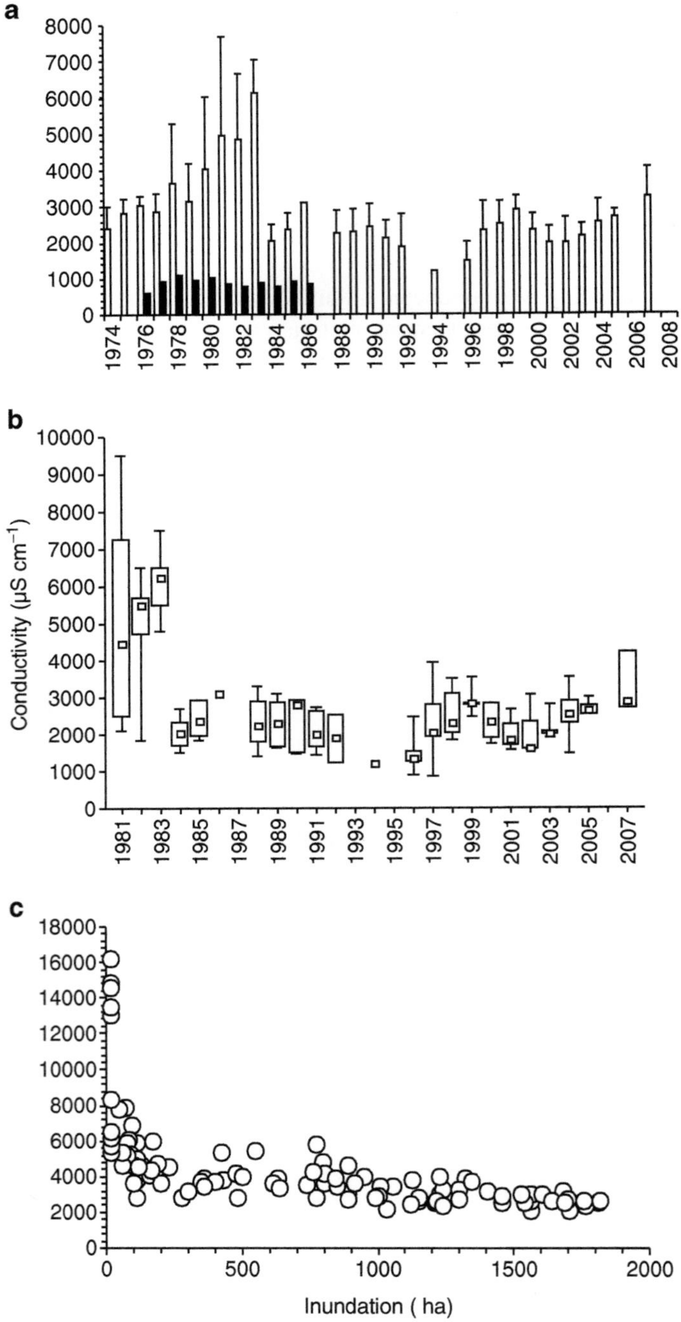

Fig. 5.1 (**a**) Electrical conductivity in waters entering the wetland from the underlying aquifer (*black bars*), an input that was discontinued in 1986, and Gigüela river (*white bars*). (**b**) Box-whisker plots of annual conductivity pooling data at the NE inlet, the central inlet until 1986

of surface discharge via the Gigüela river amounted to 2,778 ± 1,073 μS cm^{-1}. These latter inputs were higher and more variable in the early 1980s.

Inputs of varying salinity caused strong variability in wetland salinity (Fig. 5.1b). At the NE inlet, wetland salinity experienced changes associated with discharge, because salinity was lower in years of high discharge. Salinity was lower at the central inlet in the 1970s, but rose when groundwater inputs were discontinued in the mid-1980s. Finally, at the terminal outlet, salinity also depended upon water dilution, as it was higher in those years with lowest flooding, such as in 2007. Salinity ranges were very wide in all wetland sites (1,220–8,800, 710–5,360 and 2,044–16,230 μS cm^{-1} for the inlet, central inlet/outlet and terminal outlet, respectively).

Spatially, salinity distribution in this wetland changed over time, along with the changing nature of water inputs (Fig. 5.2). When surface and groundwater discharges took place, two distinct areas of salinity were observed, thus defining different habitats in the wetland (Fig. 5.2). Strong discharge via surface streams rapidly dampened out conductivity throughout the wetland when the flooding area exceeded 1,500 ha, whereas small areas of different salinity emerged once surface inputs were interrupted (Fig. 5.2).

Chemically, wetland waters are sulphate and calcium-rich, also reflecting changes in the composition of water inputs. When groundwater entered through the central inlet, as in 1975, wetland water in the inlet area was bicarbonate-sulphate- and calcium-rich, the remaining waters being sulphate and calcium-rich (Table 5.1). In 1997, when groundwater no longer flooded the wetland, a high stream discharge took place, that diminished ionic contents at the NE entrance (Table 5.1). In 2000, once surface flooding stopped, ionic contents of the wetland were roughly homogeneous with higher contents at the terminal end (Table 5.1). Sulphate enrichment has surely been the outcome of droughts in wetland sediments over time, since it is well-known that oxidation of sulphide compounds occurring in the reduced sedimentary environment are enhanced under conditions of drought (Elmers et al. 2007), which are very common in this area.

5.4 Water Pollution

The wetland harboured cyanide that sometimes resulted in some massive fish deaths. Cyanide was already recorded in 1992, with concentrations ranging 0.008–0.150 mg l^{-1}, being 20 times higher in interstitial water than in the water column.

Fig. 5.1 (continued) and outlet afterwards, and the terminal outlet. (**c**) Conductivity at the terminal outlet of the wetland as a function of flooding. Data from 1974 to 2007. Notice the hyperbolic relationship. Groundwater conductivity was measured only once a year, whereas stream data were recorded once a month and hence averages and standard deviations could be estimated. Authors' calculations from raw data of their own and those of Sáez-Royuela (1977), INYPSA (1990), Spanish Geological Survey and Guadiana Water Authority (unpublished data)

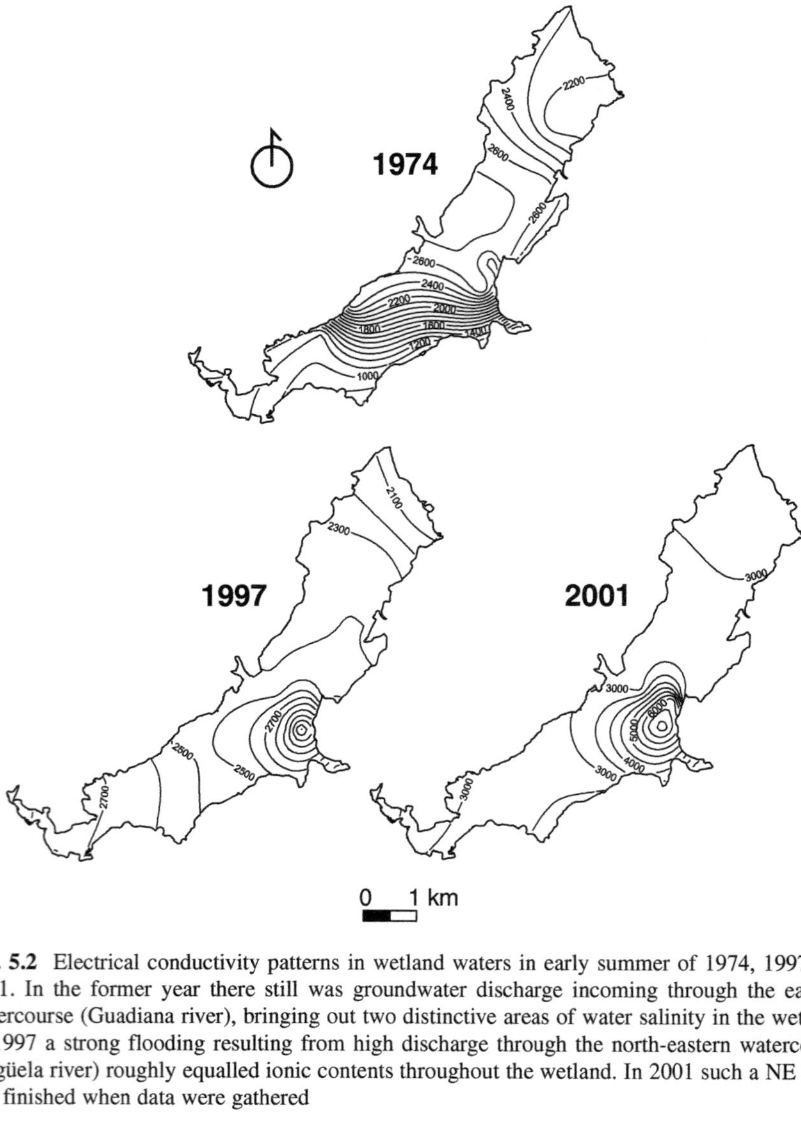

Fig. 5.2 Electrical conductivity patterns in wetland waters in early summer of 1974, 1997 and 2001. In the former year there still was groundwater discharge incoming through the eastern watercourse (Guadiana river), bringing out two distinctive areas of water salinity in the wetland. In 1997 a strong flooding resulting from high discharge through the north-eastern watercourse (Gigüela river) roughly equalled ionic contents throughout the wetland. In 2001 such a NE input had finished when data were gathered

In 2001, cyanide concentration was lower, ranging from 0.002 to 0.029 mg l^{-1}, coinciding with fish mortality. The origin of cyanide is unclear, because there are no gold mines in the catchment, which are the main source of cyanide to aquatic environments (Eisler and Wiemeyer 2004). Lethal and sub-lethal effects on freshwater fauna have been reported for similar concentrations to those found in this wetland (Eisler and Wiemeyer 2004).

The wetland does not show symptoms of lead pollution in waters and sediment despite frequent wildfowl hunting in the nineteenth century. Lead concentration

Table 5.1 Ionic content of wetland waters in 1975, 1997 and 2000. 1975 was a "normal" year before strong man-made impacts took place. In 1997 a very high discharge entered from the Upper Guadiana river basin through the NE inlet, whereas 2000 data reflected a roughly homogeneous situation once water stopped entering the wetland via NE surface stream. In 1986 the central inlet turned into outlet due to land subsidence flowing upstream as a consequence of underground peat fire. Units are meq l^{-1}

	NE inlet			Central inlet/outlet			Terminal outlet		
	1975	1997	2000	1975	1997	2000	1975	1997	2000
HCO_3^-	4.1	2.6	3.0	4.2	4.4	4.8	4.2	4.1	6.3
CO_3^{2-}	0.0	0.1	0.1	0.0	0.0	0.2	0.0	0.0	0.2
Cl^-	4.2	2.3	7.5	1.9	10.2	11.0	4.1	7.6	11.6
SO_4^{2-}	30.4	8.2	19.4	5.2	49.4	20.5	28.2	44.0	34.8
Na^+	4.1	1.9	6.5	1.7	9.2	7.1	3.6	6.6	9.8
K^+	0.5	0.1	0.3	0.1	0.5	0.3	0.5	0.4	0.5
Ca^{2+}	25.5	8.0	18.8	8.6	30.8	20.0	24.5	28.4	22.6
Mg^{2+}	11.4	4.3	5.6	4.0	22.0	11.2	11.4	18.0	19.8

ranged from 2 to 16 mg kg^{-1} of sediment. However, some pesticides such as α-HCH and lindane were recorded in water, emergent vegetation and sediments of central areas in the late 1990s, maybe arising from a recent input at the central inlet. This pesticide amounted to 0.002–0.74 mg l^{-1} in water, 0.007–0.85 mg kg^{-1} (DW) in sediment and 0.002–1.09 mg kg^{-1} (DW) in emergent vegetation.

Atmospheric deposition is negligible in the catchment of the wetland. The main impact on water quality occurred as a result of eutrophication. Heavy loads of organic carbon, nitrogen and phosphorus entered the wetland, promoting a high productivity of emergent vegetation (see Chapter 7) that resulted in high internal loading. These impacts are aggravated by a terminal dam that prevented organic matter export downstream since 1986. Nowadays, the wetland is a closed wetland, in the sense of Hopkinson (1992).

Since the catchment is very large relative to the wetland (roughly 1000:1), it is not surprising that many organic pollution episodes in the catchment have impacted the wetland. Urban and agro-industrial wastewaters have only recently started to receive treatment and have been entering untreated into the wetland for decades. Its stream course is that of the Gigüela, which mainly suffered either raw or partly treated wastewaters from Alcazar de San Juan (32,000 inhabitants, but 50 km away) and – what is more important – Villarrubia de los Ojos (10,000 inhabitants, and 3 km away). Diffuse pollution cannot be discarded either, because crops are the dominant land use in the catchment, and riparian forests are almost absent throughout (see Chapters 1 and 4). Some preliminary calculations suggest that diffuse pollution loadings can be 0–65%, 0–79% and 0–7% out of the total stream loadings at the wetland inlet for total organic carbon, total nitrogen and phosphorus, respectively. Diffuse sources are the main contributors of nitrogen pollution in crop-dominated catchments (Eriksson et al. 2007; Testa et al. 2008), and the Tablas de Daimiel watershed is mostly devoted to irrigation agriculture (see Chapter 1).

5.5 Nutrient Dynamics and Eutrophication

Organic carbon concentration in the incoming river ranged from 2.3 to 993 mg C l^{-1}, being higher in the first part of the 1980s (Fig. 5.3a), but also showing wide seasonal variation. Total nitrogen ranged from 2.1 to 46.7 mg N l^{-1}, whereas total phosphorus did from 0.05 to 71.9 mg P l^{-1} (Fig. 5.3a). In the early 1980s, the concentration of these nutrients was the expected outcome of raw wastewaters increase when municipal sewers and pipes were built in all towns and discharged into nearby rivers that drained to Las Tablas de Daimiel wetland. Later, as new wastewater treatment facilities were put into operation, the concentration of these compounds diminished. This pattern of nutrient loading reduction is well known in other European countries (Tudesque et al. 2008). All three elements co-varied somewhat in that period (organic carbon-total nitrogen variability 38%, organic carbon-total phosphorus variability 17%, total nitrogen-total phosphorus variability 26%), thus implying that nitrogen and phosphorus inputs were partially of wastewater origin. During 25% of the months recorded from 1981 to 2006 (Fig. 5.3b), the total phosphorus concentration was above the limits of the EU regulation for inputs to a National Park (Directive 91/271).

Nutrient import fluxes from the Gigüela watershed were usually high, albeit very variable (Fig. 5.4). While total organic carbon import ranged from 0 to 99,942 kg C km^{-2} $year^{-1}$, total nitrogen and total phosphorus amounted to 0–76,381 kg N km^{-2} $year^{-1}$ and 0–4,446 kg P km^{-2} $year^{-1}$, respectively. In forested wetlands impacted by municipal wastewater loadings, these figures were much lower (Day et al. 2006), only enhancing primary productivity and producing minor impaired effects on the wetland, as compared with the effects occurring in Las Tablas de Daimiel. These inputs of nutrients are higher than those in the Chinese marsh wetlands of the Xianghai Natural reserve, strongly impacted by agricultural and municipal wastewaters (Wang et al. 2004). Peaks of organic carbon and phosphorus matched, but those of nitrogen did not (Fig. 5.4), thus suggesting that phosphorus import was mainly of urban wastewater origin, but other sources were more influential on nitrogen import (see above).

Until 1986, when the Guadiana discharge was discontinued, its nutrient loadings could not be evaluated, but they may have been low, in view of the low nitrogen and phosphorus content of groundwater in those years (M. Álvarez-Cobelas, 2002). Imports from the Azuer creek, which drained 3 km upstream of the wetland, in the Guadiana river, could be more important, ranging from 0 to 177 kg N km^{-2} $year^{-1}$ and 0 to 44 kg P km^{-2} $year^{-1}$.

Organic loadings entering through the Gigüela river have often been strong enough to greatly diminish oxygen contents and kill large charophyte meadows (see Chapter 7). These strong pollution impacts occurred in 1989, 1997 and 2004; in former years they originated from the upper watershed, whereas Villarrubia de los Ojos was the source of heavy pollution in 2004. The effect of strong depletions of oxygen concentration after a heavy wastewater discharge into Tablas de Daimiel is shown in Fig. 5.5.

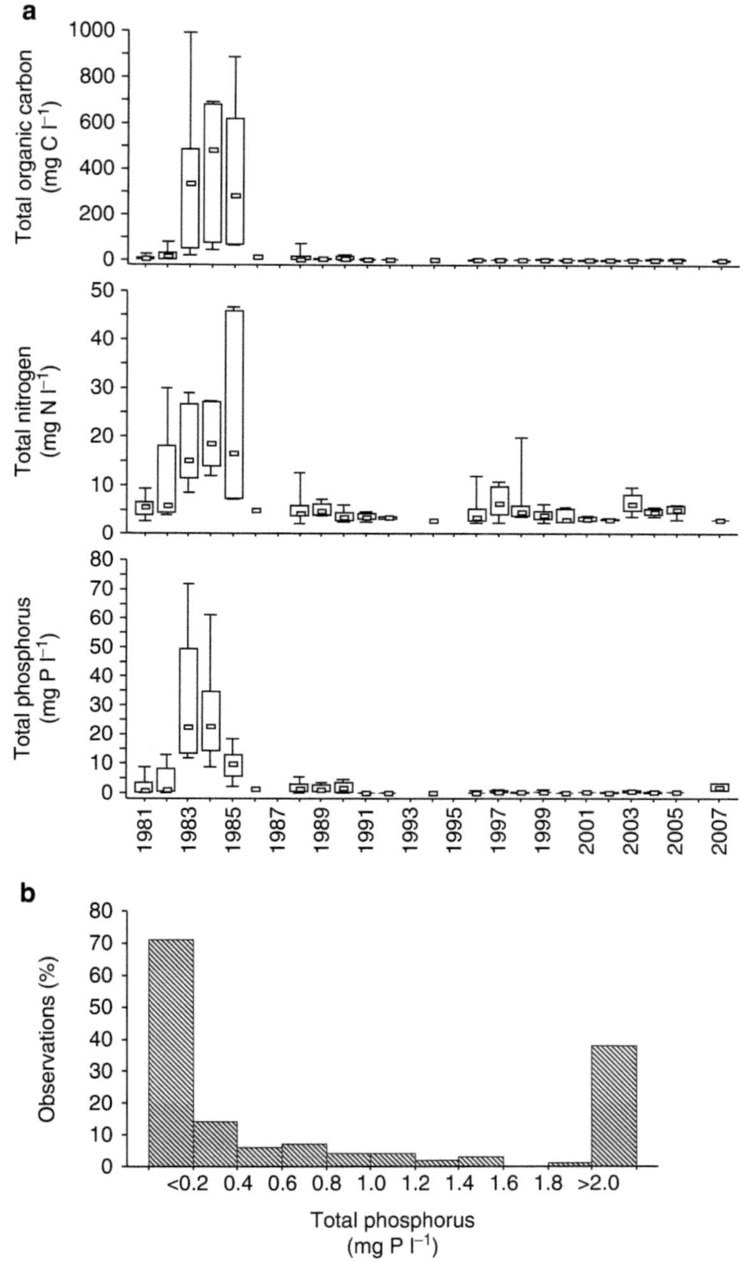

Fig. 5.3 (**a**) Box-whisker plots of total organic carbon, total nitrogen and total phosphorus in the Gigüela river at wetland entrance. (**b**) Histogram of total phosphorus concentrations at wetland entrance. EU regulations (Directive 91/271) imply that 2 mg P l^{-1} is the maximal limit to be attained by phosphorus concentrations before entering an environmentally protected area

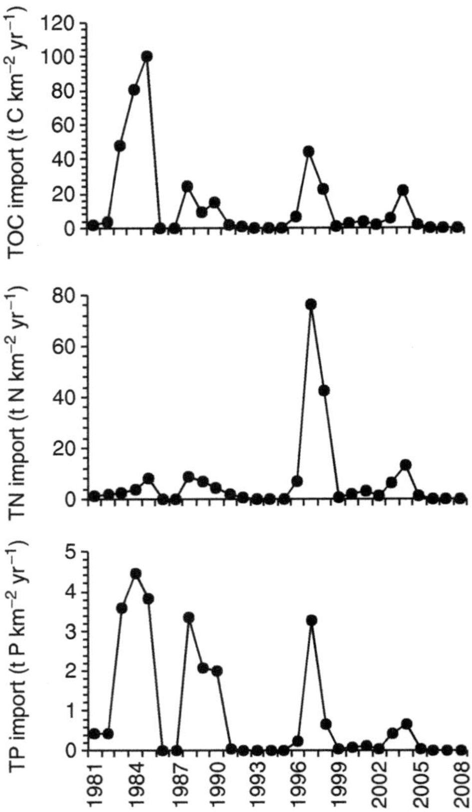

Fig. 5.4 Annual total organic carbon, nitrogen and phosphorus import to the wetland. Loadings were scaled to wetland area to estimate input fluxes

The historical record of nutrient concentrations in the wetland has always been high, but ecosystem health has been impaired since 1986, when a terminal dam closed the wetland. According to OECD (1982) standards, Las Tablas de Daimiel waters can be considered as hypertrophic. Substance concentrations have been loosely related with flooding, *i.e.* the higher the flooding, the lower the concentration. Therefore, years of higher flooding show lower average concentrations and variability (Figs. 5.6–5.9). Since 1974 (Fig. 5.6), total organic carbon concentration has ranged from 45 to 1,516 mg C l^{-1} in three wetland sites, showing more variability at the NE inlet (Gigüela river). This variable has not shown temporal coherence among sites, that is, the total organic carbon time trend has been different among sites, thus implying that site effects are more important than homogenization of water quality resulting from water connectivity. All surveyed nutrients and chlorophyll-*a* in this wetland have shown similar results as well. Dissolved organic carbon, less frequently measured, was more variable at the central and terminal sites (ranges 0.07–474 and 8–637 mg C l^{-1}, respectively) than at the entrance of the wetland (2.5–252 mg C l^{-1}). Wetland plants are important for dissolved organic carbon, because their decomposition products increase this compound in variable amounts; reed decomposition produces more coloured water than cut-sedge decomposition (Fig. 5.7).

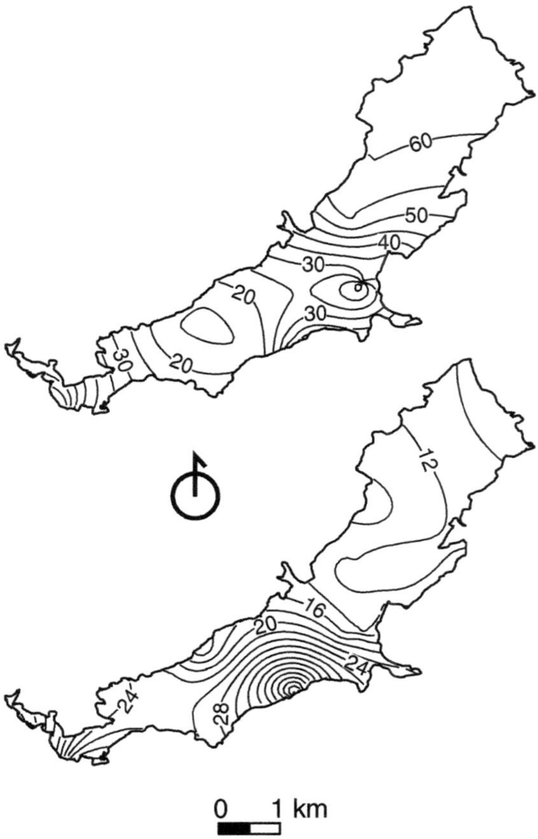

Fig. 5.5 Oxygen concentration (% saturation) in the wetland before (*upper map*) and after (*lower map*) a strong organic matter input event coming from the nearby Villarrubia de los Ojos town

Total nitrogen contents have been more variable and often higher at the NE inlet (0.48–27.57 mg N l^{-1}, Fig. 5.8). The central inlet/outlet presents a decreasing temporal trend (Fig. 5.8a). Nitrate concentrations have partly tracked those of total nitrogen, but nitrification might have been a dominant process in the wetland, because central and terminal areas show higher nitrate contents (Fig. 5.8b); this variable ranges from 0 to 66 mg l^{-1} throughout the wetland. Ammonia concentration has been higher at the terminal dam, but no temporal trend can be found in any site (Fig. 5.8c); overall, its range is between 0.005 and 6.93 mg l^{-1}.

Usually, total phosphorus contents have been higher at the NE inlet (range 0.03–5.26 mg P l^{-1}), thus revealing wastewater inputs. They can reach up to 1.8 mg P l^{-1} in the wetland, without showing any time trend (Fig. 5.9a).

Soluble reactive phosphorus concentration has ranged between 0.005–1.337, 0.005–0.399 and 0.005–0.255 mg P l^{-1} for the NE inlet, the central area and the terminal outlet of the wetland, but no temporal pattern can be detected (Fig. 5.9b).

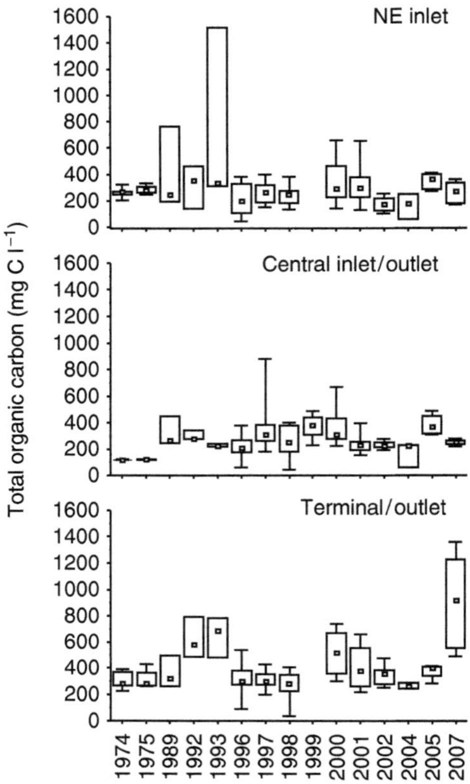

Fig. 5.6 Box-whisker plots of total organic carbon concentration in three wetland sites over time. The upper panel depicts data at the NE wetland inlet, the middle panel shows data at the central inlet/outlet (since 1986), whereas the lower panel shows carbon concentration at the terminal site

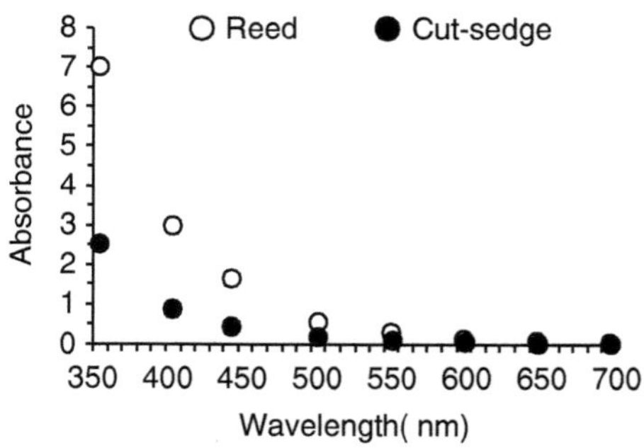

Fig. 5.7 UV–VIS absorbencies (arbitrary units) of main wetland plant extracts

Chlorophyll-*a* is usually of mixed origin, embracing phytoplankton, benthic and detritic pigments, due to the small water column involved. This may be one reason why chlorophyll-*a* variability if present, is weakly explained by nutrient variability

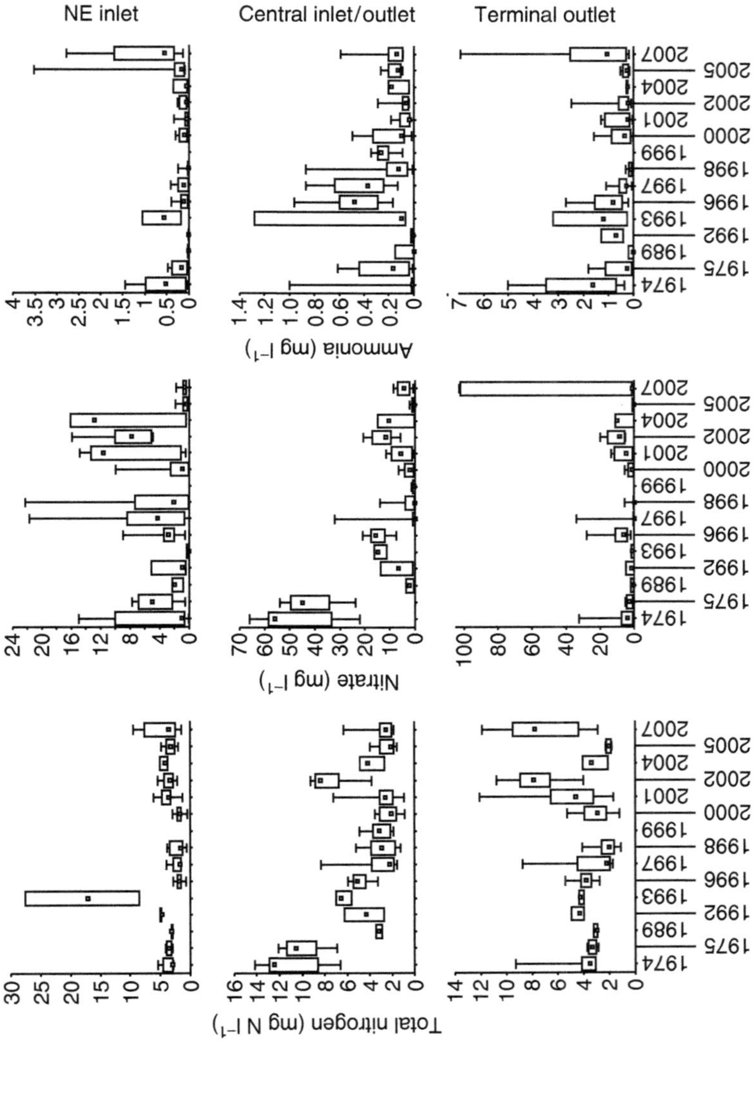

Fig. 5.8 Box-whisker plots of total nitrogen, nitrate and ammonia concentrations in three wetland sites over time. The upper panel depicts data at the NE wetland inlet, the middle panel shows data at the central inlet/outlet (since 1986), whereas the lower panel shows concentration at the terminal site

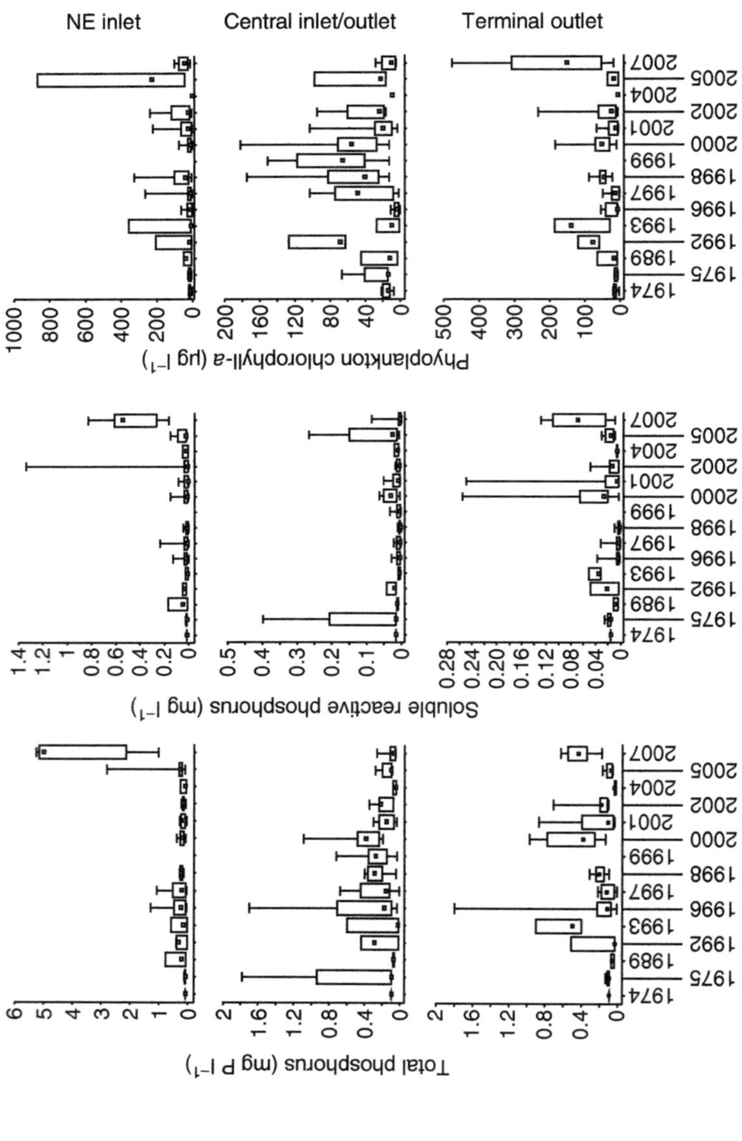

Fig. 5.9 Box-whisker plots of total phosphorus, soluble reactive phosphorus and chlorophyll-*a* concentrations at the NE wetland inlet, at the central inlet/outlet (since 1986), and at the terminal site

Table 5.2 Annual nitrogen and phosphorus budgets (g m^{-2} year^{-1}) for Las Tablas de Daimiel National Park from 1996 to 1998. Atmospheric deposition of TN and TP was negligible (Data source from Sánchez-Carrillo and Álvarez-Cobelas 2001)

	Total nitrogen			Total phosphorus		
	1996	1997	1998	1996	1997	1998
Inputs						
Inflows	8.49	66.35	25.27	0.94	2.68	1.60
Internal loading	27.22	43.32	78.44	3.69	16.58	22.81
Macrophye leaching	16.65	14.42	12.32	8.03	6.43	5.39
Waterfowl additions	0.07	0.10	0.15	0.02	0.03	0.04
Outputs						
Outflows	0	6.12	7.73	0	0.42	0.40
Sedimentation	29.65	52.94	86.40	1.34	17.20	24.45
Infiltration	7.48	28.99	5.33	0.71	2.46	0.44
Storage	15.30	36.14	16.72	10.63	5.64	4.55
Retention (%)	39	93	69	97	84	57

(see below). Its ranges are higher at the NE inlet (0.7–870 μg l^{-1}) and terminal outlet (2.8–477 μg l^{-1}), being lower at the central sites (0.6–182 μg l^{-1}; Fig. 5.9c). Because of the wide variability involved, it is difficult to envisage any temporal trend (Fig. 5.9c).

The detailed nutrient budget carried out for the period 1996–1998 (Table 5.2) pointed out that heavy mass inputs increased during wet years, mainly in the wetland, originated by runoff (Sánchez-Carrillo and Álvarez-Cobelas 2001). Sedimentation appeared to be the main nutrient removal process in TDNP (Table 5.2), as in most wetlands (Howard-Williams 1985; Moustafa et al. 1998). Nutrient inputs did not control sedimentation rates in a straightforward manner, since they appeared to be dependent upon wetland productivity and, hence, upon internal loading (Sánchez-Carrillo and Álvarez-Cobelas 2001). Nutrient deposition rates were very high in this wetland and much higher than those in the Everglades (Davis 1991) and mineral soil wetlands (Johnston 1991). TP and TN removal efficiency depended on hydrological conditions, while internal loadings were much more important (Table 5.2) than those in lakes (Ekholm et al. 1997; Knuuttila et al. 1994), increasing in wet years. Retention values at TDNP (Table 5.2) were similar to those of Moustafa et al. (1998), Ekholm et al. (1997) and Johnston (1991). TP retention fell as inflow increased in Tablas de Daimiel, a result that conflicted with that of Marion and Brient (1998). Macrophyte leaching values were similar to those of Harris et al. (1995) and increases in nutrient input did not promote leaching (Table 5.2). Contrary to the results of Manny et al. (1994) for other wetlands, waterfowl was not an important source of nutrients in TDNP (Table 5.2). Nutrient loss by infiltration, however, meant an important reduction of TN and TP content in the wetland (Table 5.2), both fractions correlating with the hydrological dynamics of TDNP (Sánchez-Carrillo and Álvarez-Cobelas 2001).

In years with high flooding, this wetland behaved as a sink for carbon and phosphorus, also suggested by their increasing content in the NE inlet and terminal outlet sediments over time (Fig. 5.10). Sedimentary nitrogen showed changing patterns

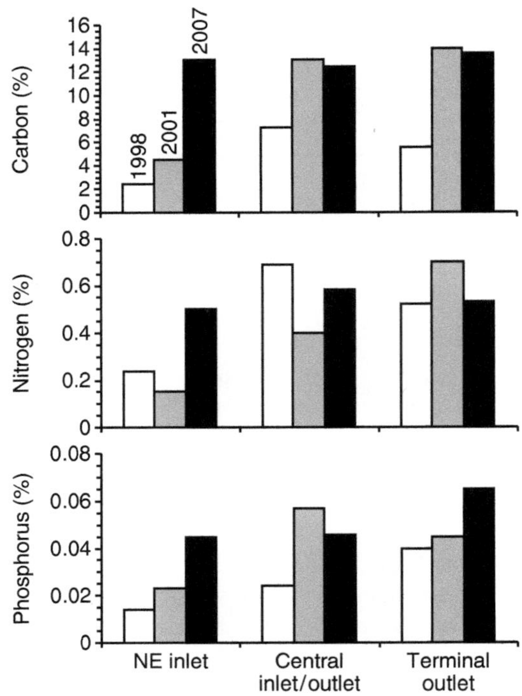

Fig. 5.10 Total carbon, nitrogen and phosphorus concentration (as a percentage of sediment dry weight) in wetland sediments in 1998–2007. PG: NE inlet, MM: central outlet, PN: terminal outlet

over time in all three studied sites, maybe arising from the more complex interplay of factors impinging on the nitrogen cycle. In the former (until 1986) central inlet, sediment nutrient dynamics, however, might experience differential processes resulting in no accumulation trend (Fig. 5.10). Sedimentary carbon ranged 1.5–13.6% of dry sedimentary matter, of which 23–99% was organic carbon. Sedimentary nitrogen was between 0.1% and 0.8%, whereas phosphorus amounted to 0.004–0.065% (Fig. 5.10). These figures were higher than others reported for municipal and crop-impacted wetlands in Nature Reserves (Wang et al. 2004).

5.6 Spatial Heterogeneity of Nutrient Controls

Environmental controls of wetland nutrients changed among sites (Table 5.3). At the wetland inlet, conductivity and total organic carbon were related, chlorophyll-*a* was related with ammonia and dissolved organic carbon, whereas total phosphorus was related with ammonia and soluble reactive phosphorus, thus suggesting a common wastewater origin (Table 5.3). At the central inlet/outlet, conductivity was also related with total organic carbon and inversely with total nitrogen and nitrate, which co-varied; chlorophyll-*a* did not appear to be controlled by any other

Table 5.3 Pearson correlations among physico-chemical variables in three areas of the wetland. Statistically significant values at $p < 0.05$ are shown in bold lettering. Floo: flooding, Con: conductivity, TOC: total organic carbon, DOC: dissolved organic carbon, TN: total nitrogen, NO_3: nitrate, NH_4: ammonia, TP: total phosphorus, SRP: soluble reactive phosphorus, TN:TP: total nitrogen-to-total phosphorus atomic ratio, Chl-a: chlorophyll-*a*

	Floo	Con	TOC	DOC	TN	NO_3	NH_4	TP	SRP	N:P
NE inlet										
Cond	−0.26									
TOC	−0.27	**0.55**								
DOC	−0.12	0.27	0.17							
TN	−0.21	0.31	0.31	0.04						
NO_3	0.08	−0.22	−0.18	−0.20	0.01					
NH_4	−0.26	0.06	0.07	**0.70**	0.27	−0.14				
TP	−0.22	−0.08	−0.03	0.23	0.18	−0.20	**0.61**			
SRP	−0.24	−0.15	−0.11	−0.02	0.09	−0.05	0.21	**0.59**		
TN:TP	−0.16	0.24	0.23	−0.17	0.38	0.06	−0.02	−0.21	−0.10	
Chl-a	−0.22	0.37	0.33	**0.90**	0.11	−0.15	**0.53**	0.22	0.08	0.03
Central inlet/outlet										
Cond	0.02									
TOC	−0.12	**0.67**								
DOC	−0.06	0.32	0.21							
TN	−0.02	**−0.51**	−0.38	0.00						
NO_3	0.06	**−0.64**	−0.48	−0.12	**0.85**					
NH_4	0.13	0.03	0.11	0.07	0.01	0.01				
TP	0.15	0.12	0.08	0.02	−0.03	0.00	0.30			
SRP	0.09	−0.13	−0.05	0.05	0.10	0.16	0.03	0.46		
TN:TP	−0.04	−0.23	−0.17	−0.09	0.33	0.29	−0.05	−0.32	−0.09	
Chl-a	0.03	0.47	0.41	0.14	−0.15	−0.27	−0.08	0.28	0.14	−0.14
Terminal outlet										
Cond	**−0.57**									
TOC	**−0.57**	**0.76**								
DOC	−0.14	0.01	−0.02							
TN	−0.36	0.43	0.34	−0.15						
NO_3	−0.21	**0.64**	**0.52**	−0.05	**0.50**					
NH_4	−0.29	**0.51**	0.43	−0.06	0.29	**0.58**				
TP	−0.23	0.28	0.34	−0.15	0.14	0.07	0.26			
SRP	−0.36	0.44	**0.55**	−0.07	0.24	0.19	0.20	**0.62**		
TN:TP	0.11	−0.12	−0.12	−0.09	0.12	0.04	−0.05	−0.31	−0.19	
Chl-a	−0.39	0.36	0.34	−0.09	0.25	−0.05	0.06	0.31	0.19	−0.15

variable (Table 5.3). At the terminal outlet, flooding appeared to inversely control conductivity and total organic carbon, conductivity being also straightforwardly related to nitrate and ammonia concentration; total nitrogen was related to ammonia and nitrate, whereas soluble reactive phosphorus co-varied with total phosphorus (Table 5.3). The nitrogen-to-phosphorus ratio showed no statistical relationship with any of the other variables tested in any wetland site.

Multiple regression models explained high variability for total organic carbon concentration, but performed poorly for the remaining nutrients and chlorophyll-*a* (Table 5.4) in the three wetland sites where long data records were available.

Table 5.4 Multiple significant ($p < 0.05$) regressions between total nutrients and chlorophyll-*a* (chl-a) and physico-chemical predictors in different sites of the wetland. Data were $\log_{10}$- transformed before analysis. The Akaike information criterion AIC has also been included to enable model comparison, but it could not be calculated for phosphorus models. TP regression at the Central inlet/outlet was non statistically significant ($p > 0.05$). Interc.: interception, Coeff.: coefficient, SE: standard error, cond.: conductivity, TOC: total organic carbon, TN: total nitrogen and TP: total phosphorus

					Independent variables										
Dependent variables	R^2	AIC		Interc.	River discharge	Incoming cond.	Incoming TOC	Flooding	NE inlet cond.	NE inlet TOC	NE inlet chl-a	Central cond.	Central TOC	Terminal cond.	Terminal TP
NE Inlet															
TOC	0.64	−68	Coeff.	−1.12					1.02						
			SE	0.34					0.1						
TN	0.26	−29	Coeff.	1.33	0.18	−0.93	0.07	−0.14	0.7						
			SE	0.83	0.06	0.29	0.02	0.06	0.25						
TP	0.06		Coeff.	−0.16				−0.19							
			SE	0.19				0.07							
Chl-a	0.13	83	Coeff.	−0.98				−0.21	0.84						
			SE	1.14				0.09	0.3						
Central inlet/ outlet															
TOC	0.58	−105	Coeff.	0.61					−0.65	0.71		0.69			
			SE	0.33					0.13	0.1		0.07			
TN	0.09	−0.2	Coeff.	1.86								−0.39			
			SE	0.38								0.11			
Chl-a	0.06	161	Coeff.	0.99							0.25				
			SE	0.14							0.09				

Terminal outlet									
TOC	0.73	−181	Coeff.	−0.45		−0.21	0.28	0.85	
			SE	0.33		0.09	0.09	0.07	
TN	0.2	−14	Coeff.	1.02	−0.18				
			SE	0.09	0.03				
TP	0.13		Coeff.	−4.27				0.96	
			SE	1.12				0.31	
Chl-a	0.37	58	Coeff.	2.6	−0.35				0.31
			SE	0.16	0.07				0.08

One month-lagged models did not improve the variability explained. Seasonally-lagged models, such as those used in eutrophication studies, did not perform better.

The spatial scale of environmental control of wetland dynamics is usually hypothesized to change as a result of water connectedness, and so, the higher the water renewal, the larger the spatial scale of environmental control. Connectivity is a great concern when addressing drought effects (Bond et al. 2008) similar to those that have been affecting Las Tablas de Daimiel. However, a previous study has demonstrated that strong site effects have prevailed on nutrient chemistry in this wetland, irrespective of the degree of wetland connectivity, since there have been very few significant correlations among different sites for a single nutrient and no significant correlations with loadings (Álvarez-Cobelas et al. 2007). Causes for this behaviour might be that water inputs to this wetland have been highly variable, and levelling ecological effects have been dampened by site processes. Other supporting evidences for this overwhelming importance of in-site processes are that 1st) no temporal trend has been observed in nutrient contents over the years, 2nd) no simultaneous seasonal patterns have been obvious in nutrient data from the same site, or for the same nutrient at different sites, and 3rd) periodicity analyses have not shown overlapping of annual rhythms. These results challenge the current view of the spatial scale of environmental control of wetland performance and suggest that the role of water availability in ecological connectivity may be more complex than previously suspected. Our study poses interesting problems to be addressed when attempting nutrient monitoring in wetlands that experience highly variable water inputs and suggests that the functioning of that type of wetland might be similar to that of wetlands created for wastewater treatment (Álvarez-Cobelas et al. 2007).

5.7 Wetland Sedimentation and Siltation

Mass accumulation rates reported for Las Tablas de Daimiel wetland (TDNP, hereafter) (Fig. 5.11 upper and middle panels) were extraordinarily high when compared with other ecosystems (Sánchez-Carrillo et al. 2001). Some similarity in the sedimentation of TP appeared at the wetland open-water areas, comparable to those of Mitsch and Reeder (1991) from a freshwater coastal wetland of Lake Erie. Similarly, sedimentation rates on particulate matter measured at the wetland inflow were somewhat higher than those of Hensel et al. (1998), obtained from riverine sites of the Rhône River Delta (France) and those of Meeker (1996) from riverine zones of a coastal wetland of Lake Superior. Finally, TN and TP sedimentation rates in the middle macrophyte-dominated zone were also slightly higher than those reported by Johnston (1991) for a riparian forest levee in Wisconsin. Accumulation rates in TDNP concurred more with shallow lakes records, *i.e.*, those of the large shallow Lake Vortjärv in Estonia (Noges et al. 1999) and of the small shallow Lake Tämnaren in Sweden (Bengtsson and Hellström 1992).

Accretion rates in TDNP (Fig. 5.11 lower panel) were also higher (annual averages during 1997 and 1998: 2.22 ± 0.96 cm year^{-1} and 2.46 ± 0.99 cm year^{-1}, respectively) than in many other wetlands. Accretion measured at the inlet was similar to that in

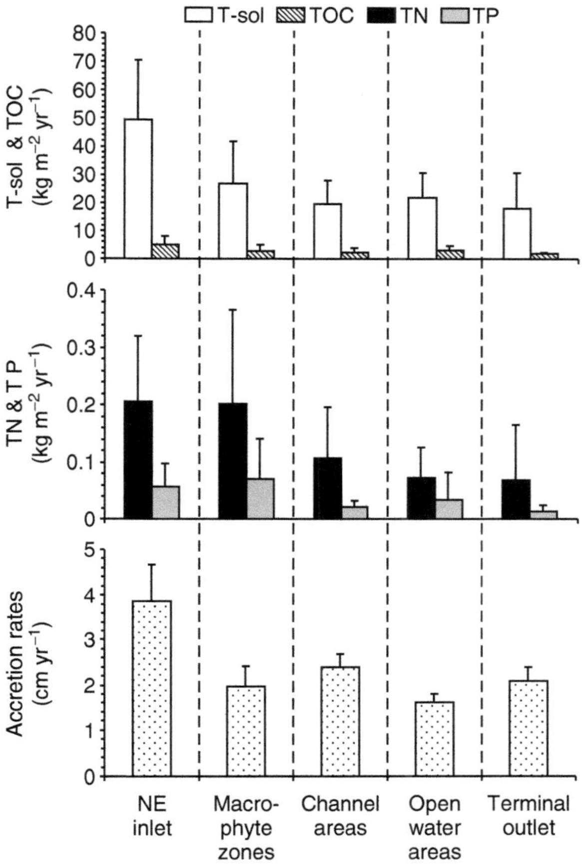

Fig. 5.11 Annual accumulation rates of total solids (T-sol), total organic carbon (TOC), total nitrogen (TN), total phosphorus (TP) as well as accretion in geomorphologically distinctive zones of TDNP during 1997 and 1998 (Data source from Sánchez-Carrillo et al. 2001)

wetlands of Central Pennsylvania (Wardrop and Brooks 1998). At the remaining sites, accretion rates were comparable to the data reported by Thoms (1998) from floodplain wetlands in Australia, and those of Johnston et al. (1984) and Eckblad et al. (1977) from a wetland in Illinois and a floodplain lake in the Mississippi river, respectively.

High sedimentation may be associated with activities such as agriculture, hydrologic modification, urban runoff, unsatisfactory wastewater treatment and erosion from uplands due to vegetation losses (Wardrop and Brooks 1998). At present, the TDNP wetland is threatened by all of them.

Seasonal sedimentation rates peaked at the end of Summer and in Fall, which matched the senescent stage of macrophytes (Sánchez-Carrillo et al. 2001). Johnston et al. (1984) observed similar relationships in a Wisconsin wetland, even though its values were higher in Autumn during litterfall. High TP sedimentation rates recorded at TDNP in Spring (Sánchez-Carrillo et al. 2001) concurred with the seasonality observed by Meeker (1996) in a wetland of Lake Superior (Wisconsin), where higher rates were reported during the initial vegetation growth periods than for those in the senescent stage.

Spatial heterogeneity in Tablas de Daimiel sedimentation rates (Fig. 5.11) is similar to that reported by Hupp and Bazemore (1993) for a forested freshwater wetland in Tennessee (USA), with numerous channels and dams. These authors indicated that the variations of accumulation rates were related to changes in topography (water-depth) that generated different flow velocities on the floodplain. In TDNP, with similar features, the explanation for the effect of water level variance on sedimentation was not high (R^2 = 0.13; Sánchez-Carrillo et al. 2001). Sedimentation rates were lower at the deeper sites in TDNP (outlet dam area) and higher at the shallower sites dominated by emergent macrophytes (Fig. 5.11), concurring with the inverse relationship between depth and sedimentation rates found by Janus et al. (1990). Reddy et al. (1993) reported from the Everglades that there seemed to be an inverse relation between yearly TP sedimentation and the distance from the inflow as occurred in TDNP (Fig. 5.12), although the decrease in sedimentation rates in the first kilometer after inflow is greater than in the Everglades. The exponential reduction of sedimentation might explain the TDNP sedimentation spatial pattern, with the inlet zone operating as a huge sediment sink in which allochthonous material was stored. In fact, some areas close to the inlet channel are so silted nowadays that they no longer support flooding (personal observation).

Although sedimentation depends to a high degree on external inputs, as has been reported by Marion and Brient (1998) from a wetland in Grand-Lieu (France), such a relationship was weakly resolved at TDNP (Sánchez-Carrillo et al. 2001). Sedimentation of autochthonous organic and inorganic matter generated in the wetland prevails in TDNP (Sánchez-Carrillo et al. 2000). The proportion of settling autochthonous organic matter oscillated from 45% to 65%, while TN and TP sedimentation accounted for over 80% when inflows decreased (Sánchez-Carrillo et al. 2000). Our results are consistent with Meeker's (1996) observations on a wetland of Lake Superior which showed increased sedimentation of organic matter in zones dominated by emergent macrophytes.

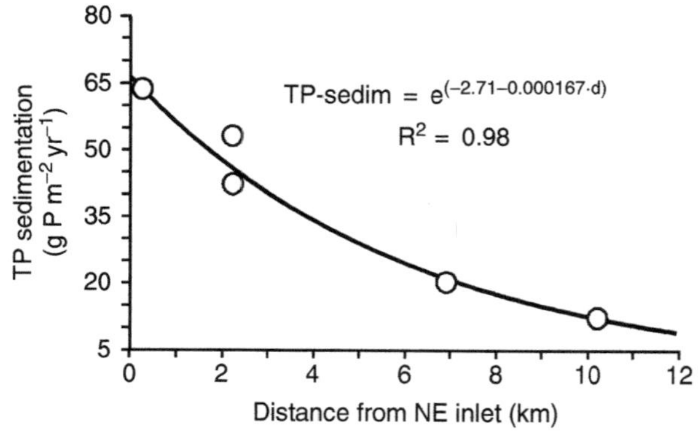

Fig. 5.12 Relationship between distance (d in the equation) from the NE inlet and total annual phosphorus accumulation in TDNP during 1997–1998 (Data source from Sánchez-Carrillo et al. 2001)

In many cases, organic matter accumulating in wetlands originates from emergent macrophytes. Davis (1991) has shown that vegetation accumulates P and N in detritus, which contributes to accretion of organic sediments and nutrients in the Everglades. The net effect of macrophyte production and decomposition is detritus accumulation (Davis 1991). Reed vegetation occupies vast areas in the wetland (see Chapter 7), and its annual reproduction cycle contributes to periodic accumulations of detritus in the sediment (Mitsch and Gosselink 2000). A large fraction of accumulated organic matter may chiefly be derived from emergent vegetation, as shown by a marked statistical relationship between emergent macrophyte covers and total sedimentation rates of organic matter (Reed cover $R^2 = 0.78$, cut-sedge $R^2 = 0.57$; Sánchez-Carrillo et al. 2001). This finding is in agreement with observations made by Butler and Malanson (1995) in Montana ponds (USA) and by Reddy et al. (1993) in the Everglades.

Alluvial wetlands are depositional features in the geomorphological sense, since all appear as a result of sediment deposition (Richards 1982). However, high sediment deposition, as is the case in TDNP, may cause siltation and hence significant alterations in the biological communities (Vargo et al. 1998). As in other wetlands, sedimentation in TDNP depends on a variety of factors, such as hydrology, external inputs, mineral precipitation, and production of detritus derived from the biota. Mass accumulation and high mineralization of organic matter dictates the annual development of vegetation, via nutrient recycling (Wetzel 1990), this resulting in continuously increasing detritus accumulation. In a previous study (Sánchez-Carrillo et al. 2000) we demonstrated through a simple sedimentation model that, if the currently high annual organic matter accumulation rates are maintained, wetland siltation could occur within the next 100 years. Mass accumulation must, therefore, be taken into account in management strategies, in order to preserve this important ecosystem (see Chapters 10 and 11).

References

Álvarez-Cobelas M, Cirujano S (2003) Las Tablas de Daimiel National Park. http://www.humedalesibericos.com/basic/finished projects

Álvarez-Cobelas M, Cirujano S (eds) (1996) Las Tablas de Daimiel: ecología acuática y sociedad. Ministerio de Medio Ambiente, Madrid

Álvarez-Cobelas M, Cirujano S, Sánchez-Carrillo S (2001) Hydrological and botanical man-made changes in the Spanish wetland of Las Tablas de Daimiel. Biol Conserv 97:89–97

Álvarez-Cobelas M, Sánchez-Carrillo S, Cirujano S (2007) Strong site effects dictate nutrient patterns in a Mediterranean floodplain. Wetlands 27:326–336

APHA (1989) Standard methods for the examination of water and wastewater, 17th edn. AWWA, Washington, DC

Bachmann RW, Canfield DE (1996) Use of an alternative method for monitoring total nitrogen concentrations in Florida lakes. Hydrobiologia 323:1–8

Bengtsson L, Hellström T (1992) Wind-induced resuspension in a small shallow lake. Hydrobiologia 241:163–172

Bloesch J (1996) Towards a new generation of sediment traps and a better measurement/understanding of settling particle flux in lakes and oceans: a hydrodynamical protocol. Aquat Sci 58:283–296

Bond NR, Lake PS, Arthington J (2008) The impacts of drought on freshwater ecosystems: an Australian perspective. Hydrobiologia 600:3–16

Breitburg DL, Baxter JW, Hatfield C, Howarth RW, Jones CG, Lovett GM et al (1998) Understanding effects of multiple stressors: ideas and challenges. In: Pace ML, Groffman M (eds) Successes, limitations and frontiers in ecosystem science. Springer Verlag, New York, pp 416–431

Butler DR, Malanson GP (1995) Sedimentation rates and patterns in beaver ponds in a mountain environment. Geomorphology 13:255–269

Davis SM (1991) Growth, decomposition, and nutrient retention of Cladium jamaicense Crantz and Typha domingensis Pers. in the Florida Everglades. Aquat Bot 40:203–224

Day JW Jr, Westphal A, Pratt R, Hyfield E, Rybczik J, Kemp GP, Day JN, Marx B (2006) Effects of long-term municipal effluent discharge on the nutrient dynamics, productivity, and benthic community structure of a tidal freshwater forested wetland in Louisiana. Ecol Eng 27:242–257

Eckblad JW, Peterson NL, Ostlie K (1977) The morphometry, benthos and sedimentation rates of a floodplain lake in Pool 9 of the Upper Mississippi River. Am Midl Nat 97:433–443

Eisler R, Wiemeyer SN (2004) Cyanide hazards to plants and animals from gold mining and related issues. Rev Environ Contam Toxicol 183:21–54

Ekholm P, Malve O, Kirkkala T (1997) Internal and external loading as regulators of nutrient concentrations in the agriculturally loaded Lake Pyhäjärvi (southwest Finland). Hydrobiologia 345:3–14

Elmers MC, Watmough SA, Buttle JM, Dillon PJ (2007) Drought-induced sulphate release from a wetland in south-central Ontario. Environ Monit Assess 127:399–407

Eriksson H, Pastuszak M, Löfgren S, Mörth CM, Humborg C (2007) Nitrogen budgets of the Polish agriculture 1960–2000: implications for riverine nitrogen loads to the Baltic Sea from transitional countries. Biogeochemistry 85:153–168

Gunatilaka A (1991) Nutrient cycling in a freshwater reed marsh. Verhandlungen der Internationalen Vereinigung für Limnologie 24:980–983

Harris TT, Williges KA, Zimba PV (1995) Primary productivity and decomposition of five emergent macrophyte communities in the Lake Okeechobee marsh ecosystem. Arch Hydrobiol 45:63–78

Hensel PF, Day JW Jr, Pont D, Day JN (1998) Short-term sedimentation dynamics in the Rhône River Delta, France: the importance of riverine pulsing. Estuaries 21:52–65

Hopkinson CS jr (1992) A comparison of ecosystem dynamics in freshwater wetlands. Estuaries 15:549–562

Howard-Williams C (1985) Cycling and retention of nitrogen and phosphorus in wetlands: a theoretical perspective. Freshw Biol 15:391–431

Hupp CR, Bazemore DE (1993) Temporal and spatial patterns of wetland sedimentation, West Tennessee. J Hydrol 141:179–196

INYPSA (1990) Estudio de seguimiento de los resultados del Plan de Regeneración Hídrica del Parque Nacional Las Tablas de Daimiel. Final Report, Madrid

Janus LL, Scybala DM, Jones BA (1990) Nutrient budget analyses and phosphorus loading goal for Lake Okeechobee, Florida. Verhandlungen der Internationalen Vereinigung für Limnologie 24:538–546

Johnston CA (1991) Sediment and nutrient retention by freshwater wetlands: effects on surface water quality. Crit Rev Environ Control 21:491–565

Johnston CA, Bubenzer GD, Lee GB, Madison FW, Mc Henry JR (1984) Nutrient trapping by sediment deposition in a seasonally flooded lakeside wetland. J Environ Qual 13:283–290

Knuuttila S, Pietiläinen OP, Kauppi L (1994) Nutrient balances and phytoplankton dynamics in two agriculturally loaded shallow lakes. Hydrobiologia 275(276):359–369

Maltby E, Hogan DV, Immirzi CP, Tellam JH, van der Peijl MJ (1994) Building a new approach to the investigation and assessment of wetland ecosystem functioning. In: Mitsch WJ (ed) Global wetlands: old world and new. Elsevier Science, Amsterdam, pp 637–658

Manny BA, Johnson WC, Wetzel RG (1994) Nutrient additions by waterfowl to lakes and reservoirs: predicting their effects on productivity and water quality. Hydrobiologia 279(280):121–132

Marion L, Brient L (1998) Wetland effects on water quality: input-output studies of suspended particulate matter, nitrogen (N) and phosphorus (P) in Grand-Lieu, a natural plain lake. Hydrobiologia 373(374):217–235

Marker AFH, Nusch EA, Rai H, Rieman B (1980) The measurement of photosynthetic pigments in freshwater and standardisation of methods: conclusions and recommendations. Arch Hydrobiol 14:91–106

Meeker JE (1996) Wild-rice and sedimentation processes in a Lake Superior coastal wetland. Wetlands 16:219–231

Mitsch WJ, Gosselink JG (2000) Wetlands, 3rd edn. Van Nostrand Reinhold, New York

Mitsch WJ, Reeder BC (1991) Modelling nutrient retention of a freshwater coastal wetland: estimating the roles of primary productivity, sedimentation, resuspension and hydrology. Ecol Model 54:151–187

Moustafa MZ, Fontaine TD, Guardo M, James RT (1998) The response of a freshwater wetland to long-term "low level" nutrient loads: nutrients and water budget. Hydrobiologia 364:41–53

Noges P, Tuvikene L, Noges T, Kisand A (1999) Primary production, sedimentation and resuspension in large shallow Lake Vortsjärv. Aquat Sci 61:168–182

OECD (1982) Eutrophication of waters. Monitoring, assessment, control. OCDE, Paris

Öquist MG, Svensson BH (1996) Non-tidal wetlands. In: Watson RT, Zinyowera MC, Moss RH (eds) Climate change 1995, impacts, adaptations and mitigation of climate change: scientific technical analyses. Cambridge University Press, Cambridge, pp 215–239

Phillips JD (1989) Fluvial sediment storage in wetlands. Water Resour Bull 25:867–873

Phillips JD (1996) Wetland buffers and runoff hydrology. In: Mulamoottil G, Warner BG, McBean EA (eds) Wetlands: environmental gradients, boundaries, and buffers. CRC Press, New York, pp 207–220

Reddy CN, DeLaune RD, DeBusk WF, Koch MS (1993) Long-term nutrient accumulation rates in the Everglades. Soil Sci Soc Am J 57:1147–1155

Richards KS (1982) Rivers: form and process in alluvial channels. Methuen, London

Sáez-Royuela R (1977) Contribución al estudio ecológico de Las Tablas de Daimiel. III. Las aguas. Anales INIA 3:101–149

Sánchez-Carrillo S, Álvarez-Cobelas M (2001) Nutrient dynamics and eutrophication patterns in a semiarid wetland: the effects of fluctuating hydrology. Water Air Soil Pollut 131:97–118

Sánchez-Carrillo S, Álvarez-Cobelas M, Angeler DG, Vizcayno-Muñoz C (2000) Tasas de acreción y características de los sedimentos actuales en el parque nacional Las Tablas de Daimiel (Ciudad Real). Estudios Geológicos 56:239–250

Sánchez-Carrillo S, Álvarez-Cobelas M, Angeler DG (2001) Sedimentation in the semi-arid freshwater wetland Las Tablas de Daimiel (Spain). Wetlands 21:67–79

Testa JM, Kemp WM, Boynton WR, Hagy JD III (2008) Long-term changes in water quality and productivity in the Patuxent River Estuary: 1985 to 2003. Estuaries Coasts 31:1021–1037

Thoms MC (1998) Floodplain–wetlands: transient storage areas of sediment and pollutants. In: Williams WD (ed) Wetlands in a dry land: understanding for management. Environment Australia, Biodiversity Group, Canberra, Australia, pp 205–215

Tudesque L, Gevray M, Grenouillet G, Lek S (2008) Long-term changes in water physicochemistry in the Adour–Garonne hydrographic network during the last three decades. Water Res 42:732–742

Vargo SM, Neely RK, Kirkwood SM (1998) Emergent plant decomposition and sedimentation: response to sediment varying in texture, phosphorus content and frequency of deposition. Environ Exp Bot 40:43–58

Wang GP, Liu JS, Tang J (2004) The long-term nutrient accumulation with respect to anthropogenic impacts in the sediments from two freshwater marshes (Xianghai Wetlands, Northeast China). Water Res 38:4462–4474

Wardrop DH, Brooks RP (1998) The occurrence and impact of sedimentation in Central Pennsylvania wetlands. Environ Monit Assess 51:119–130

Wetzel RG (1990) Land-water interfaces: metabolic and limnological regulators. Verhandlungen der Internationalen Vereinigung für Limnologie 24:6–24

Part III
Ecological Communities

Chapter 6
Plankton Ecology and Diversity

C. Rojo and M.A. Rodrigo

Abstract In this chapter we describe plankton responses induced by fluctuating hydrology and eutrophication process in the semi-arid TDNP wetland. We have followed the planktonic community since 1992, at seasonal and interannual scales, covering as well the spatial heterogeneity of the wetland. The studied planktonic components were bacterioplankton, autotrophic picoplankton, nano and microphytoplankton and zooplankton, including ciliates. Plankton has been studied in terms of species composition (diversity), functional groups, spatial heterogeneity, population dynamics. The present data from this wetland allow us to determine which factors (resources and conditions) are relevant for each group and how their specific responses affect the food web. Bacterioplankton biovolume ranged between 0.1 and 5.0 $mm^3\ l^{-1}$, autotrophic picoplankton (0–1.7 $mm^3\ l^{-1}$), nano- and microphytoplankton (0.03–774 $mm^3\ l^{-1}$, more than 90% nanoplankton), zooplankton (0.01–280 $mm^3\ l^{-1}$, more than 50% ciliates).

Plankton community shows an evident spatial heterogeneity, with the seasonal and inter annual dynamic of all groups differing between wetland areas. Shallowness and hydrologic conditions impinge strongly on the seasonal dynamics of plankton communities; patterns during drought and wet periods were different. Therefore, a lack of pattern among plankton assemblages is the most remarkable feature. We suggest that alternative states of plankton taxonomic structure in this eutrophic semiarid wetland occur as a sequence of substitutions, due to close coupling with sediment and hydrological disturbance. The autogenic trajectory is not attained and the trophic relationships or top-down control are possibly irrelevant.

C. Rojo (✉)
Cavanilles Institute of Biodiversity and Evolutionary Biology, University of Valencia, P.O. Box 2085, E-46071 Valencia, Spain
e-mail: carmen.rojo@uv.es

S. Sánchez-Carrillo and D.G. Angeler (eds.), *Ecology of Threatened Semi-Arid Wetlands: Long-Term Research in Las Tablas de Daimiel*, DOI 10.1007/978-90-481-9181-9_6,

6.1 Introduction

Wetland systems are characterized by remarkable environmental heterogeneity, caused by the complex interactions between surface water, groundwater and riparian systems; this sets the stage for high biodiversity (Ward and Tockner 2001; Rodrigo et al. 2002; Pithart et al. 2007). Moreover, in semiarid wetlands, matter and energy flows occur as pulses in the form of high and low water periods, when the environments become isolated from the inlet channels and between them (see Chapters 3 and 5). This behaviour also allows for the presence of more co-existent competitors than in stable systems and, thus, higher biodiversity than would be expected (Levins 1979; Rojo and Álvarez-Cobelas 2003).

However, the drought or hydrological stress of these environments during a period of several months or years, favours only the abundance of acclimating species (*sensu* Reynolds 1997), limiting richness. In addition, Chase (2007) found much higher species similarity among communities in water bodies than in those experiencing drought, and this, possibly, was the result of filtering out species from the regional pool which are unable to withstand such environmental harshness. As a result, the species pool is homogenised, reducing diversity at regional scales. Therefore, aquatic biodiversity in a semiarid wetland is a balance between richness reached through environmental heterogeneity and species loss, due to stressing conditions.

On the other hand, from a dynamic point of view, a lower richness in a wetland may be the result of similar seasonal assemblages (alternate states, Allen et al. 1977) in different places, and higher global richness is produced when alternative states occur depending on local conditions (Scheffer and van Nes 2007). Therefore, the challenge when studying plankton communities of a semiarid wetland is to broach their spatial and temporal variability, taking into account different time scales and environmental disturbances which affect stability and connectivity of different parts of wetlands; these are the goals of this study. In this chapter we report results about planktonic communities in Las Tablas de Daimiel wetland from 1996 to 2002 (monthly samplings), and compare them with the results obtained during the period 1992–1993. The study also embraces the spatial heterogeneity of this wetland, which faces severe hydrological changes (see Chapter 3).

Another interesting ecological issue is the response of different hierarchical levels (populations, taxonomical or functional groups) to environment factors and their changes (Reynolds et al. 2002; Marquet et al. 2005; Wagner and Adrian 2009). The resulting diversity of plankton groups acts as the dynamics of primary and secondary producers and the food web function (Beckerman et al. 2006; Angeler et al. 2000; Ortega-Mayagoitia et al. 2002).

Plankton has been studied in terms of species composition, diversity, population dynamics, functional and taxonomical groups, spatial heterogeneity, and factors controlling the planktonic food web. The planktonic components studied were bacterioplankton, autotrophic picoplankton, nano- and microphytoplankton and zooplankton, including ciliates. The present data from this wetland allow us to determine which factors (resources and conditions) are relevant for each group and how their specific responses affect the food web. We consider this study a contribution to a better understanding of the semiarid aquatic ecosystems, less studied than temperate ones and so different from them (Álvarez-Cobelas et al. 2005).

6.2 Materials and Methods

6.2.1 Field Sampling and Counting Techniques

Five sampling sites were chosen for this study (Fig. 6.1): (a) The shallow input site of the Gigüela river, with emergent vegetation which acts as a natural wastewater treatment (input site). (b) The main water entrance to the wetland, a very shallow (usually lower than 0.5 m deep) flow-through channel located at the NE area of the wetland, which often receives wastewater and silt from a nearby village (NE-inlet). (c) A deeper pond area (1.0–1.5 m deep) located in the middle of the wetland (Central inlet). This site was formerly a water input site from a groundwater-fed stream, but aquifer overexploitation caused this site to dry-up permanently since 1986. Dark grey peat serves as a source of dissolved colour and frequent winds contribute to sediment re-suspension and high turbidity. (d) The central site is 0.5–1 m

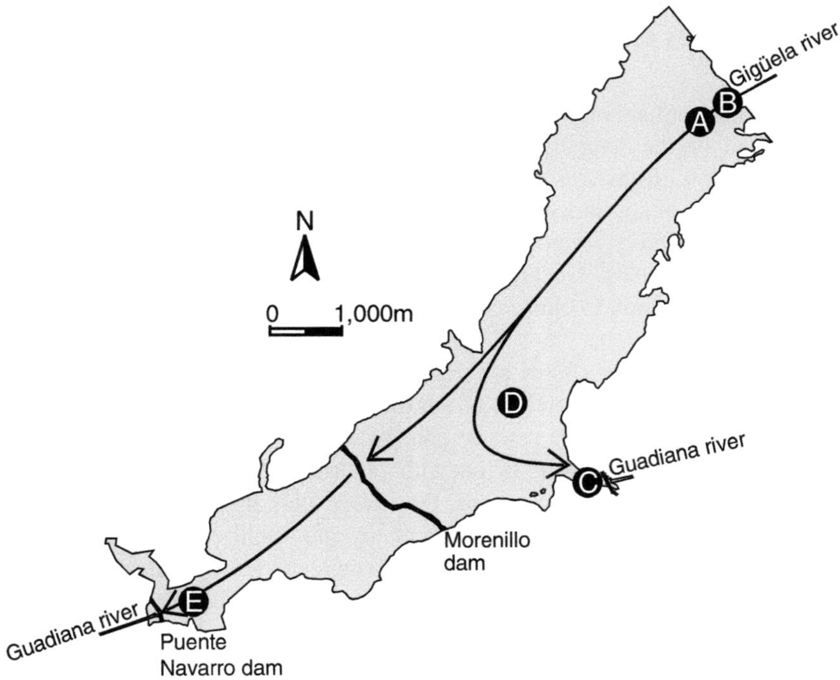

Fig. 6.1 Location of the sites studied at Las Tablas de Daimiel National Park: A: inlet site, B: NE-inlet, C: central inlet, D: central site and E: outlet (see text for a detailed explanation). *Arrows* show the main flooding pattern in the wetland

deep, is located at the centre of the wetland and characterised by extensive *Chara hispida* meadows. (e) The terminal south-western outlet site is a 4 m deep area during high floods (Outlet). It lies at a distance of 14 km downstream from the Inlet site. All sampling sites were hydrologically connected from January to June in 1997 and 1998 (see Chapters 1 and 3). More details about spatial heterogeneity in the wetland are described by Alvarez-Cobelas et al. (2001 and 2007).

Data used for this study correspond to 1992–1993 and 1996–2002 (in monthly intervals). Bacterioplankton diversity was determined by the use of the 16S rRNA gene as molecular marker (see more details in D'Auria et al. 2010). Picoplankton was counted using fluorescence microscopy: autofluorescence was used for autotrophic picoplankton (APP) and acridine orange staining for bacterioplankton (see more details in Ortega-Mayagoitia et al. 2002; Rodrigo et al. 2003). The remaining plankton fractions were identified and counted using an inverted microscope (Rojo et al. 2000a; Ortega-Mayagoitia and Rojo 2000a, 2000b) obtaining counting errors of 5% (Lund et al. 1958). Biomass was calculated on a fresh weight basis, using geometrical volumes in the case of phytoplankton, ciliates and rotifers (Rott 1981; McCauley 1984), and the allometric equations reported by McCauley (1984) for crustaceans. More details about methods are described by Ortega-Mayagoitia and Rojo (2000a, 2000b), Rojo et al. (1999, 2000a), Conforti et al. (2005) and Rodrigo et al. (2003).

6.2.2 Indexes and Statistical Methods

The well known Shannon–Wiener formula (Shannon and Weaver 1963) has been used to evaluate the diversity; the data set for calculation was the biovolume of each population (Rojo and Álvarez-Cobelas 1993) and calculated on clusters obtained at 97% of similarity for bacterioplankton gene sequences (D'Auria et al. 2010). The equitability (evenness, E) as potential diversity index was also calculated (Magurran 1988). Jaccard similarity index and average linkage between groups were calculated to cluster years and sites as a function of the presence of plankton species.

Pearson's correlations (Sokal and Rholf 1981) were calculated among flooding area, conductivity, concentrations of nutrients (total nitrogen, total phosphorus, nitrate, ammonia and soluble reactive phosphorus), TN:TP ratio, organic material, total organic carbon and Chlorophyll-a, with total fresh weight and percentage of different taxonomic or functional groups. Data were log-transformed (x + 1) prior to analyses, to meet statistical criteria for normality and stabilize variances. Backward, stepwise, multiple regression analyses were employed for detecting relationships between (i) primary and secondary producers (food web effects) and (ii) groups of secondary producers and abiotic variables (independent from food web). Cross-correlation analyses were carried out, with a month's delay, to relate total biomass of different planktonic groups with other biotic and abiotic variables.

Analyses of variance (ANOVAs, Sokal and Rholf 1981) were used to test spatial heterogeneity, interannual differences and the effect of some relevant disturbances on plankton abundance. As intra-group variance can be high in such fluctuating environments, the homogeneity of variance (Levene test) was taken into account, to prevent its masking the statistical significance of results. Effects were considered significant at $p < 0.05$. Time series trends were calculated by regression fit, and autocorrelation analyses were performed to detect periodicity (SPSS statistics 17.0).

6.3 Diversity of Heterotrophic and Autotrophic Picoplankton

The first results regarding bacterioplankton diversity in the central part of the wetland show a high diversity (H = 4.9; E = 0.95) during summer. This diversity value is higher than in other types of environments where bacterial diversity has been described by similar methodology, for example the monimolimnion of lake Pavin in France (H = 3.1–3.6; Lehours et al. 2007) or the water column of a freshwater pond, also in France (H = 4.0; Briée et al. 2007). In the Korean Woopo wetland, Baik et al. (2008) reported a value of H = 4.4. The high bacterial diversity in the Tablas de Daimiel wetland may be due, among other factors, to the influence of the submerged vegetation where many other organisms (epiphytic algae, zooplankters, macroinvertebrates, fish, etc.) associated with them, can provide nutrient sources, attachment surfaces, microenvironments, etc. for certain types of microorganisms (Basu et al. 2000).

The phylum Proteobacteria was the most abundant, with 46% of retrieved clones (as in the Korean Woopo wetland; Baik et al. 2008) followed by Bacteroidetes (14%) (Fig. 6.2). A very important and striking feature is that certain amounts (11%) of the retrieved clones belong to sequences without close references in the public sequence repository. Among Proteobacteria, classes Deltaproteobacteria and Betaproteobacteria were the most abundant (Fig. 6.2). Although certain general patterns typical of freshwater lakes and rivers (Zwart et al. 2002), such as the dominance of the phyla Proteobacteria and Bacteroidetes, which can also be observed in this brackish wetland waters, this system has a particular microbiota. The new phylogroups belonged mainly to phylum Proteobacteria (specifically to families Desulfobacteraceae, Hydrogenophilaceae and Chromatiaceae), phylum Bacteroidetes (Sphingobacteria), Chloroflexi (Anaerolineae) and Acidobacteria. Thus, the Tablas de Daimiel wetland may be considered a reservoir of bacterial biodiversity, characterized by a high number of new or previously not well described bacterial taxa (D'Auria et al. 2010).

6.4 Bacteria and Autotrophic Picoplankton Dynamics

The range of bacterial density (Table 6.1) in the wetland was wide and, on most occasions, nearer to values typical for eutrophic–hypertrophic waters (Sommaruga and Robarts 1997). There was an increasing tendency of bacterial abundances throughout time in the Central and Outlet sites (Fig. 6.3, Table 6.1). The annual averages were statistically different (ANOVA $p < 0.001$). Some seasonal trends in

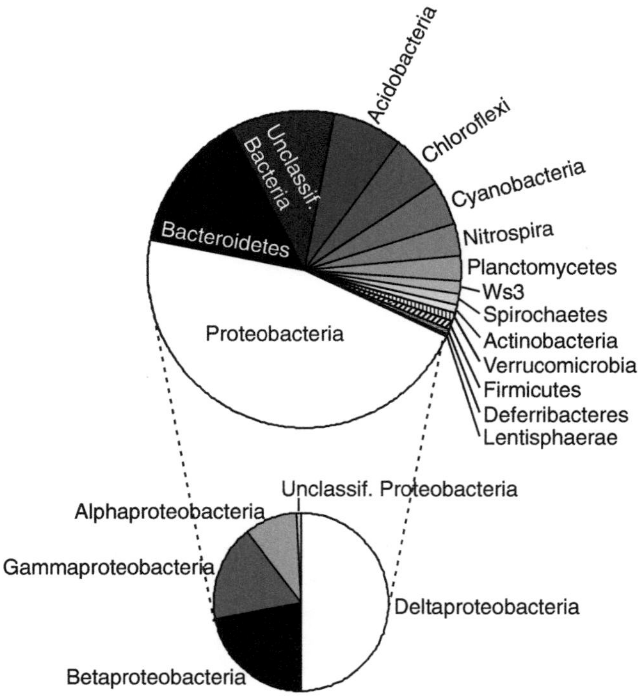

Fig. 6.2 Distribution of clone percentages among Phylum levels for bacterioplankton in summer 2007 (*upper circle*) at Central site. Detailed distribution of clone percentages among Class levels for Proteobacteria (*lower circle*)

Table 6.1 Annual averages (minimum and maximum between brackets) of bacterial and autotrophic picoplankton abundances (10^5 cells/mL). nd: no data

	Bacterial		Autotrophic	
	Central site	Outlet	Central site	Outlet
1996	11.0 (2.4–20.2)	14.9 (4.3–22.1)	0.26 (0.001–1.5)	0.02 (0.001–0.05)
1997	15.6 (8.3–23.5)	23.5 (7.4–101.1)	0.26 (0.001–2.5)	1.8 (0.001–7.4)
1998	14.2 (4.5–47.1)	16.9 (4.7–64.7)	3.96 (0.001–2.5)	4.6 (0.001–14.8)
1999	37.7 (2.3–102.1)	nd	17.02 (0.003–90.0)	nd
2000	75.0 (18.3–168.7)	92.5 (4.6–170.0)	8.26 (0.009–49.5)	20.2 (0.001–70.9)
2001	nd	nd	2.53 (0.001–13.1)	47.9 (0.02–335)

bacterial dynamics can be observed. During 1996 and 1997, there was no sharp seasonal variation in bacterial abundance, however, a clear summer maximum (June–July) was observed during the other 3 years. The water level may have affected bacterial dynamics, since there was little variation between the cool and warm seasons in the dry year (1996). By contrast, during more humid years, there were clear peaks in summer, although the abundance peaks in winter

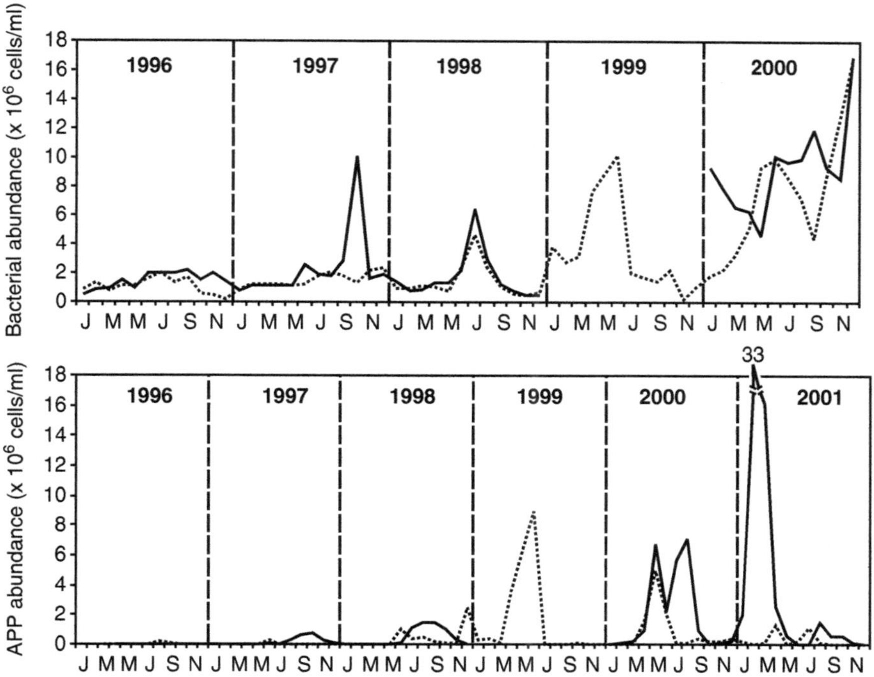

Fig. 6.3 Bacterial (*upper graph*) and APP (*lower graph*) abundance dynamics at Central (*dotted line*) and Outlet (*solid lines*) sites of TDNP

(when temperature was low) had no statistically significant correlation to temperature. There was no significant correlation of organic matter with bacterial abundance, as reported by other authors (Jeppesen et al. 1997).

The autotrophic picoplankton community was mainly represented by phycocyanin-containing coccoid cyanobacteria (0.9–1.5 μm in diameter) that fluoresced red under green excitation. No aggregates of picophytoplankton were observed. The APP abundance range was also wide (Table 6.1), varying from 100 to 33 × 10^6 cells/mL. The APP dynamics also exhibited the same increasing trend, but only until 1999, decreasing later on (Fig. 6.3). The annual averages were statistically different (ANOVA $p < 0.01$). In 1999, a large population growth was observed, starting in April and with a large peak in June. An almost identical pattern was observed during 2000, although with lower values. Two smaller peaks were observed in 2000. In 2001, APP biomass was very different at the outlet: the largest peak recorded in the wetland was measured at the end of February, with a value as high as 33 × 10^6 cells/mL. The abundance decreased considerably later. On several occasions, APP maximum development was prior to the maximum of larger phytoplankton (see Section 6.6 Phytoplankton). However, the above mentioned APP peak in February coincided with another maximum of phytoplankton biomass, consisting mainly of *Ochromonas* sp. and *Anabaena sphaerica*. The highest APP abundance reported in the literature (Sommaruga and Robarts 1997: 3.4 × 10^6 cells/mL; Stockner et al. 2000: 1.3 × 10^7 cells/mL) is far from the maximum abundance recorded at the Las Tablas de Daimiel wetland. In summary, APP exhibits seasonal patterns in this wetland, although differing between sites, a circumstance caused by spatial heterogeneity (Rojo et al. 2000a; Ortega-Mayagoitia et al. 2000). Some of the patterns are similar to those reported for lakes, but other features, such as large winter peaks, are unique.

6.5 Species Composition and Diversity of Phytoplankton

Two hundred and twenty three species of phytoplankton were identified in the wetland, including all sites sampled from 1992 to 2002 (Table 6.2): 45 taxa were Cyanobacteria, 10 were Dinophyceae, 37 Euglenophyceae, 1 Xantophyceae, 11 Cryptophyceae, 12 Chrysophyceae, 25 Bacillarophyceae, 5 Zygnematales, 77 Chlorophyceae. Moreover, a group of unidentified small flagellates was also recorded.

Table 6.2 Total phytoplankton taxa found from 1992 until 2002

Cyanophyceae
Anabaena bergii Ostenfeld
Anabaena sphaerica Bornet et Flahault
Anabaena sp.
Anabaenopsis circinalis (C. S. West) Wolosynka and Miller
Anabaenopsis elenkinii Miller
Aphanizomenon flos-aquae (Linn.) Ralfs ex Bornet and Flahault
Aphanizomenon sp.
Aphanocapsa delicatissima W. et G. S. West
Aphanocapsa holsatica (Lemmermann) Cronberg et Komárek
Aphanothece minutissima (W. West) Komárková-Legnerová et Cronberg
Chroococcus microscopicus Komárková-Legnerová et Cronberg
Chroococcus minor (Kützing) Nägeli
Chroococcus minimus (Keissler) Lemmermann
Chroococcus minutus (Kützing) Nägeli
Cyanodictyon planctonicum Meyer
Geitlerinema cf. *amphibium* (Agardh ex Gomont) Anagnostidis
Geitlerinema cf. *nematodes* (Skuja) Anagnostidis
Geitlerinema cf. *tenue* (Anisimova) Anagnostidis
Glocapsa turgida (Kützing) Nägeli
Lemmemaniella cf. *Flexa* Hindák
Limnothrix amphigranulata (Can Goor) Meffert
Limnothrix redekei (Van Goor) Meffert
Lingbya sp.
Merismopedia punctata Meyen
Merismopedia tennuissima Lemmermann

(continued)

Table 6.2 (continued)

Microcystis aeruginosa (Kützing) Kützing
Microcystis flos-aquae (Wittrock) Kirchner
Oscillatoria lanceaformis Kalbe
Oscillatoria tenuis Agardh ex Gomont
Phormidium formosum (Bory ex Gomont) Anagnostidis
Phormidium tenue (Bory ex Gomont) Anagnostidis
Planktolyngbya brevicellularis Cronberg et Komárek
Planktolyngbya contorta (Lemmermann) Anagnostidis et Komárek
Planktolyngbya limnetica (Lemmermann) Komárkova-Legnerová et Cronberg
Planktothirx agardhii (Gomont) Anagnostidis et Komárek
Pseudanabaena catenata Lauterborn
Pseudanabena cf. *frigida* (Fritsh) Anagnostidis
Pseudanabena galeata Böcher
Pseudanabaena limnetica Lemmermann
Pseudanabena cf. *minima* (G.S.An) Anagnostidis
Pseudanabaena raphidioides (Geitler) Anagnostidis et Komárek
Spirulina abreviata Lemmermann
Spirulina meneghiniana Zanardini ex Gomont
Synechocystis aquatilis Sauvageau
Synechococcus sp.
Dinophyceae
Ceratium hirundinella (O. F. Müller) Dujardin
Glenidinium cf. *inaquale* Chodat
Gymnodinium eurytopum Skuja
Gymnodinium mitratum Schiller
Gymnodinium cf. *wawrikae* Schiller
Katodinium fungiforme (Anisimova) Loeblish III
Peridinium cinctum (O. F. Müller) Ehrenberg
Peridinium inconspicurum Lemmermann
Peridinium umbonatum Stein
Peridinium willei Huitfeld-Kaas
Euglenophyceae
Astasia sp.
Euglena acus Ehrenberg
Euglena agilis Schmarda
Euglena anabaena var. *minima* Mainx
Euglena clara Skuja
Euglena clavata Skuja
Euglena deses var. intermedia Klebs
Euglena cf. *elastica* Prescott
Euglena gracilis Klebs
Euglena mutabilis Schmitz
Euglena oxyuris Schmitz
Euglena polymorpha Dangeard
Euglena cf. *proxima* P. A. Dang
Euglena sanguinea Ehrenberg

(continued)

Table 6.2 (continued)

Euglena spiroides Lemmermann
Euglena splendens Dangeard
Euglena tripteris (Dujardin) Klebs
Euglena viridis Ehrenberg
Lepocinclis acuta Prescott
Lepocinclis fusiformis (Carter) Lemmermann
Lepocinclis ovum (Enrenberg) Lemmermann
Phacus aenigmaticus Drezepolski
Phacus agilis Skuja
Phacus brachykenton Pochmann
Phacus brevicaudatus (Klebs) Lemmermann
Phacus costatus Conrad
Phacus inflexus (Kisselev) Pochmann
Phacus granum Drezepolski
Phacus polytrophos Pochmann
Phacus pyrum (Ehrenberg) Stein
Phacus splendens Pochmann
Phacus sp.
Trachelomonas abrupta Swirenko emend. Deflandre
Trachelomonas armata (Ehrenberg) Stein
Trachelomonas intermedia P. A. Dang
Trachelomonas sculpta Balech
Trachelomonas volvocinopsis Swirenko
Xantophyceae
Centritractus sp.
Cryptophyceae
Chloromonas sp.
Chroomonas sp.
Chroomonas caudata L. Geitler
Cryptomonas erosa Ehrenberg
Cryptomonas erosa var. *reflexa* Marsson
Cryptomonas marsonii Skuja
Cryptomonas obovata Skuja
Cryptomonas phaseolus Skuja
Cryptomonas rostratiformis Skuja
Plagioselmis lacustris Pascher and Ruttner
Plagioselmis nannoplanctica Skuja
Chrysophyceae
Chrysochromulina parva Lackey
Chrysococcus cf. *cordiformis biconicus* Ettl
Chrysococcus sp.
Desmarella cf. *brachycalyx* Skuja
Desmarella moniliformis Kent
Desmarella sp.
Kephyrion sp.
Monosiga ovata Kent

(continued)

Table 6.2 (continued)

Ochromonas cf. *vischeri* Bourrelly
Ochromonas sp.
Ophiocytium capitatum Wolle
Primnesium cf. *saltans* J. Massart
Bacillariophyceae
Asterionella formosa Hassall
Aulacoseira granulata (Ehrenberg) Simonsen
Chaetoceros muelleri Lemmermann
Cyclotella atomus Hustedt
Cyclotella meneghiniana Kützing
Cyclotella ocellata Pantocsek
Fragilaria ternera (W. Smith) Lange-Bertalot
Fragilaria ulna (Nitzsh) Lange-Bertalot
Fragilaria cf. *brevistriata* Grun. In Van Heurck
Navicula cryptocephala Kützing
Navicula cuspidata (Kützing) Kützing
Navicula cf. *Halophila* (Grunow) Cleve
Navicula oblonga Kützing
Navicula sp.
Nitzschia acicularis (Kützing) W. Smith
Nitzschia amphibia Grunow
Nitzschia closterium (Enrenberg) E. Smith
Nitzschia gracilis Hantzsch
Nitzschia cf. *pusilla* Grunow
Nitzschia palea (Kützing) W. Smith
Nitzschia reversa W. Smith
Nitzschia sigmoidea (Kützing) W. Smith
Nitzschia tryblionella Hantzsh in Rabenh.
Synedra ulna Ehrenberg
Rhizosolenia sp.
Zignematales
Closterium cynthia De Notaris
Closterium ehrenbergii Meneghini ex Ralfs
Mougeotia sp.
Spirogyra parvua (Hassall) Kützing
Spirogyra protecta Wood
Chlorophyceae
Aktinastrum hantzschii Lagerheim
Ankistrodesmus spiralis (Turn.) Lemmermann
Ankistrodesmus falcatus (Corda) Ralfs
Ankyra judayi (Smith) Fott
Ankyra paradoxidoides (Smith) Fott
Aulacomonas hyalina HL Skuja
Pyramimonas cf. *nanella* Conrad and Kufferath
Carteria sp.
Chlamydomonas acuta Korshikov

(continued)

Table 6.2 (continued)

Chlamydomonas cf. altera Skuja
Chlamydomonas ehrenbergii Goroschankin
Chlamydomonas fusus Ettl
Chlamydomonas reindhardtii P. A. Dang
Chlamydomonas sp.
Chlorella sp.
Chlorogonium elongatum PA Dangeard
Chlorogonium sp.
Closteriopsis acicularis (G. M. Smith) Belcher and Swale
Closteriopsis longissima Lemmermann
Coelastrum astroideum De Notaris
Coelastrum microporum Nägeli in A. Braun
Crucicengia tetrapedia (Kirchn.) W. and G. S. West
Dictyosphaerium ehrenbergianum Nägeli
Dictyosphaerium tetrachotomum Printz
Eudorina elegans Ehrenberg
Golenkinia radiata Chodat
Golenkiniopsis parvula (Woronichin) Korshikov
Golenkiniopsis solitaria Korsikov
Golenkiniopsis cf. *varians* Thérèz and Couté
Golenkiniopsis sp.
Kirchneriella irregularis var. spiralis (Smith) Korschikov
Korshikoviella gracilipes (Lambert) Silva
Lagerheimia baltonica (Scherff. In Kol) Hnd.
Lagerheimia genevensis (Chodat) Chodat
Lagertheimia longiseta (Lemmerman) Wille
Lagerheimia marssonii Lemmermann
Monoraphidium arcuatum (Korschikoff) Hindák
Monoraphodium circinale (Nyg.) Nyg.
Monoraphidium contortum (Thur.) Komárkova-Legnerová
Monoraphidium convolutum (Corda) Komárkova-Legnerová
Monoraphidium dybowskii (Woloszynska) Hindák and Komárková-Legnerová
Monoraphidium griffithii (Berk.) Komárkova-Legnerová
Monoraphidium komarkovae Nygaard
Monoraphidium minutum (Nägeli) Komárkova-Legnerová
Monoraphidium pusillum (Printz) Komárková-Legnorová
Monoraphidium skujae Fott
Monoraphidium subclavatum Nyg.
Monoraphidium tortile (W. West and G.S. West) Komárkova-Legnerová
Oocystis lacustris Chodat
Oocystis sp.
Pandorina morum (O. F. Müller) Bory
Pediastrum boryanum (Turpin) Meneghini
Pteromonas angulosa (Carter) Lemmermann
Pteromonas limnetica Hortobágyi
Pteromonas cf. *robusta* (Gicklhorn)

(continued)

Table 6.2 (continued)

Scenedesmus acuminatus (Lagerheim) Chodat
Scenedesmus acutus Meyen
Scenedesmus bernardii G. M. Smith
Scenedesmus dimorphus (Turpin) Kuetzing
Scenedesmus ecornis (Ehrenberg) Chodat
Scenedesmus hortobagy Hortobagyi
Scenedesmus intermedius Chodat
Scenedesmus obliquus (Turpin) Kützing
Scenedesmus oocystis Chodat
Scenedesmus ovalternus Chodat
Scenedesmus quadricauda (Turpin) Brébisson sensu Chodat
Scenedesmus spinosus Chodat
Schroederia setigera (Schroeder) Lemmermann
Scourfieldia cf. *quadrata* Pascher
Spermatozopsis exultans Korschikoff
Tetraedron minimum (Bréb.) Hansg.
Tetraedron triangulare Korschikoff
Tetraedron cf. *trigonum* (Nägeli) Hansgirg
Tetraselmis cf. *arnoldi* (Proshkina-Lavrenko) Norris et al.
Tetraselmis suecica (Kylin) Butcher
Tetraselmis sp.
Tetrastrum staurogeniaforme (Schröder) Lemmermann

Table 6.3 Number of new species (+), lost species (–) and net change of richness () of phytoplankton at each study site, comparing the 1992–1993 list of algae with 1997–1998 and this with the 2001–2002 list. The lists of algae found in the wetland in each different period is also shown

	NE inlet	Central inlet	Outlet	TDNP
92–93/97–98	+36, –18 (18)	+40, –18 (22)	+15, –8 (7)	+50, –41 (9)
97–98/01–02	+68, –5 (63)	+63, –15 (48)	+66, –8 (58)	+11, –3 (8)

To study the changes of richness throughout different years, lists of taxa reported three different periods were compared (Table 6.3): 1992–1993 (Rojo 1996), 1997–1998 (Rojo et al. 2000a) and 2001–2002 (Lionard et al. 2005). In the first studied period, between drought stressing circumstances and a hydrological disturbance with an increase of flooding area, 50 new taxa appeared and 41 of those originally recorded were lost. Small local changes implied large changes in the overall algal list. From 1997 to 2002, when the state of low flooding was recovered, great changes in local algal composition only resulted in 11 new species in the wetland and only three species disappeared. Then, when flooding was maintained in the wetland, an exchange between local algal compositions was observed. Richness dynamics in the Central inlet (Fig. 6.4) showed a mean richness of 17

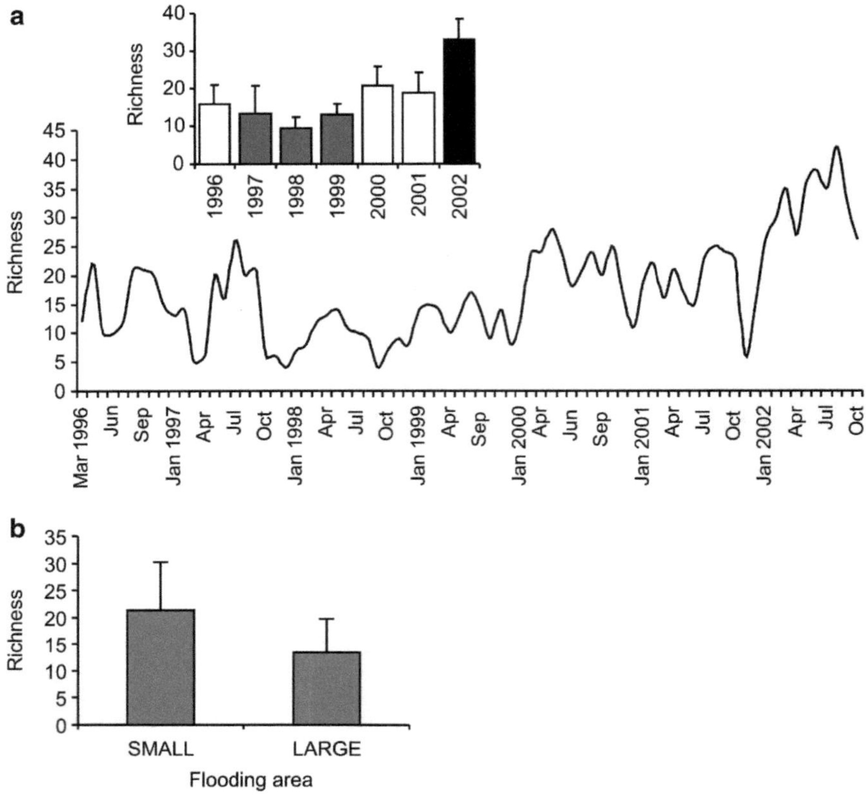

Fig. 6.4 (**a**) Dynamics of phytoplankton richness during the 7 years study at Central inlet. The annual richness averages are shown in histogram. Different colours indicate similar years clustered by a post-hoc Bonferroni test. (**b**) Average richness observed when flooding area was small (<800 ha) or large (>800 ha). Bars represent standard deviations

species per sample (standard deviation = 8 species). A statistically significant trend towards richness increase ($r = 0.58$, $p < 0.001$) occurred during the undisturbed period (1997–2002). Lower richness was recorded during 1997–1999 and higher richness from 2000 to 2002. An ANOVA test confirmed the different richness between years (Fig. 6.4; $p < 0.0001$): richness average was significantly higher when the flooding area was smaller in comparison to a lower richness during the hydrological disturbance in 1997–1998, when the flooding area experienced an increase (Fig. 6.4).

The Shannon–Wiener diversity index, calculated on the biovolumes of populations (Fig. 6.5), shows a range between 0.4 bits in October 2002 and 3.9 bits in July 1997 (mean = 2.4 bits, standard deviation = 0.7 bits). High differences between index values occurred even in consecutive samples, and when the diversity trajectories of different areas is compared, it is evident that there is an obvious lack of similarity (Rojo et al. 2000b). Evenness average was 60% (standard deviation = 14%), which

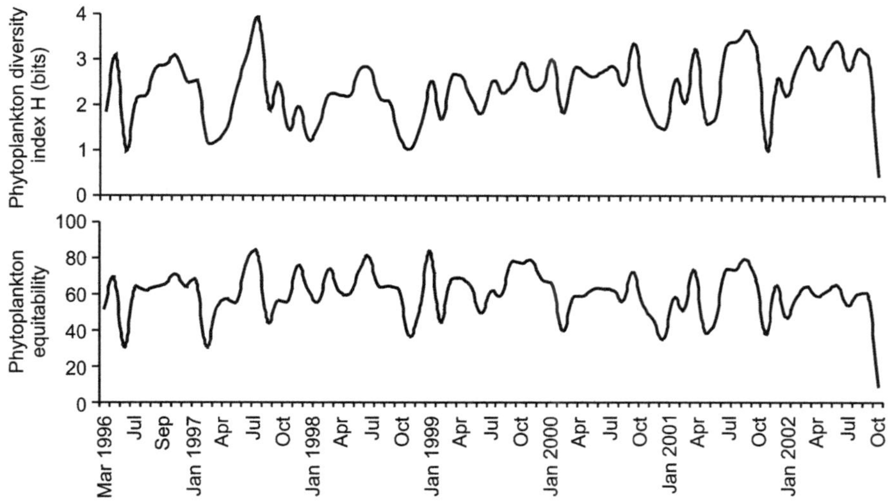

Fig. 6.5 Dynamics of Shannon–Wiener index, calculated on phytoplankton bio volume, and equitability during 7 years study at Central inlet

means communities are far from their best possible location (they have frequent dominant species). Evenness was also lowest in October 2002 (Fig. 6.5) due to a *Cyclotella meneghiniana* bloom (70% on total biomass). Temporal trends in diversity index or evenness were not statistically significant. Mean diversity was similar between years, ranging between 2.0 bits in 1999 and 2.7 bits in 2002. Their coefficient of variation implies seasonal variability from 1996 to 1998 (CV = 29%, 40%, 29% respectively), quite stationary diversity from 1999 to 2000 (CV = 16%, 20%) and highly irregular but not clearly seasonal diversity from 2001 to 2002 (CV = 36%, 33%).

6.6 Phytoplankton Abundance Dynamics

From 1996 to 2002, abundance of phytoplankton varied irregularly, inter annually and among sites. In this sense, the average density and standard deviation in the NE-inlet site were $22 \times 10^3 \pm 39 \times 10^3$ ind mL^{-1} (minimum value 183 ind mL^{-1}, maximum 231×10^3 ind mL^{-1}, CV = 177%). In the Central inlet, mean and variation were $38 \times 10^3 \pm 56 \times 10^3$ ind mL^{-1} (minimum value 162 ind mL^{-1}, maximum 367×10^3 ind mL^{-1}, CV = 145%) and in the Outlet, mean density was $57 \times 10^3 \pm 153 \times 10^3$ ind mL^{-1} (min 88 ind mL^{-1}, max $1{,}200 \times 10^3$ ind mL^{-1}, CV = 270%). The average densities showed a geographical gradient (Lionard et al. 2005), increasing almost a third from the NE-inlet site to the Central inlet and Outlet sites. The peaks of total density were achieved through species blooms, rather than by a

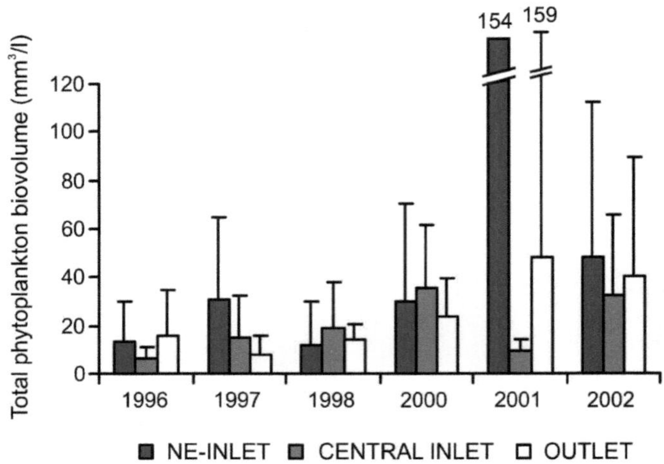

Fig. 6.6 Annual average of total phytoplankton biovolume at NE-inlet, Central inlet and Outlet sites. Bars are standard deviation values

simultaneous increase of several species, this fact being highlighted by the diversity indices. Moreover the peaks were not simultaneous between sites and the same population never peaked in two sites at a time (Lionard et al. 2005).

From 1996 to 2002, the dynamics of phytoplankton biovolume show that the minimum and maximum values of biovolume were in the NE-inlet site (0.03 mm^3 L^{-1} in spring 1996 and 774 mm^3 L^{-1} in summer 2001, *Planktothrix agardhii* bloom). As for annual averages (Fig. 6.6), both the lowest and most conservative total biovolumes were registered in the Central inlet site (coefficient of variation range = 60–100%). The highest value was measured in the NE-inlet (CV range = 110–180%) and the most irregular biovolume dynamics were at the Outlet site (CV range = 50–230%). However, an ANOVA test did not find any significant differences between sites (throughout years) or between years, using the biovolume of all samples. Intrannual variability of biovolume in all sites is high enough to mask the interannual differences between them.

The dominance of taxonomic groups (Fig. 6.7) varied throughout time, without any evident spatial or temporal pattern. No percentage of taxonomic groups was statistically different between years. An ANOVA test on the percentage of taxonomic groups, carried out during the 6 studied years, showed their spatial heterogeneity (Fig. 6.8). Only Cyanobacteria had both different mean biovolume and percentage between sites, being higher in the NE-inlet ($p < 0.0001$). Cryptophyceans had lowest percentages in that site, and Bacillariophyceae were lower at the Outlet site. There, again, we have an example of a lack of assemblage pattern.

A lack of seasonality and alternative dominance of species were detected when focussing on the density and biovolume dynamics of phytoplankton at the Central inlet (Fig. 6.9). An autocorrelation analysis did not reveal any periodicity. Biovolume showed a positive trend during the 7 studied years

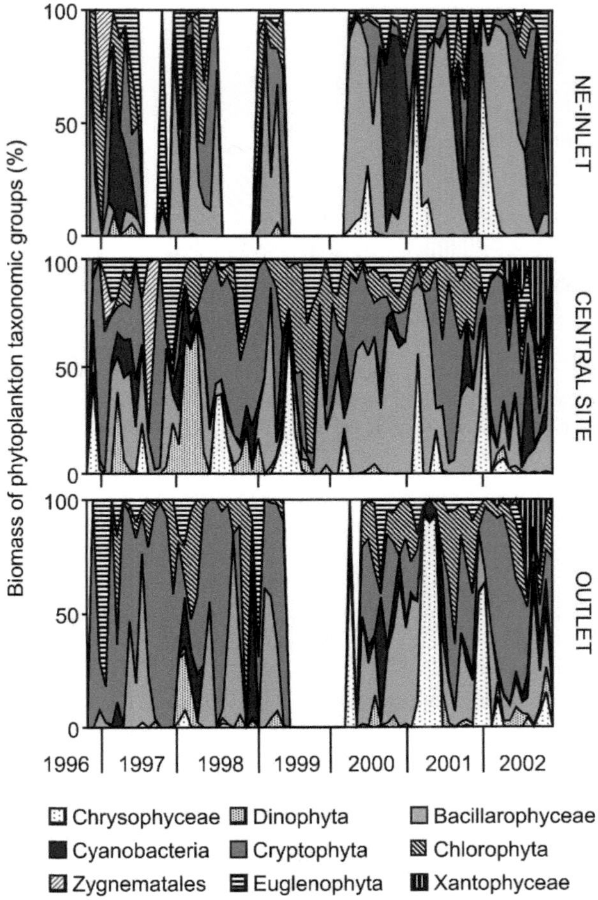

Fig. 6.7 Percentage of biomass of phytoplankton taxonomic groups at NE-Inlet, Central inlet and Outlet, from 1996 to 2002

($r = 0.26$, $p = 0.022$); moreover, when the time series was analysed from 1996 to 2000 (excluding the extremely low values of 2001), the increase of biovolume was expressed by a regression coefficient of 0.5 mm^3 L^{-1} per month ($p < 0.0001$). Despite this statistically significant trend, the interannual fluctuation was also evident (Fig. 6.10).

The dynamics of taxonomic groups point out fluctuating alternate phytoplankton assemblages (Fig. 6.7) with a weak pattern: Cryptophyceans in Winter–Spring, Cyanobacteria and Dinophyceae appeared together in summer, and diatoms alternating with them from spring to autumn. The statistical value of their correlation coefficients (Tables 6.4 and 6.5) corroborated these apparent negative relationships. Relationships between structural features of phytoplankton assemblages (Tables 6.4 and 6.5) indicate that the total biomass, which co-varied with chlorophyll-a, was

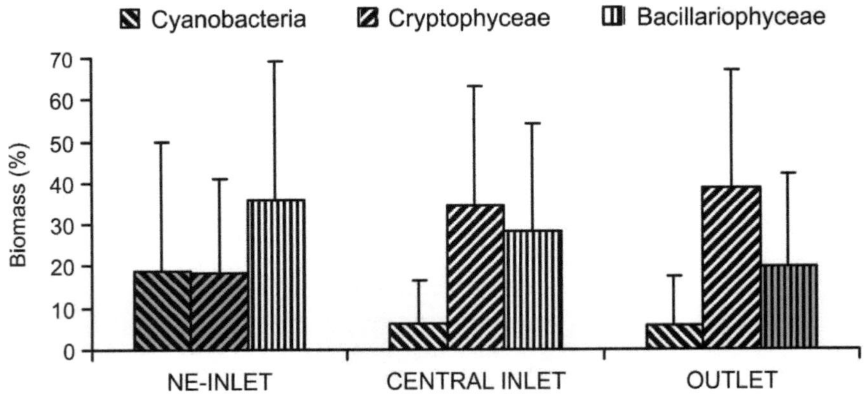

Fig. 6.8 Average of percentage of Cyanobacteria, Cryptophyceae and Bacillariophyceae bio volume found in each site from 1996 to 2002. Bars are standard deviation values. Similar background highlights no statistically significant differences (ANOVA, post hoc Bonferroni test)

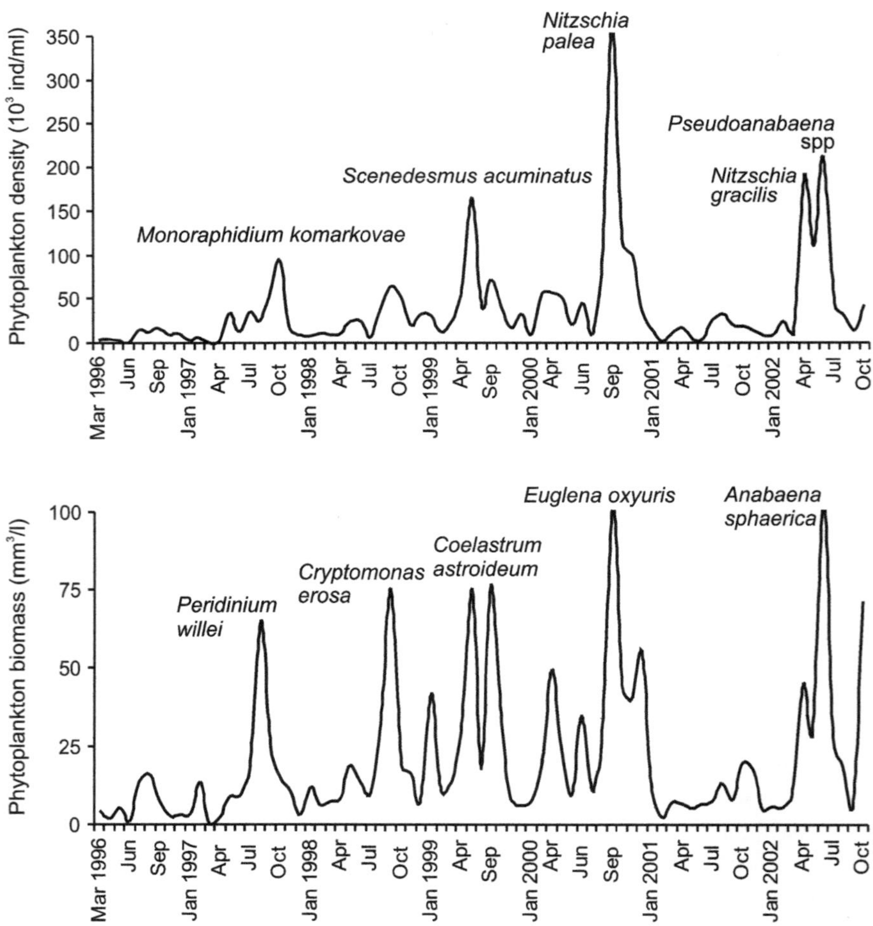

Fig. 6.9 Dynamics of total density and total bio volume of phytoplankton in Central inlet of TDNP during 7 years monthly sampling. The dominants species, which produced peaks, are indicated

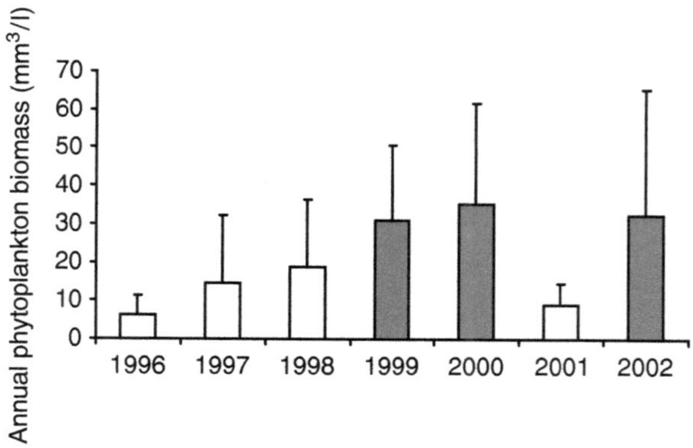

Fig. 6.10 Annual changes of mean phytoplankton biomass in Central inlet. Bars indicate standard deviation. Statistically different data are shown in grey (Bonferroni *post* hoc test)

achieved thanks to the biomass of all taxonomic groups (Table 6.4). Thus, the amount of planktonic primary producers can not be related to a single taxonomic group (Table 6.5).

6.7 Species Composition and Diversity of Zooplankton

From 1992 to 2002, a total richness of 115 zooplankton species has been found; of them, 84 were Rotifers, 22 Cladocerans and 9 Copepods (Table 6.6). Based on the available data on spatial heterogeneity of zooplankton richness (1996–1998) a similar richness of the different groups is observed between sites in the wetland (Fig. 6.11). Zooplankton richness can be similar between wetlands in a trophic or disturbance gradient, even when zooplankton biomass is clearly different between them (Beaver et al. 1999).

The most common taxa were bdelloid rotifers and *Lecane closterocerca* (present in 93% and 79% of samples, respectively), as well as *Testudinella patina* (58%) and *Lepadella patella* (55%). *Ceriodaphnia dubia, Daphnia curvirostris* and *Acanthocyclops robustus* were the most frequent crustaceans (Ortega-Mayagoitia et al. 2000). However, most species are not common in the whole area; a similarity analysis based in species composition revealed that the Central inlet is similar to the Outlet site, and they are weakly related with the NE-inlet; the Central site, however, was not related to this cluster (Fig. 6.11). Richness in the wetland, therefore, is achieved thanks to different zooplankton assemblages between areas.

Variability of richness throughout years can be observed in the Central inlet site from 1996 to 2001. The lowest numbers of rotifer species occurred in 1996 (16 populations) and maximum in 1999 (27 populations). 22% of species appeared in, at least, 5 of 6 of the years studied and 60% of species were only present in 1 or 2 years. A hierarchical analysis based on similarity of rotifers presence, clustered years 1997–1998 with 2001 and then with 1999–2000, while 1996 remained

Table 6.4 Pearson correlations between structural variables of phytoplankton assemblage at Central inlet (1996–2002). TB: total phytoplankton biomass, Ch-a: Chlorophyll a, RICH: number of species, DIV: Shannon–Wiener diversity index, and biomass percentages of CIA: Cianophyceae, DIN: Dinophyceae, EUG: Euglenophyceae, CRY: Cryptophyceae, CHR: Chrysophyceae, BAC: Bacillariophyceae, CHL: Chlorophyceae. Correlation significance at 0.01 (**) or 0.05 levels (*)

	TB	Ch-a	RICH	DIV	CIA	DIN	EUG	CRY	CHR	BAC	CHL
TB	1										
Ch-a	0.60**	1									
RICH	0.24*		1								
DIV			0.63**	1							
CIA				0.33**	1						
DIN						1					
EUG		0.29*		0.28*			1				
CRY	–0.47**	–0.36**	–0.25*	–0.26*	–0.25*	–0.26*	–0.28*	1			
CHR									1		
BAC					–0.23*	–0.28*		–0.40**	–0.24*	1	
CHL		0.30**						–0.34**			1

Table 6.5 Pearson correlations between structural variables of phytoplankton assemblage (% considered) at Central inlet (1996–2002). TB: total phytoplankton biomass; Ch-a: Chlorophyll a; RICH: number of species; DIV: Shannon–Wiener diversity index; and percentage of total biomass contributed by CIA: Cianophyceae, DIN: Dinophyceae, EUG: Euglenophyceae, CRY: Cryptophyceae, CHR: Chrysophyceae, BAC: Bacillariophyceae, CHL: Chlorophyceae. Correlation is significant at 0.01 (**) or 0.05 levels (*)

	TB	Ch-a	RICH	DIV	CIA	DIN	EUG	CRY	CHR	BAC	CHL
TB	1										
Ch-a	0.60**	1									
RICH	0.24*		1								
DIV			0.63**	1							
CIA	0.31**		0.30**		1						
DIN	0.23*				0.24*	1					
EUG	0.49**	0.29*			0.56*		1				
CRY	0.49**	0.36**	–0.25*					1			
CHR				–0.25*					1		
BAC	0.54**	0.36**					0.26*	0.35**		1	
CHL	0.40**	0.30**					0.38**				1

Table 6.6 List of zooplankton species found at Tablas de Daimiel wetland from 1992 to 2002

Rotifera
Anuraeopsis fissa Gosse 1851
Asplachna sp.
Asplachna brightwelli Gosse 1850
Asplachna girodi De Guerne 1888
Bdelloideae
Brachionus angularis Gosee 1851
Brachionus bidentata Anderson 1889
Brachionus calyciflorus Pallas 1776
Brachionus calyciflorus anuraeiformis Brehm 1909
Brachionus cf. *ibericus* Ciros-Pérez, Gómez and Sierra 2001
Brachionus leydigi Cohn 1862
Brachionus plicatilis O.F.M. 1786
Brachionus quadridentatus brevispinus Ehrenberg 1832
Brachionus rotundiformis Tschugunoff 1921
Brachionus cf. *rubens* Ehrenberg 1838
Brachionus urceolaris O.F.M. 1773
Brachionus variabilis Hempel 1896
Cephalodella catellina O.F.M. 1786
Cephallodela cf. *forficula* Ehrenberg 1838
Cephalodella gibba Ehrenberg 1838
Cephalodella sp.
Collotheca sp.
Colurella cf. *adriatica* Ehrenberg 1831
Colurella cf. *obtusa* Gosse 1886
Colurella uncinata O.F.M. 1773
Collotheca sp.
Encentrum sp.
Epiphanes senta Müller 1773
Euchlanis dilatata Ehrenberg 1832
Euchlanis sp.
Hexarthra fennica Levander 1892
Hexarthra oxyuris Sernov 1903
Keratella cochlearis var. tecta Lauterborn 1900
Keratella cochlearis Gosse 1851
Keratella quadrata O.F.M. 1786
Keratella tropica tropica Apstein 1907
Keratella tropica brehmi Klausener 1908
Keratella tropica reducta Fadeew 1927
Lecane aculeata Jaubski 1912
Lecane batillifer Murray 1913
Lecane bulla Gosse 1886
Lecane closterocerca Schmarda 1859
Lecane decipiens Daday 1913
Lecane furcata Murray 1913
Lecane hamata Stockes 1896

(continued)

Table 6.6 (continued)

Lecane hastata Murray 1913
Lecane hornemanni Ehrenberg 1834
Lecane lamellata Daday 1893
Lecane cf. *lateralis* Sharma 1978
Lecane luna O.F.M. 1776
Lecane lunaris Ehrenberg 1832
Lecane nana Murray 1913
Lecane ohioensis Herrick 1885
Lecane pyriformis Daday 1905
Lecane quadridentata Ehrenberg 1832
Lecane tenuiseta Harring 1914
Lecane cf. *scutata*
Lecane sp.
Lepadella patella O.F.M. 1786
Lepadella sp.
Liliferotrocha sp.
Lindia torulosa Dujardon 1841
Lindia sp.
Lophocaris salpina Ehrenberg 1834
Mytilina ventralis Ehrenberg 1832
Notholca acuminata Ehrenberg 1832
Notholca salina Focke 1961
Notholca squamula O.F.M. 1786
Notommata glyphura Wulf 1961
Pompholyx sulcata Hudson 1885
Platyas quadricornis Ehrenberg 1832
Polyarthra dolichoptera Idelson 1925
Polyarthra sp.
Proales sp.
Ptygura sp.
Synchaeta oblonga Ehrenberg 1831
Synchaeta sp.
Testudinella patina Hermann 1783
Testudinella patina intermedia Anderson 1889
Trichocerca cf. agnatha Wulf 1939
Trichocerca cf. elongata Gosse 1886
Trichocerca cf. rattus O.F.M. 1776
Trichotria pocillum O.F.M. 1776
Tripleuchlanis plicata Levander 1894
Cladocera
Alona costata Sars 1862
Alona quadrangularis O.F.Müller 1785
Alona rectangula Sars 1862
Alona rustica Scott 1895
Alona sp.

(continued)

Table 6.6 (continued)

Bosmina longirostris O.F.Müller 1785
Ceriodaphnia laticaudata P.E.M. 1867
Ceriodaphnia reticulata Jurine 1820
Ceriodaphnia cf. dubia Richard 1894
Ceriodaphnia sp.
Chydorus sphaericus O.F.Müller 1785
Daphnia curvirostris Eymann 1887
Daphnia magna Strauss 1820
Daphnia longispina
Daphnia parvula Fordyce 1901
Daphnia sp.
Ilyocryptus sordidus Liévin 1848
Leydigia sp.
Oxyurella tenuicaudis Sars 1862
Pleuroxus aduncus Jurine 1820
Simocephalus exspinosus DeGeer 1778
Simocephalus vetulus O.F.Müller 1776
Copepoda
Acanthocyclops robustus Sars 1863
Acanthocyclops vernalis setiger Fischer 1853
Canthocampus saphylinus Jurine 1820
Cyclops abysorum Sars 1863
Eucyclops cf. *serrulatus* Fischer 1851
Macrocyclops albidus Jurine 1820
Megacyclops viridis Jurine 1820
Tropocyclops prasinus Fischer 1860
Paracyclops fimbriatus Fisher 1853

separated (Fig. 6.11). Therefore, the presence of rotifers segregates dry, flooding and stable years and contributes to increase the global biodiversity.

6.8 Zooplankton Abundance Dynamics

A lack of pattern was evident for zooplankton assemblages in the wetland, similar to what was observed for phytoplankton spatial heterogeneity and dynamics (Ortega-Mayagoitia et al. 2000).

Zooplankton was most abundant in the Central inlet and Outlet sites (2 ± 5 mm^3 L^{-1} and 4 ± 6 mm^3 L^{-1} respectively) whereas in the Central and NE-inlet sites it was 0.3 ± 1 mm^3 L^{-1} (Fig. 6.12). Biomass distribution of taxonomic classes showed differences between years and places. Zooplankton biomass in the NE-inlet and Outlet sites was higher in the last dry year (1996); whereas in the Central site, maximum was reached in 1997 and in 1998, which were both wetter years (Fig. 6.12 and Ortega-Mayagoitia et al. 2000). Throughout these years, a lack of spatial pattern

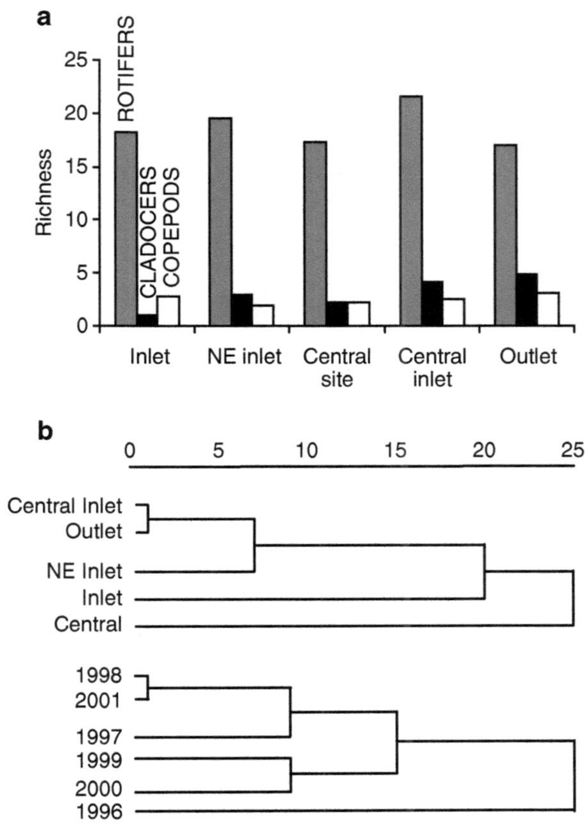

Fig. 6.11 (**a**) Total number of species of zooplankton taxonomic groups observed in different sites in the wetland from 1996 to 1998. (**b**) Dendrograms showing similarity between sites or years based on zooplankton species composition (rescaled distance cluster in the scale)

was found, and each part of the wetland had a different behaviour in relation with zooplankton changes between a dry year and after flooding. Rotifer biomass increased after flooding, possibly as a result of eutrophication, cladoceran populations decreased after flooding, likely due to macrophyte loss and, consequently, copepod biomass increased to the stabilized water level (Ortega-Mayagoitia et al. 2000). With respect to seasonality, and taking into account the mean between sites throughout the years, copepods peaked mostly in late winter and spring, cladocerans appeared in spring, alternating with copepods and, finally, this latter group was substituted by rotifers (Fig. 6.13).

As an average for five sites, ciliate biomass (Fig. 6.14) was 2 ± 3 mg/L in 1996, 3 ± 4 mg/L in 1997 and 18 ± 47 mg/L in 1998. Therefore, ciliates increased their biomass after flooding. The highest variability throughout the years was in the NE-inlet (17 ± 56 mm^3 L^{-1}) and the most stable was the Central inlet (8 ± 11 mm^3 L^{-1}). A multivariate correlation showed similar statistically significant dynamics of the

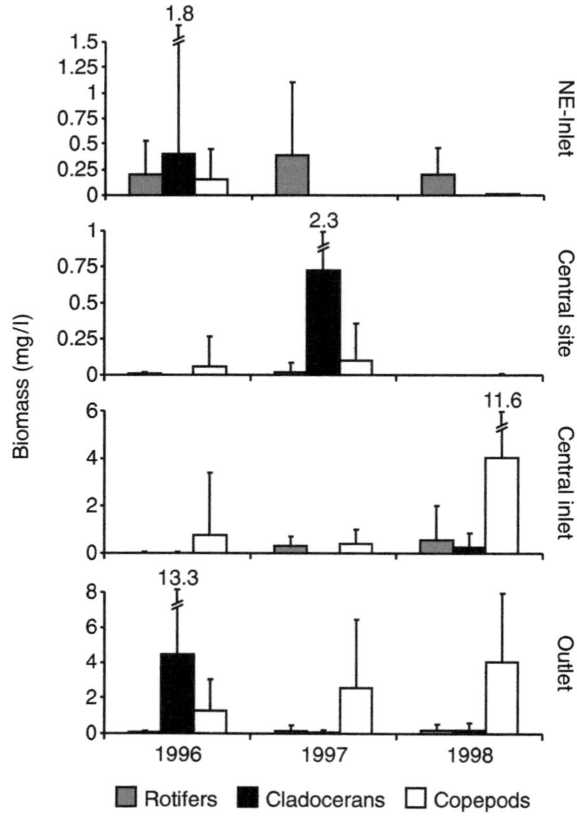

Fig. 6.12 Annual average of zooplankton biomass (taxonomic groups) reached in different sites in TDNP in 3 consecutive years. Bars are standard deviation

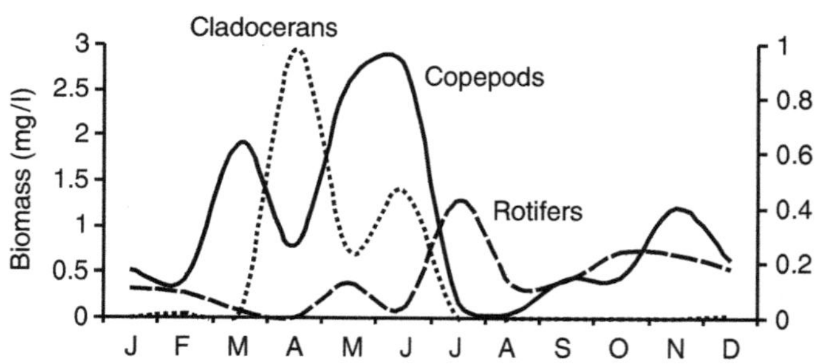

Fig. 6.13 Monthly zooplankton biomass dynamics of averaged data for sites and years. Secondary Y-axis is for rotifers

Central inlet with the Central site and the NE inlet ($r = 0.66$ and $r = 059$, respectively; $p < 0.0001$). Ciliates contributed more than 50% to total zooplankton biomass in 56% of observed samples. Seasonally, ciliates were the most important

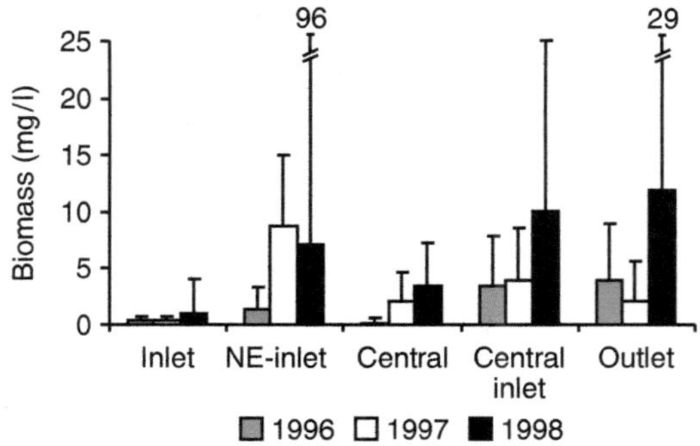

Fig. 6.14 Annual average of ciliates biomass reached in different sites of the wetland in 3 consecutive years. Bars are standard deviation

component of zooplankton in terms of biomass (Ortega-Mayagoitia et al. 2000), except in spring, when cladocerans were dominant (Fig. 6.14).

The time-course of zooplankton group abundance, from 1996 to the most stable period (1996–2001) in the Central inlet (Fig. 6.15), showed that all taxonomic groups peaked in the summer of 1998, the first year after the flooding event. Despite the difference between years, the monthly variability (test of homogeneity of variances $p < 0.001$) does not allow to confirm interannual differences for any taxonomic group. The lack of seasonality of taxonomic groups is also evident and supported by an autocorrelation analysis.

The analyses presented here show a clear difference in the zooplankton community structure in different places of the same wetland and no unidirectional temporal trend was observed; community structure is, to a considerable degree, affected by different features in each wetland site. This lack of pattern is observed in other long-term dynamics (Gislason et al. 2009).

6.9 Bacterioplankton and APP Relationship with Abiotic and Biotic Factors

The relationship between bacteria and chlorophyll has been extensively investigated in lakes and in the ocean, but remains poorly documented for wetlands. Data from this wetland support the pattern of decreasing ratio of bacterial biomass to chlorophyll-a concentration (see Chapter 5) across a trophic gradient in wetlands too, as in other systems (Currie 1990). This relationship was statistically significant and had a slope of 0.24, similar to those reported by Jeppesen et al. (1997) for hypertrophic Lake Sobygard (0.29; range of chlorophyll: 3–1,313 $\mu gChl/L^{-1}$), or the slopes of multi-lake equations, like those reported by Currie (1990) (0.2–0.4; 0.2–100 $\mu gChl/L^{-1}$), and by del Giorgio and Peters (1993) (0.27; 0.7–38 $\mu gChl/L^{-1}$).

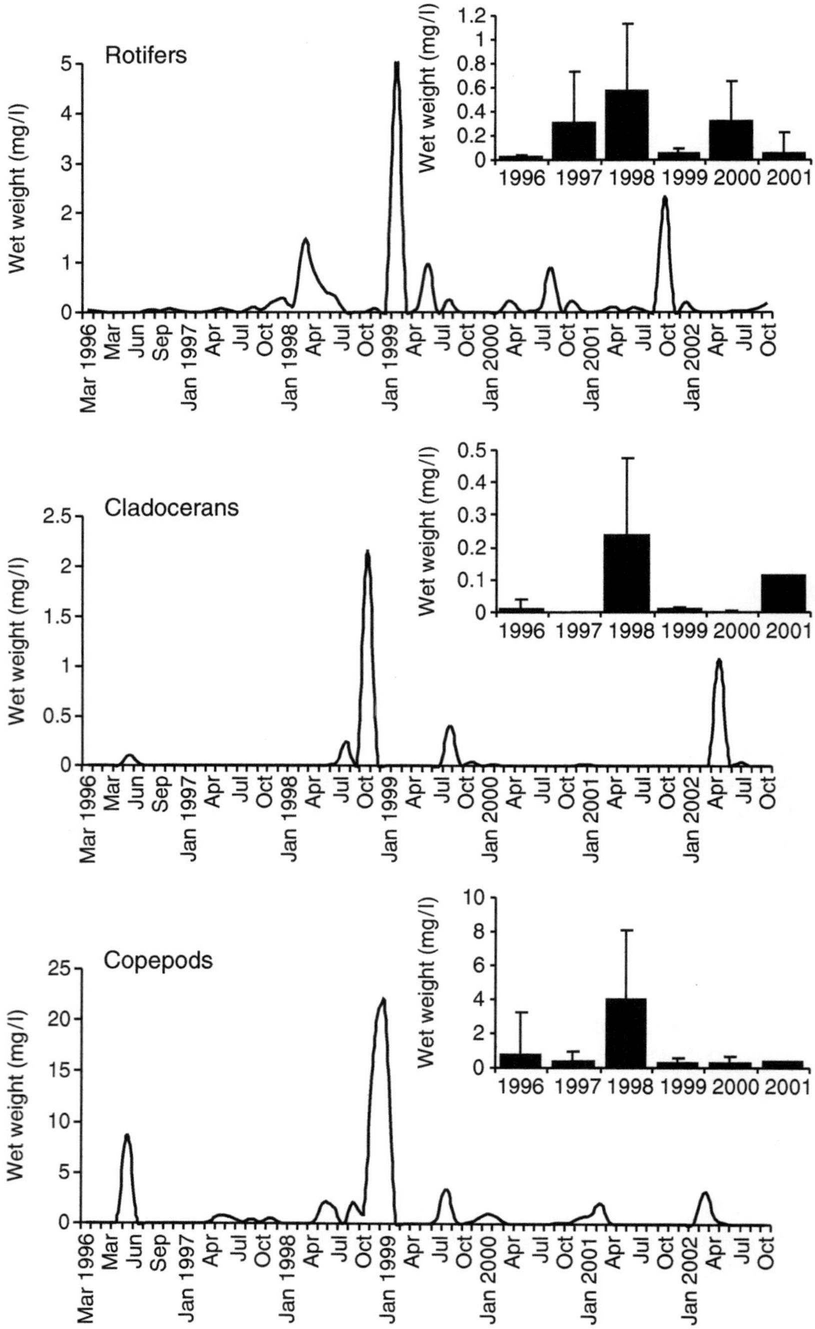

Fig. 6.15 Dynamics of different taxonomic groups at Central inlet site from 1996 to 2001. Annual average (right corners) is also shown with standard deviation in bars

Table 6.7 Pearson's significant correlation coefficients between bacterial and APP bio volumes, and environmental variables at Central inlet and Outlet sites. The number of data pairs used in the analysis (n) is indicated between brackets. Data, when necessary, were log-transformed. There were no significant correlations with dissolved oxygen nor organic matter.*: $p < 0.05$; **: $p < 0.01$

	Bacteria		APP	
	Central inlet	Outlet	Central inlet	Outlet
Temperature			0.32* (60)	
Conductivity		0.56** (36)	0.41** (60)	0.39** (36)
pH			0.40** (60)	
NO_2			−0.38** (60)	
NO_3			−0.33** (70)	−0.31* (57)
NH_4	−0.31* (60)		−0.37** (70)	
Total Nitrogen	−0.30* (60)		−0.25* (68)	
SRP	0.50** (60)	0.49** (48)	0.29* (70)	0.31* (57)
Total phosphorus		0.33* (48)		
TN/TP molar ratio		−0.35* (48)		−0.28* (48)
Suspended solids	0.47* (24)	0.72** (12)		
DOC		0.66** (15)		
Chlorophyll-*a*	0.42** (60)	0.33* (48)	0.49** (70)	0.32* (57)
Phytoplank. biovol.	0.48** (58)	0.32* (45)	0.53** (70)	0.43** (57)
APP bio volume	0.64** (58)	0.53** (45)	1	1

There was no significant correlation between bacterial biomass and temperature (Table 6.7). Bacteria were correlated with soluble reactive phosphorus (SRP) both in the Central inlet and Outlet sites and only with total phosphorus in the Central inlet site. Dissolved organic carbon presented a highly significant correlation with bacteria in the Outlet site. Bacteria also correlated significantly with chlorophyll *a* (and phytoplankton biomass) and APP biomass in both sites. There were negative correlations with ammonium and total nitrogen at the Central inlet site. APP was significantly correlated with temperature at the Central inlet site (Table 6.7). Horn and Horn (2008) also found correlation of APP with temperature in two reservoirs, since this small autotrophic fraction only developed from the onset of stratification. In both sites of the wetland water conductivity was related to APP, and nitrate also correlated negatively in both sites (Table 6.7).

Multiple regression analyses showed that APP biomass and SRP concentration were the two predictive variables of bacterial abundance at the Central inlet site (Table 6.8), explaining 50% of the variance. In the APP analysis, bacteria, phytoplankton, pH and temperature explained 61% of the variance. In the Outlet site, when SRP concentrations were used as a predictive variable, the bacterial biomass is explained in 43% by APP biomass and SRP concentration. For APP, only bacterial biomass was selected as a predictive variable (Table 6.8).

Table 6.8 Equations of the stepwise multiple regression analyses for bacterial and APP biomass at Central inlet and Outlet sites ($p < 0.01$). Relationship between bacteria and planktonic primary producers (estimated as chlorophyll *a* and phytoplanktonic biomass) with pooled data from all sites. TP: total phosphorus; SRP: soluble reactive phosphorus; CHL: chlorophyll-a

Central inlet		
logBACT = 5.0 + 0.16 log APP + 7.0 log SRP	$r^2 = 0.50$	n = 58
logAPP = −19.5 + 1.2 log BACT + 0.72 log PHYTO + 1.3pH + 0.06TEMP	$r^2 = 0.61$	n = 56
Outlet		
logBACT = 5.22 + 0.15 log APP (TP used as predictive variable)	$r^2 = 0.25$	n = 45
logBACT = 5.27 + 0.12 log APP + 3.8 log SRP (SRP used as predictive variable)	$r^2 = 0.43$	n = 45
logAPP = −5.25 + 1.6 log BACT	$r^2 = 0.25$	n = 45
Pooled data		
logBact = 5.96 + 0.24logCHL	$r^2 = 0.11$	n = 135
logBact = 4.33 + 0.28logPHYTO	$r^2 = 0.17$	n = 129

The relationship between bacteria and planktonic primary producers (estimated as chlorophyll *a* and phytoplanktonic biomass with pooled data from all sites) is shown in the equations in Table 6.8. The pooled data of APP abundance did not show any relationship to total phosphorus, neither did the relative contribution of APP to total phytoplankton.

The APP biovolume contribution to total phytoplankton (Fig. 6.16) increased throughout time, with some monthly values as high as 15–31%. There was no relationship between the proportion of APP in total phytoplankton biomass. Bell and Kalff (2001) concluded that, in freshwater lakes, APP typically contributes more than 50% of the total biomass in ultraoligotrophic waters (0.54 μgChl/L), declining in hypertrophic waters to 3% on average at concentrations of 100 μgChl/L. Vörös et al. (1998) found that, below 10 μgChl/L, the percentage contribution of picocyanobacteria biomass to total phytoplankton biomass exceeded 70% in some cases, whereas above 10 μgChl/L the highest contribution was only of 10%. Conversely, at the Las Tablas de Daimiel wetland, in the months that chlorophyll-a was below 10 μg/L, the percentage contribution of APP never exceeded 5%, and the highest percentages were found at higher chlorophyll *a* concentrations. APP peaks preceded the maximum development of larger phytoplankton (nano- and microphytoplankton). This is also a characteristic of phytoplankton succession in eutrophic lakes (Sommer et al. 1986). Vörös et al. (1998) reported declines in APP abundance coincident with N depletion and the commencement of blooms of colonial N_2 fixing cyanobacteria. There is evidence that the N:P ratio may determine the dominance of APP (prokaryotes) over nano- or microphytoplankton in lakes (Stockner et al. 2000) and that high N:P molar ratios (>25:1), and not necessarily the trophic state, can control the APP abundance. However the N:P molar ratio in the wetland was

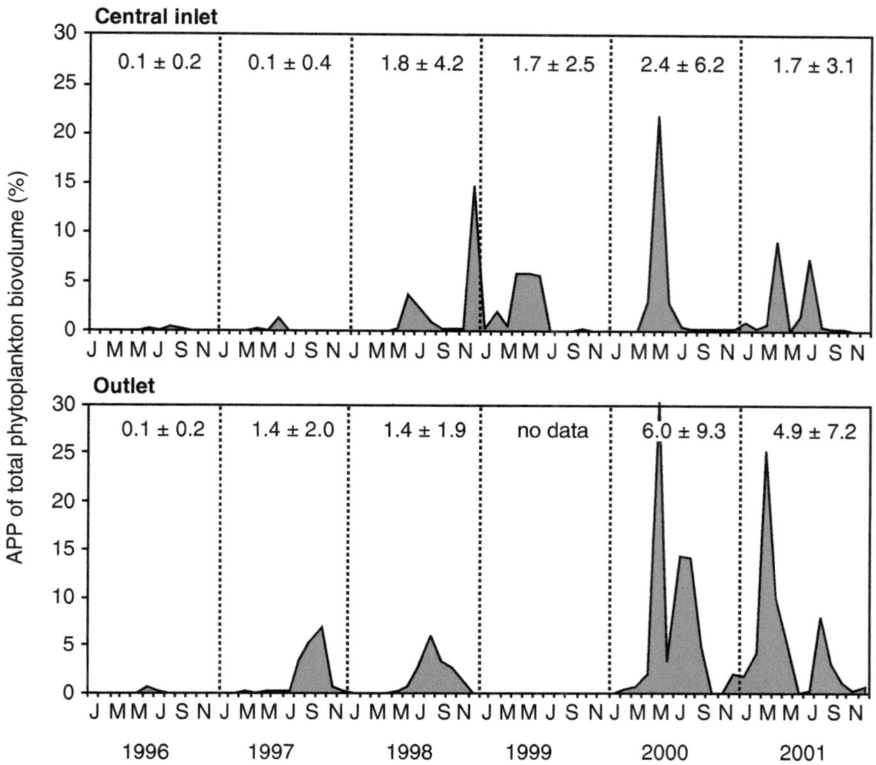

Fig. 6.16 Seasonal distribution of APP contribution to total phytoplankton bio volume at Central inlet and Outlet sites from 1996 to 2001 (the numbers indicate the annual averaged values ± standard deviations)

not significantly correlated with APP abundances. A lack of competition for nutrients between the different sized fractions of planktonic primary producers is expected in hypereutrophic system like Las Tablas de Daimiel (Rojo et al. 2000a).

Nutrient quantity and quality, grazing by microzooplankters (Weisse 1988) and water temperature (the same as for bacteria) are among the environmental factors thought to affect autotrophic picoplankton abundance, production and growth in natural systems, but also competition with larger phytoplankton and the light regime. In this wetland there is a weak positive correlation between temperature and APP biomass. It seems clear that temperature is the triggering factor that initiates the rapid growth of APP, but its populations decreased considerably when temperature was still high (end of summer), due to possible biotic interactions (grazing, competition). Horn and Horn (2008) provided clear evidence for top-down control of APP, mainly by daphnids, based on a study of long-time data series in two German reservoirs. We have also observed that APP was negatively affected by zooplankton grazing in microcosm experiments from the Tablas de Daimiel wetland (Ortega-Mayagoitia et al. 2002). On the other hand, outdoor enclosure experiments performed in Las Tablas de Daimiel wetland (Angeler et al. 2005) suggest that water depth may strongly mediate APP abundances, through variations in light quality and quantity (particularly under drawdown conditions, when the refuge against damaging radiation decreases with falling water depth; Angeler and Rodrigo 2004).

Bacterial and APP abundances were well correlated. This could be interpreted in the sense that factors affecting one of these communities also influence the other. Moreover, it is difficult to find clear-cut dichotomies (top-down versus bottom-up) in the control of plankton communities of wetlands (Keddy 2000). It is possible that both operate simultaneously, that neither operates except on rare exceptions or that other factors, such as habitat productivity, habitat heterogeneity or omnivory may override the apparent dichotomy. Omnivory exerted by zooplankton, which may mask trophic relationships, has also been suggested to be at work in the Las Tablas de Daimiel wetland (Ortega-Mayagoitia et al. 2002).

6.10 Environmental Factors Related to Phytoplankton

Resource dynamics and changes of environmental conditions throughout the 6 studied years were related to the main phytoplankton assemblage descriptors in the Central inlet site. Phytoplankton biomass is related to conductivity and inversely with ammonia concentration and TN:TP ratio (Table 6.9). Phytoplankton biomass was higher when conductivity increased (lower water level, see Chapter 5) and was more limited by nitrogen compounds than by phosphorus. These results agree with those of Bovo-Scomparin and Train (2008) reported for a floodplain river area (Brazil).

However, richness is related to nitrate and total nitrogen concentration and shows a negative relationship with flooding area, whereas variability of diversity index is not explained by any environmental factors studied. This is in accordance

Table 6.9 Pearson's correlations (r) between phytoplankton variables (abbreviations as in Tables 4 and 5; prefix % indicates biomass percentages, while no prefix indicates total biomass) and physical and chemical percentages at Central inlet from 1996 to 2002. COND: Conductivity; TN: total nitrogen; TP: total phosphorus; NO_3: nitrate, NH_4: ammonia; SRP: soluble reactive phosphorus; MAT: organic material; FLOO: flooding area; TOC: total organic carbon and TN:TP ratio. Table shows only variables with statistically significant r-values at $p < 0.01$

	TB	RICH	DIN	%DIN	%CRY	BAC	%BAC	%CHL
COND	0.34			0.30	−0.30			0.35
TN		0.42						
TP								
NO_3		0.30						
NH_4	−0.37	−0.38						
SRP							0.39	
MAT				0.30				
FLOO		−0.41			0.31	−0.34	−0.34	

with the weak relation between phytoplankton biomass and diversity measures (Tables 6.4 and 6.5).

More specifically, Dinophyceae are related to conductivity, organic material and total organic carbon, which means summer conditions with high evaporation and metabolism. Conversely, Cryptophytes are well related with flooding area and, consequently, inversely to conductivity, as Chlorophytes percentage (they are algae which withstand water flux disturbance, Bovo-Scomparin and Train 2008). Only diatoms were related to nutrients (phosphors concentration) and negatively to flooding area, as occurred in the Yangtze floodplain (Yang et al. 2008). Therefore, environmental conditions related to water evaporation or flooding seem to be the best related with algae, although in different directions. This is a possible explanation why no group of algae significantly explains the total biomass variance.

6.11 Environmental Factors Related to Zooplankton and the Food Web

When comparing dry 1996 with wet 1997, there was evidence that inflows in the form of flooding act as a bottom-up structuring force of total plankton biomass (Angeler et al. 2000), in the sense that flooding had a deleterious effect on the food web cascade (crustacean effect). However, when several more consecutive years are analysed, the conclusion is a lack of phytoplankton relationship with others planktonic groups: bacteria, APP (competitor) or predators (rotifers and crustaceans). Therefore, the pattern seems that, throughout the years, both structuring factors tested (food web and flooding) overlap their actions on plankton, highlighting the high impact of hydrological changes on zooplankton structure, as observed in other very different aquatic ecosystems (Gislason et al. 2009). The hydrological variability in Mediterranean shallow waterbodies makes it difficult to establish reliable predictions on the dynamics and relationships of plankton in these ecosystems (Moreno-Ostos et al. 2007).

Bacteria and autotrophic picoplankters have been described as an available food source for many micrograzers (Sherr et al. 1991), both with similar growth rates. This makes them an important component of the microbial food web in many aquatic ecosystems (Sommaruga and Robarts 1997). APP can be an important contributor (more than 50% of picoplanktonic biomass) at the base of the microbial food web. Protozoa are the most important APP consumers, but rotifers and several Cladocera genera are also able to ingest APP (Stockner et al. 2000). Moreover, the fact that the APP observed in the Las Tablas de Daimiel wetland is mainly composed of single cells, means that there is no reason for the material and energy flow from picocyanobacteria to grazers to be interrupted. The annual averaged APP contribution to total picoplankton biovolume ranged from 0.1% to 22%, with monthly contributions as high as 90%. In spite of this, no taxonomic group of zooplankton seems to depend (statistically) on bacteria, APP or phytoplankton. Multiple regression analyses only revealed significant relationship between copepods biomass and environmental factors (normalized variables): Copepod biomass = –0.43 organic materials +0.38 inundation area ($p < 0.0001$). This is in accordance with the above-mentioned results marking the stable years after flooding as more appropriate for copepods. Cross correlations between prey and predators were not statistically significant with a month lag. Top-down control on phytoplankton is not evident in shallow water bodies with macrophytes (Muylaert et al. 2006; Goldyn and Szelag-Wasielevoska 2005), during hydrological disturbances (Moreno-Ostos et al. 2007; Angeler et al. 2002) when omnivorous copepods, rather than cladocerans, dominate zooplankton communities (Ortega-Mayagoitia et al. 2002; Claps et al. 2004).

Plankton communities show an evident spatial heterogeneity and site-specific seasonal and interannual dynamic across the wetland. Shallowness and hydrologic conditions impinge strongly on the seasonal dynamics of plankton communities; the weak patterns during drought and wet periods were different. Therefore, a lack of pattern among plankton assemblages is the most remarkable feature. We suggest that alternative states of plankton taxonomic structure in this eutrophic semiarid wetland occur as a sequence of substitutions, caused by its close coupling with sediment (Ortega-Mayagoitia et al. 2002) and communities are persistently altered by disturbances. The autogenic trajectory is unattained and the trophic relationships or top-down control are possibly irrelevant.

References

Allen TFH, Bartell SM, Koonce JF (1977) Multiple stable configurations in ordination of phytoplankton community changes rates. Ecology 58:1076–1082

Álvarez-Cobelas M, Cirujano S, Sánchez-Carrillo S (2001) Hydrological and botanical man-made changes in the Spanish wetland of Las Tablas de Daimiel. Conserv Biol 97:89–97

Álvarez-Cobelas M, Rojo C, Angeler D (2005) Mediterranean limnology: current status, gaps and the future. J Limnol 64:13–29

Álvarez-Cobelas M, Sánchez-Carrillo S, Cirujano S (2007) Strong site effects dictate nutrient patterns in a Mediterranean floodplain. Wetlands 27:326–336

Angeler DG, Rodrigo MA (2004) Ramp-disturbance-ramp response: a simple model for wetland disturbance ecology. Mar Freshw Res 55:33–37

Angeler DG, Álvarez-Cobelas M, Rojo C, Sánchez-Carrillo S (2000) The significance of water inputs to plankton biomass and the trophic relationships in a semi-arid freshwater wetland. J Plankton Res 22:2075–2093

Angeler DG, Álvarez-Cobelas M, Sánchez-Carrillo S, Rodrigo MA (2002) Assessment of exotic fish impacts on water quality and zooplankton in a degraded semi-arid floodplain wetland. Aquat Sci 64:76–86

Angeler DG, Sánchez-Carrillo S, Rodrigo MA, Viedma O, Álvarez-Cobelas M (2005) On the importance of water depth, macrophytes and fish in wetland picocyanobacteria regulation. Hydrobiologia 549:23–32

Baik K, Park S, Kim E, Bae K, Ann JH, Ka JO, Chun J, Seong C (2008) Diversity of bacterial community in freshwater of Woopo wetland. J Microbiol 46:647–655

Basu BK, Kalff J, Pinel-Alloul B (2000) The influence of macrophyte beds on plakton communities and their export from fluvial lakes in the St Lawrence River. Freshw Biol 45:373–382

Beaver JR, Miller-Lemke AM, Acton JK (1999) Midsummer zooplankton assemblages in four types of wetlands in the Upper Midwest, USA. Hydrobiologia 380:209–220

Beckerman A, Petchey OL, Warren PH (2006) Foraging biology predicts food web complexity. Proc Natl Acad Sci USA 103:13745–13749
Bell T, Kalff J (2001) The contribution of picophytoplankton in marine and freshwater systems of different trophic status and depth. Limnol Oceanogr 46:1243–1248
Bovo-Scomparin VM, Train S (2008) Long-term variability of the phytoplankton community in an isolated floodplain lake of the Ivinhema River State Park, Brazil. Hydrobiologia 610:331–344
Brieé C, Moreira D, López-García P (2007) Archaeal and bacterial community composition of sediment and plankton from a suboxic freshwater pond. Res Microbiol 158:213–227
Chase JM (2007) Drought mediates the importance of stochastic community assembly. Proc Natl Acad Sci USA 104:17430–17434
Claps MC, Gabellone NA, Benitez HH (2004) Zooplankton biomass in an eutrophic shallow lake (Buenos Aires, Argentina): spatio-temporal variations. Ann Limnol Int J Limnol 40:201–210
Conforti V, Lionard M, Rojo C (2005) Euglenophyes in Tablas de Daimiel Nacional Park as example of biotic indices limitations. Limnetica 62:163–179
Currie DJ (1990) Large-scale variability and interactions among phytoplankton. Bacterioplakton and phosphorus. Limnol Oceanogr 35:1437–1455
D'Auria G, Barón-Rodríguez MM, Durbán-Vicente A, Moya A, Rojo C, Latorre A, Rodrigo MA (2010). Unravelling the bacterial diversity to be found in the semiarid Tablas de Daimiel National Park wetland. Aquat Microb Ecol 59:33–44
Del Giorgio PA, Peters RH (1993) The influence of DOC on bacteria-chlorphyll relationships in lakes. Verh Int Verein Limnol 25:359–362
Gislason A, Petursdottir H, Astthorssonn OS, Gudmundsson K, Valdimarsson H (2009) Inter-annual variability in abundance and community structure of zooplankton south and north of Iceland in relation to environmental conditions in spring 1990–2007. J Plankton Res 31:541–551
Goldyn R, Szelag-Wasielewska E (2005) The effects of two shallow reservoirs on the phyto- and bacterioplankton of lowland river. Polish Journal of Environmental Studies 14:437–444
Horn H, Horn W (2008) Bottom-up or top-down – How is the autotrophic picoplankton mainly controlled? Results of long-term investigations from two drinking water reservoirs of different trophic state. Limnologica 38:302–312
Jeppesen E, Erlandsen M, Sondegaard M (1997) Can simple empirical equations describe the seasonal dynamics of bacterioplankton in lakes?: An eight-year study in shallow hypertrophic and biologically dynamic lake Sobygard, Denmark. Microb Ecol 34:11–26
Keddy P (2000) Wetland ecology. Principles and conservation. Cambridge University Press, Cambridge
Lehours AC, Evans P, Bardot C, Joblin K, Gerard F (2007) Phylogenetic diversity of Archaea and Bacteria in the anoxic zone of a meromictic lake (Lake Pavin, France). Appl Environ Microbiol 73:2016–2019
Levins R (1979) Coexistence in a variable environment. Am Nat 114:765–783
Lionard M, Ortega-Mayagoitia E, Rojo C (2005) Evolución del fitoplancton en el parque Nacional de Las Tablas de Daimiel (Ciudad Real, 1996–2001). Limnetica 24:171–182
Lund JWG, Kipling C, LeCen ED (1958) The inverted method of estimating algal number the statistical basis of estimations and counting. Hydrobiologia 11:143–170
Magurran AE (1988) Ecological diversity and its measurement. Princeton University Press, Princeton, NJ
Marquet PA, Quiñones RA, Abades S, Labra F, Tognelli M, Arim M, Rivadeneira M (2005) Scaling and power-laws in ecological systems. J Exp Biol 208:1749–1769
McCauley E (1984) The estimation of the abundance and biomass of zooplankton in samples. In: Downing JA, Rigler FH (eds) A manual on methods for the assessment of secondary production in freshwater. Blackwell, Oxford
Moreno-Ostos E, da Silva SLR, de Vicente I, Cruz-Pizarro L (2007) Interannual and between-site variability in the occurrence of clear water phases in two shallow Mediterranean lakes. Aquat Ecol 41:285–297

Muylaert K, Declerck S, Van Wichelen J, De Meester L, Vyverman W (2006) An evaluation of the role of daphnids in controlling phytoplankton biomass in clear water versus turbid shallow lakes. Limnologica 36:69–78
Ortega-Mayagoitia E, Rojo C (2000a) Fitoplancton del Parque Nacional Las Tablas de Daimiel. II Las cianofíceas, los dinoflagelados, las criptofíceas, las crisofíceas y las xantofíceas. Anal Jar Bot Madrid 57:251–266
Ortega-Mayagoitia E, Rojo C (2000b) Fitoplancton del Parque Nacional Las Tablas de Daimiel. III Diatomeas y clorofitas. Anal Jar Bot Madrid 58:17–37
Ortega-Mayagoitia E, Rojo C, Armengol J (2000) Structure and dynamics of zooplankton in a semi-arid wetland, the National Park Las Tablas de Daimiel (Spain). Wetlands 20:629–638
Ortega-Mayagoitia E, Rojo C, Rodrigo MA (2002) Factors masking the trophic cascade in shallow eutrophic wetlands: evidence from a microcosm study. Archiv Hydrobiol 155:43–63
Pithart D, Pichlová R, Bílý M, Hrbácek J, Novotná K, Pechar L (2007) Spatial and temporal diversity of small shallow waters in river Luznice floodplain. Hydrobiologia 584:265–275
Reynolds CS (1997) Vegetation processes in the pelagic: a model for ecosystem theory. Ecology Institute, Oldendorf-Luhe
Reynolds CS, Huszar VLM, Kruk C, Naselli-Flores L, Melo S (2002) Towards a functional classification of the freshwater phytoplankton. J Plankton Res 24:417–428
Rodrigo MA, Rojo C, Armengol X (2002) Plankton biodiversity in a landscape of shallow water bodies (Mediterranean coast, Spain). Hydrobiologia 506(509):317–326
Rodrigo MA, Rojo C, Álvarez-Cobelas M (2003) Autothrophic and heterotrophic picoplankton in wetlands: differences with lake patterns. Int Rev Hydrobiol 88:464–481
Rojo C (1996) Fitoplancton. In: Alvarez-Cobelas M, Cirujano S (eds) Las Tablas de Daimiel: ecología acuática y sociedad. Ministerio de Medio Ambiente, Madrid
Rojo C, Álvarez-Cobelas M (1993) Hypertrophic phytoplankton and the intermediate disturbance hypothesis. Hydrobiologia 249:43–57
Rojo C, Álvarez-Cobelas M (2003) Are there steady-state phytoplankton in the field? Hydrobiologia 502:3–12
Rojo C, Ortega-Mayagoitia E, Conforti V (1999) Fitoplancton del Parque Nacional de las Tablas de Daimiel. I Las Euglenofitas. Anal Jar Bot Madrid 57:15–23
Rojo C, Ortega-Mayagoitia M, Álvarez-Cobelas M (2000a) Lack of pattern among phytoplankton assemblage. Or, what does the exception to the rule mean? Hydrobiologia 424:133–139
Rojo C, Ortega-Mayagoitia M, Rodrigo MA, Álvarez-Cobelas M (2000b) Phytoplancton structure and dynamics in a semiarid wetland, the National Park "Las Tablas de Daimiel" (Spain). Archiv Hydrobiol 148:397–419
Rott E (1981) Some results from phytoplankton counting intercalibrations. Schweizerische Zeitschrift für Hydrologie 43:34–62
Scheffer M, van Nes EH (2007) Shallow lakes theory revisited: various alternative regimes driven by climate, nutrients, depth and lake size. Hydrobiologia 584:455–466
Shannon CE, Weaver W (1963) The mathematical theory of communication. University of Illinois Press, Urbana, IL
Sherr EB, Sherr BF, Berman T (1991) High abundance of picoplankton-ingesting ciliates during late fall in lake Kinneret, Israel. J Plankton Res 13:789–799
Sokal RR, Rholf FJ (1981) Biometry, the principles and practice of statistics in biological research. W.H. Freeman, San Francisco, CA
Sommaruga R, Robarts RD (1997) The significance of autotrophic and heterotrophic picoplankton in hypertrophic ecosystems. FEMS Microbiol Ecol 24:187–200
Sommer U, Gliwicz ZM, Lampert W, Duncan A (1986) PEG-model of seasonal succession of planktonic events in freshwaters. Archiv für Hydrobiologie 106:433–471
Stockner J, Callieri C, Cronberg G (2000) Picoplankton and other non-bloom-forming cyanobacteria in lakes. In: Whitton BA, Potts M (eds) The ecology of cyanobacteria. Kluwer, The Netherlands
Vörös L, Callieri C, Balogh KV, Bertoni R (1998) Freshwater picocyanobacteria along a trophic gradient and light quality. Hydrobiologia 369(370):117–125

Wagner C, Adrian R (2009) Exploring lake ecosystems: hierarchy responses to long-term change? Global Change Biol 15:1104–1115

Ward JV, Tockner K (2001) Biodiversity: towards a unifying theme for river ecology. Freshw Biol 46:807–819

Weisse T (1988) Dynamics of autotrophic picoplankton in Lake Constance. J Plankton Res 10:1179–1188

Yang X, Anderson ND, Dong X, Shen JI (2008) Surface sediment diatom assemblages and epilimnetic total phosphorus in large, shallow lakes of the Yangtze floodplain: their relationships and implications for assessing long-term eutrophication. Freshw Biol 53:1273–1290

Zwart G, Crump BC, Agterveld M, Hagen F, Han SK (2002) Typical freshwater bacteria: an analysis of available 16S rRNA gene sequences from plankton of lakes and rivers. Aquat Microb Ecol 28:141–155

Chapter 7
Macrophyte Ecology and Its Long-term Dynamics

S. Cirujano, M. Álvarez-Cobelas, and R. Sánchez-Andrés

Abstract Dynamics of hydrophytic and helophytic vegetation is analyzed in relation to environmental changes from 1956 until the present day. These changes are mainly related to hydrological characteristics and water quality. In the case of submerged vegetation, community changes are related to alterations of the salinity and eutrophication regime, manifested in a decreased coverage and the extinction of some species. In the case of the helophytic vegetation, changes are associated with the fragmentation of the original vegetation patches, originally dominated by *Cladium mariscus*. Nowadays emergent vegetation is dominated by *Typha domingensis*, *Phragmites australis*, and, in the last years, by annual vegetation, nitrophilous taxa and woody species (*Tamarix canariensis*, *T. gallica*). This vegetation type indicates wet and saline conditions in soils. Despite these changes at the structural level, there is also evidence of a considerable increase in the biomass that accumulates every year in the wetland and which accelerates wetland siltation in the long term.

7.1 Introduction

Macrophytes in wetlands are important biotic elements that usually function as ecosystem engineers. Some of them, mostly helophytes, provide the characteristics of the wetland (Keddy 2000), *i.e.* they contribute to shape the wetland landscape. They also play other environmental roles, without which wetland performance would be quite different. Among them we can point out the following:

1. Helophytes greatly increase local transpiration (Sánchez-Carrillo et al. 2004).
2. Helophytes limit water evaporation because their patches reduce wind fetch and, hence, wind effects on the water surface (Anderson and Idso 1987).

S. Cirujano (✉)
Spanish National Research Council (CSIC), Royal Botanic Garden,
Pza. Murillo 2, E-28014 Madrid, Spain
e-mail: santos@rjb.csic.es

S. Sánchez-Carrillo and D.G. Angeler (eds.), *Ecology of Threatened Semi-Arid Wetlands: Long-Term Research in Las Tablas de Daimiel*, DOI 10.1007/978-90-481-9181-9_7,

3. Surface water motion is limited by helophyte patch distribution, focusing it through preferential paths. The distribution and flow of particulate and dissolved material in the wetland is, therefore, somewhat related to helophyte distribution (Kadlec and Knight 1996).
4. Helophytes modify the sedimentary chemical environment through their root activity. Submerged vegetation also takes up dissolved nutrients and increases dissolved oxygen in water tables (Duarte et al. 1994; Mitsch and Gosselink 2001; Kufel and Kufel 2002).
5. Macrophyte decomposition increases nutrient contents when decaying (Mason and Bryant 1974).
6. Macrophyte decay increases organic matter accumulation if there is no material export downstream, contributing to wetland siltation (Cronk and Fennessy 2001).
7. They provide refuge for many invertebrate and vertebrate animals, including mainly waterfowl (Dykyjová and Kvet 1978, www.ramsar.org).

All these ecological services can be observed in Las Tablas de Daimiel wetland. The structural and dynamic changes of flora and aquatic vegetation in the last 40 years are related to the same impacts that affect all wetlands in Mediterranean Europe: diminishing water availability as a result of agricultural practices, drying, changing morphometry of their basin, water pollution, salinization, invasive species and fires (Cirujano et al. 2002). Similar changes have also been found in tropical wetlands, such as in the Everglades of Florida (Ogden 2005). In this chapter we will describe the changes that submerged and emergent vegetation of this wetland has experienced since 1956.

The first studies relating water quality to macrophytes were carried out after the area was declared a National Park in 1970s (Pascual 1976; Sáez-Royuela 1977). These studies clearly state that aquifer overexploitation and surface water pollution were the main problems threatening the aquatic flora of this wetland, a finding which was also sustained by more recent research. The aim of this chapter is to review this previous research, building especially on the papers of Pascual (1976), Álvarez-Cobelas and Cirujano (1996, 2007), Cirujano et al. (1996, 2002), Camargo and Cirujano (1996), Álvarez-Cobelas et al. (2001, 2008), Ribeiro (2005) and Ribeiro et al. (2004).

7.2 Materials and Methods

7.2.1 Surveys and Reconstructions

To survey the flora we used common botanical methods, with seasonal surveys over most years since 1974. The long-term reconstructions were based on emergent vegetation maps of cut-sedge and reed and were drawn from aerial photographs taken in 1956, 1977, 1984, 1993, 1997, 2001 and 2007. In 1993 one member of the team (Cirujano) carried out field surveys which allowed to ascribe photographic patterns to either cut-sedge or reed vegetation in all maps. In 2001, along with *Typha* increase, a special effort was made to separate *Phragmites* and *Typha* covers in aerial photographs, through intensive field work, but this proved to be very difficult (see also Maheu-Giroux and de Blois 2005), and, hence, both plant species had to be considered together for the 2001 plant cover quantification. All maps were rasterized using the ARCVIEW-GIS software package (ESRI 2006). Maximum flooded area of the wetland (1,675 ha) was divided into a 1,675 (100 × 100 m) cells grid. The aerial cover of the two major emergent species was recorded in each cell, using the ARCVIEW planimetric tool, and results were reported as percentage cover.

7.2.2 Estimation of Growth and Decay

To measure the individual growth of helophytes, ten plants of each species were tagged in four stands distributed in the wetland to cover environmental variability, and their main leaf (*Cladium* and *Typha*) or stem (*Phragmites*) growth was recorded monthly. Growth rates of these elongating shoots were calculated assuming an exponential growth mode for all plant species, applying the commonly used relative growth rate equation (*e.g.* Hutchings 1997).

The decay of helophyte biomass (*Cladium* and *Phragmites*) was studied during 18 months in triplicate bags, using Polunin's (1982) methods to assess nutrient leaching and overall plant decomposition.

Water level and the content of nitrogen and phosphorus in sediments and the water column were related with instantaneous growth rates of each species. The duration of flooding was correlated with yearly-averaged growth rates of each species, using non-parametric Spearman correlations (Conover 1999). The STATISTICA6.0 statistical package was used for most tests.

7.2.3 Biomass and Primary Production Estimations

Biomass and plant cover were recorded in smaller patches less than 1 ha. During late summer in 2000, 2001, 2002 and 2008, maximum aboveground biomass was cut in 1-m^2 plots distributed throughout the wetland in 5–20 well developed patches. For charophytes, the biomass in 0.25-m^2 plots were gathered in 1996, 1997, 2001 and 2008. After counting all plants in each plot to estimate species density, fresh biomass was dried in an oven at 65°C up to constant weight to obtain dry weight. For patch biomass, it was assumed that 1-m^2 plots were roughly representative of biomass in the 1-ha (or smaller) patches; preliminary replicates of biomass measurements within 1-ha plots showed coefficients of variation lower than 30% for *Cladium mariscus*, *Phragmites australis*, and *Thypha domingensis* involved (Cirujano et al. 2003). Belowground biomass was measured only in 2002 and 2008; using a retrodigger, square plots of 1 m^2 were excavated down to the deeper layer of roots (usually 1.0–1.5 m deep). All roots, rhizomes and stems were gathered, sorted, sent to the laboratory, dried and weighted separately for each species. Drawings of underground organs, reported elsewhere (Conway 1936; Fiala 1978; Rodewald-Rudescu 1974), enabled us to separate rhizomes from roots of each species prior to belowground biomass quantification. Primary production of *Cladium* and *Phragmites* in 2008 was determined using the methods reported in Cronk and Fennessy (2001).

7.2.4 Spatial Distribution of Helophytes

Spatial patterns were estimated by means of spatial correlograms (Legendre and Fortin 1989). Correlograms have been calculated on plant cover using Moran's *I* autocorrelation statistic (Moran 1950). Because we dealt with sites on a surface (*i.e.* a grid of 1,675 cells), we expected a high number of values that could have complicated computations. Therefore, we reduced distance values by grouping and recoding them into distance classes by means of the R-Package (Casgrain and Legendre 2001), considering distance classes of equal length (141 m) (equidistant classes). Critical values for the autocorrelation values at each distance class were determined by means of a randomization method, followed by a Bonferroni correction (Legendre and Legendre 1998), whereas the overall significance of correlograms was tested following the Oden's (1984) method, which also used the Bonferroni correction.

Cluster analysis, using a UPGMA algorithm (Legendre and Legendre 1998), was used to search for similarity of spatial patterns among correlograms of both emergent species. Euclidean distance was the metric used to calculate each correlation matrix for cluster analyses, because this metric more readily detects differences among correlograms and, hence, facilitated the identification of spatial structure. The overall correlogram was interpreted as a surrogate of spatial structure of combined environmental and biological traits.

Water depth and quality appear to be the main variables responsible for controlling cover of emergent plants (Cronk and Fennessy 2001). We reconstructed earlier data of flooding using gauge data in the catchment at the inlets of the wetland, recorded by the Guadiana Water Authority since 1940. Using both flooding data and a digital terrain model (Sánchez-Carrillo 2000), we calculated average water depth in 1-ha cells in the same years of aerial photographs (see above).

Surface water always entered the wetland through the NE Gigüela stream, and the main outlet was located in the SW part of the wetland; hence that NE–SW direction comprised the main wetland gradient. Since no data on water quality prior to 1993 were available, we used the coordinates of each cell in the NE–SW gradient as a surrogate of water quality, because most pollution entered the wetland through the Gigüela stream. The rationale for such an approach was that water pollution always impacted the wetland through the Gigüela river, and that this pollution effect decreased with distance to the Gigüela inlet (Sánchez-Carrillo and Álvarez-Cobelas 2001). Therefore, the closer a given site was from that inlet, the more impaired was its water quality.

To outline the main environmental factors explaining the spatial scale of plant cover, we regressed the principal coordinates selected in the principal coordinates of neighbour matrices with either water depth or water quality. Furthermore, to relate spatial patterns and controlling factors, four variables were examined, namely, reed- and cut-sedge cover in 1-ha plots, space as the Euclidean distance of the given 1-ha cell from the main water input to the wetland, and average water level in the preceding year of plant cover assessment in that 1-ha cell. Water depth and plant cover were measured in cm and percentage of the 1-ha cell, respectively.

Total phosphorus is usually considered as the key factor controlling primary productivity in many freshwater environments (Vollenweider 1968) and, more recently, in wetlands (Keddy 2000). Because most water pollution entered the wetland through the NE stream, water quality effects were more important the closer a given wetland area was to the NE stream. We calculated that effect as the Euclidean distance to the input site. Total available phosphorus data showed a high correlation with those distances, and they were considered as a proxy for phosphorus distribution in the wetland.

Pearson correlation among the four variables (helophyte cover of both species, hydrolevel, total phosphorus) was calculated to relate spatial pattern and their controlling factors. Its significance test was corrected using the CRH procedure to account for underlying spatial autocorrelation (Clifford et al. 1989), by means of the PASSAGE package (Rosenberg 2002). Briefly, this test adjusts degrees of freedom depending upon autocorrelation of spatial variables, based on the amount of autocorrelation in the data. Moran's I statistic estimates the spatial autocorrelation in the plant cover and environmental data sets. Space (as a surrogate for water quality), water depth and plant cover of the other species (as an index of competition between emergent species) could be considered the main environmental controlling factors other than the biological traits of a given species. Therefore, the unexplained variability of a plant species cover by those factors in a given year can be considered as a proxy for biological traits of that species (Seabloom et al. 2001). This approach was used to reveal controlling factors of the spatial structure of both emergent plants at the whole wetland scale.

7.3 Aquatic Flora

As highlighted in previous chapters, Las Tablas de Daimiel wetland was flooded by water from different sources, having also specific chemical features, which resulted in two different habitats: one having permanent, fresh (860 ± 125 μS cm^{-1}) and deeper (1–6 m) waters, and another having seasonal, oligohaline (2,778 ± 1,073 μS cm^{-1}) and shallower (0–1 m) waters. While the former was due to the upwelling of the Llanura Manchega groundwater (see Chapter 1) and entered the wetland through the Guadiana River, the latter habitat was the result of flooding by the temporary Gigüela River (Fig. 7.1). The aquatic flora of these habitats also showed somewhat different composition and structure.

Large hydrophytes grew (*Chara hispida* var. *major*, *Nuphar luteum*, *Nymphaea alba*, *Potamogeton lucens*, *Hydrocharis morsus-ranae*, Fig. 7.1, Table 7.1) in the permanent habitat on peat substrate arising from the accumulation of decaying wetland vegetation lying on calcareous rock. In the shallow areas, and growing on sulfate-carbonate-rich limestone that were transported by the Gigüela river, there were compact submerged meadows comprising several charophyte species (*Chara canescens*, *C. hispida*, *C. aspera*, *C. hispida* var. *major*, etc.), locally termed "ovas" (Morillo and González 1996; Álvarez-Cobelas et al. 2001; Cirujano et al. 2002; Fig. 7.1, Table 7.1).

Helophytic vegetation was very homogenous and included large formations of cut-sedge (or giant sword grass, *Cladium mariscus*), locally called "masegones".

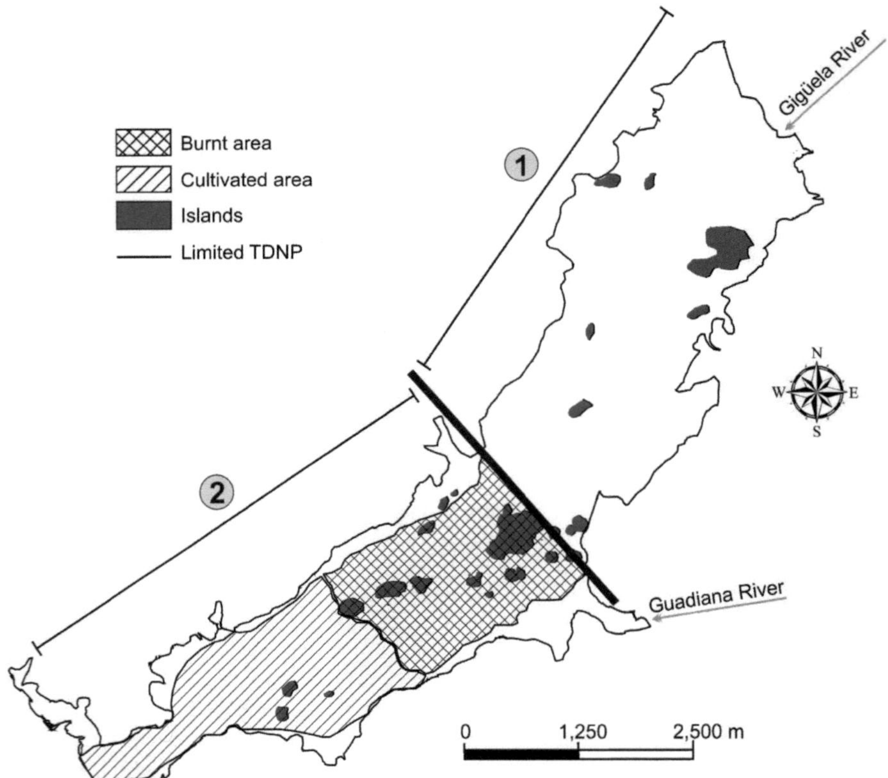

Fig. 7.1 Map showing ecosystems dependent on Gigüela (1) and Guadiana (2) rivers; the latter were agricultural areas in 1969–1977, were flooded and burned by the 1987 fire

Table 7.1 Changes in hydrophyte species richness of Las Tablas de Daimiel from 1956 until present in the areas flooded by the Gigüela (1) and Guadiana (2) rivers. See Fig. 7.1 for location. •: occurrence

Taxa	Zone 1 1956	Zone 1 Present	Zone 2 1956	Zone 2 Present
Charophytes				
Chara aspera	•	•	•	
Chara canescens	•	•		
Chara connivens		•		
Chara hispida	•	•		
Chara hispida var. *major*	•	•	•	•
Chara vulgaris	•	•	•	•
Nitella tenuissima			•	
Tolypella glomerata	•		•	•
Tolypella hispanica				•
Vascular plants				
Ceratophyllum demersum	•		•	
Ceratophyllum submersum	•		•	•
Hydrocharis morsus-ranae			•	
Lemna gibba		•		•
Lemna minor		•		•
Lemna trisulca			•	
Myriophyllum spicatum			•	
Nuphar luteum			•	
Nymphaea alba			•	
Potamogeton fluitans			•	
Potamogeton lucens			•	
Potamogeton pectinatus	•	•	•	•
Ranunculus peltatus			•	•
Ranunculus trichophyllus				•
Ruppia maritima				•
Utricularia australis			•	
Zannichellia pedunculata	•	•	•	•
Species richness	10	10	18	12

Littoral areas were covered by the remaining helophytes, mostly *Phragmites australis*, *Typha domingensis*, *T. latifolia*, *Carex riparia* and *C. hispida* (Table 7.2).

The pristine landscape structure was based upon the equilibrium between helophyte patches and helophyte-free areas, the latter being more frequent and extensive in the eastern, shallower area of the wetland, whose bottom areas were mostly covered by charophytes (Fig. 7.1, Zone 2). During the last 40 years, such a structure gradually changed, as a result of many anthropogenic impacts that started with partial desiccation and changing morphometry of the basin in the late 1960s, followed by corn and sunflower crops of the western part (25.5% out of the whole wetland). Crop cultivation within the wetland was discontinued in the 1980s, but

Table 7.2 Changes in helophyte and marginal species richness of Las Tablas de Daimiel from 1956 until the present day in the areas flooded by the Gigüela (1) and Guadiana (2) rivers. See Fig. 7.1 for location. •: occurrence

	Zone 1	Zone 1	Zone 2	Zone 2
Taxa	1956	Present	1956	Present
Vascular plants				
Aeluropus littoralis				•
Carex divisa	•	•	•	•
Carex hispida	•	•	•	•
Carex riparia			•	
Cladium mariscus	•	•		•
Iris pseudacorus			•	•
Juncus acutus	•	•	•	•
Juncus gerardi			•	•
Juncus maritimus	•	•	•	•
Juncus subulatus			•	•
Lythrum salicaria	•	•	•	•
Phragmites australis	•	•	•	•
Puccinellia fasciculata				•
Rumex palustris		•		•
Schoenus nigricans	•	•	•	•
Scirpus tabernaemontani			•	•
Scirpus lacustris			•	•
Scirpus litoralis				•
Scirpus maritimus	•	•	•	•
Tamarix canariensis	•	•	•	•
Tamarix gallica	•	•	•	•
Typha domingensis	•	•	•	•
Typha latifolia	•		•	
Veronica anagallis-aquatica	•	•	•	•
Species richness	14	14	19	22

fires that burnt the central area in 1987 (20% out of the total) caused new ecosystem-level alterations. Diminishing water inputs and the ongoing water pollution that has taken place since the 1970s to the present day have contributed to disrupt the natural hydrological regime in the wetland.

The main change was towards a chemically more homogeneous environment, but more fluctuating and somewhat stochastic as well, as a result of strong changes in water availability and ongoing eutrophication (see Chapter 6). This influenced plant species richness in a variety of ways. While the absence of fresh and permanent waters was the cause for the disappearance of large macrophytes (33.3% of species lost), water pollution, lower hydrolevels and hydroperiods and salt contents increasingly enhanced the occurrence of plant species that were better adapted to these conditions (*Lemna minor, L. gibba* and *Chara connivens*; *Ranunculus peltatus, R. trichophyllus*; *Ruppia maritima, Tolypella hispanica*, respectively). The final outcome was that the numbers of species associated in the area with the temporary Gigüela river did not change, but species composition did (Table 7.1).

The wetland is gradually reducing its submerged flora in favour of helophytes, and is even characterized by the invasion of terrestrial vegetation (see below). This phenomenon has also been reported in other agriculturally-impacted wetlands, such as Clear Lake in Iowa (Egertson et al. 2004).

Submerged meadows of charophytes are the best index of ecological conservation for this wetland. Its occurrence and abundance are dependent upon hydroperiod, water quality and the pressure of two invasive species that strongly feed on them, the common carp (*Cyprinus carpio*) and, more specifically, the American red swamp crayfish (*Procambarus clarkii*). An assay on crayfish food has demonstrated that it feeds selectively on charophytes rather than on other aquatic plants (*Ceratophyllum submersum*), to that charophyte consumption enhances crayfish weight from 0.27 ± 0.06 to 0.78 ± 0.12 g DW (Cirujano et al. 2004; Fig. 7.2).

Water pollution impaired growth of charophyte meadows, total phosphorus being a measure of such impairment (Blindow 1988, 1992; Kufel and Kufel 2002). In this wetland, charophyte meadows normally grow better at 0.17 ± 0.11 mg P l^{-1}, starting to decline when total phosphorus concentration reaches 1.77 ± 0.41 P l^{-1}. Charophyte cover has fluctuated as a result of both impaired water quality and reduced flooding; nowadays it is lower than 1 ha, caused by a recent drought (Fig. 7.3), but it is likely that increased flooding might promote recolonization of charophytes from propagule banks in former areas of growth.

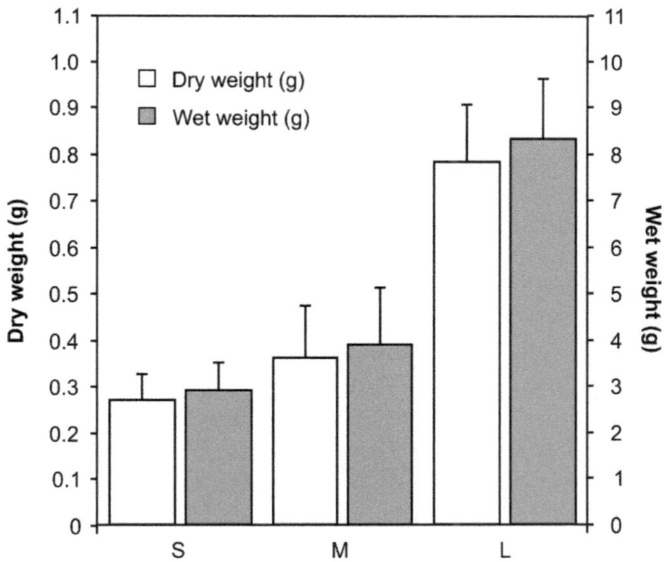

Fig. 7.2 Mean values (n = 5) of dry- and wet weight (g) of daily consumption of *Chara hispida* var. *major* by crayfish specimens of different size (small: 15.3 ± 1.25 g; medium: 25.9 ± 2.8 g and large: 40.8 ± 5.72 g) from Las Tablas de Daimiel in a mesocosm experiment (Cirujano et al. 2004)

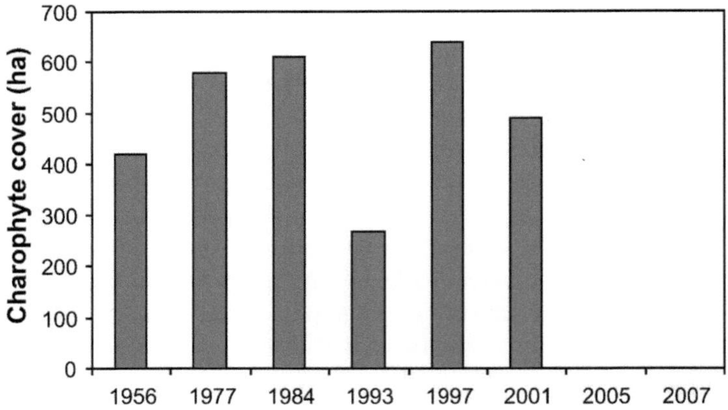

Fig. 7.3 Changes in charophyte cover from 1956 until the present day

7.4 Helophyte Cover and Growth

From a spatial perspective, changes in emergent vegetation resulted in strong fragmentation and cover loss of cut-sedge patches (*Cladium mariscus*; see below, Álvarez-Cobelas et al. 2008), with a decrease of 89% of the original cover, and a variable increase of reed (*Phragmites australis*) and cattail (*Typha domingensis*) cover, up to 24% and 237%, respectively. Such a substitution was caused by the elimination of 405 ha of cut-sedge cover in the period 1969–1977, the subsequent burning of 110 ha in 1987, and the stochastic flooding occurring later, that enhanced reed growth (Figs. 7.4 and 7.5). A key factor for maintaining biological richness of this wetland was the balance between helophyte-free and helophyte covered areas. Thus, helophyte-free area in 1956, when the wetland was in pristine condition, amounted to 22.6% of the whole wetland and since then it fluctuated between 14.5% and 38% without any clear time trend. This did not imply that those areas were colonized by charophyte meadows, since either water scarcity or pollution prevented or limited submerged plant growth (Fig. 7.6).

In recent years, water scarcity changed vegetation dynamics into terrestrial plant communities in which helophyte-free areas were colonized by invasive nitrophilous species (*Erygeron canadensis*, *E. bonariensis*, *Aster squamatus*), that occupied 79% and 95% of total helophyte-free areas in 2007 and 2008, respectively, with a growing contribution of woody species (*Tamarix canariensis*, *T. gallica*) that comprised the potential forest vegetation of wet and saline soils (Fig. 7.7).

Growth rates of cut-sedge were higher at NE sites, ranging from 0.001 to 0.022 d-1. Reed growth was similar at all sites, its overall range being 0.001–0.045 d^{-1}. In 2001, cattail growth rates were markedly higher at SW sites than at NE sites, its range being 0–0.040 d^{-1} (Álvarez-Cobelas and Cirujano 2007).

Sedimentary nutrients displayed significant relationships with plant growth rates. While *Cladium* growth rate was positively related with nitrogen, *Phragmites* growth rate was negatively related with the N:P ratio, with the critical value around 4.

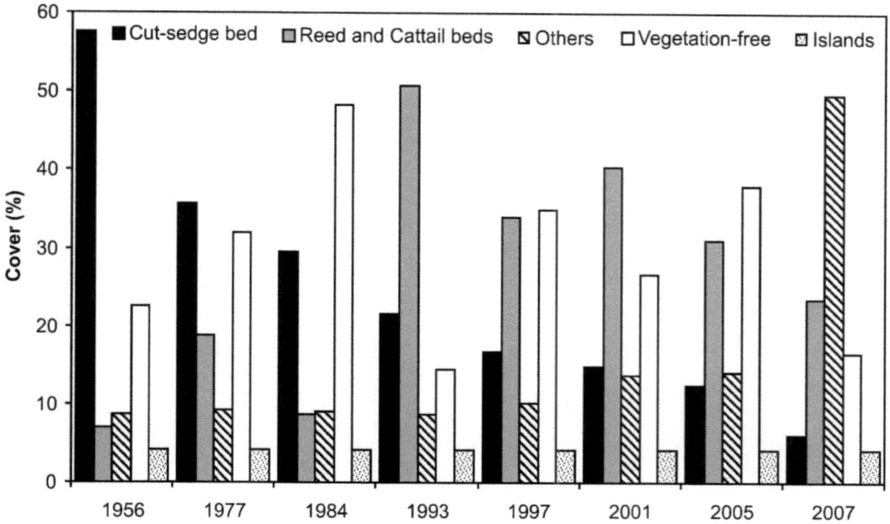

Fig. 7.4 Changes in percentage of terrestrial vegetation cover (others) and emergent vegetation cover (cut-sedge, reed and cattail) from 1956 until the present day

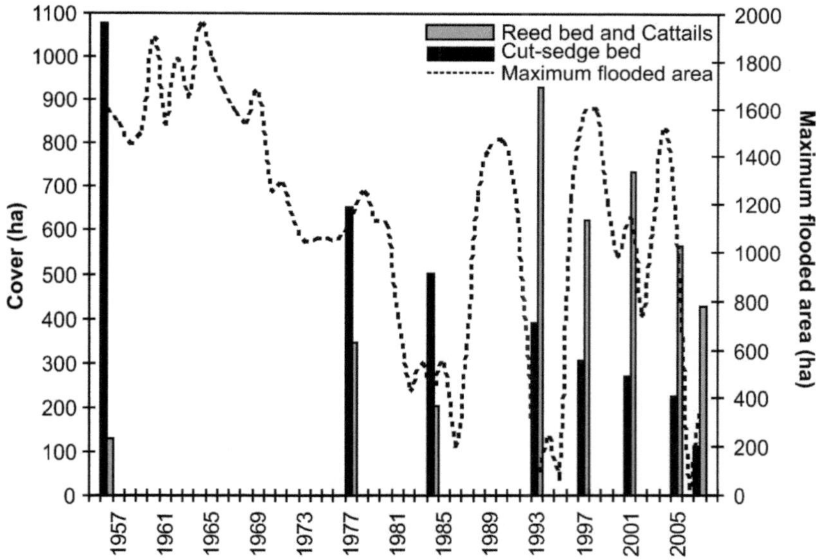

Fig. 7.5 Changes in helophyte cover as related with maximal flooding from 1956 until present

Typha growth rate was negatively related with sedimentary nitrogen and phosphorus (Álvarez-Cobelas and Cirujano 2007). Water level was weakly but positively related with growth rates of *Cladium* and *Typha* (Álvarez-Cobelas and Cirujano 2007).

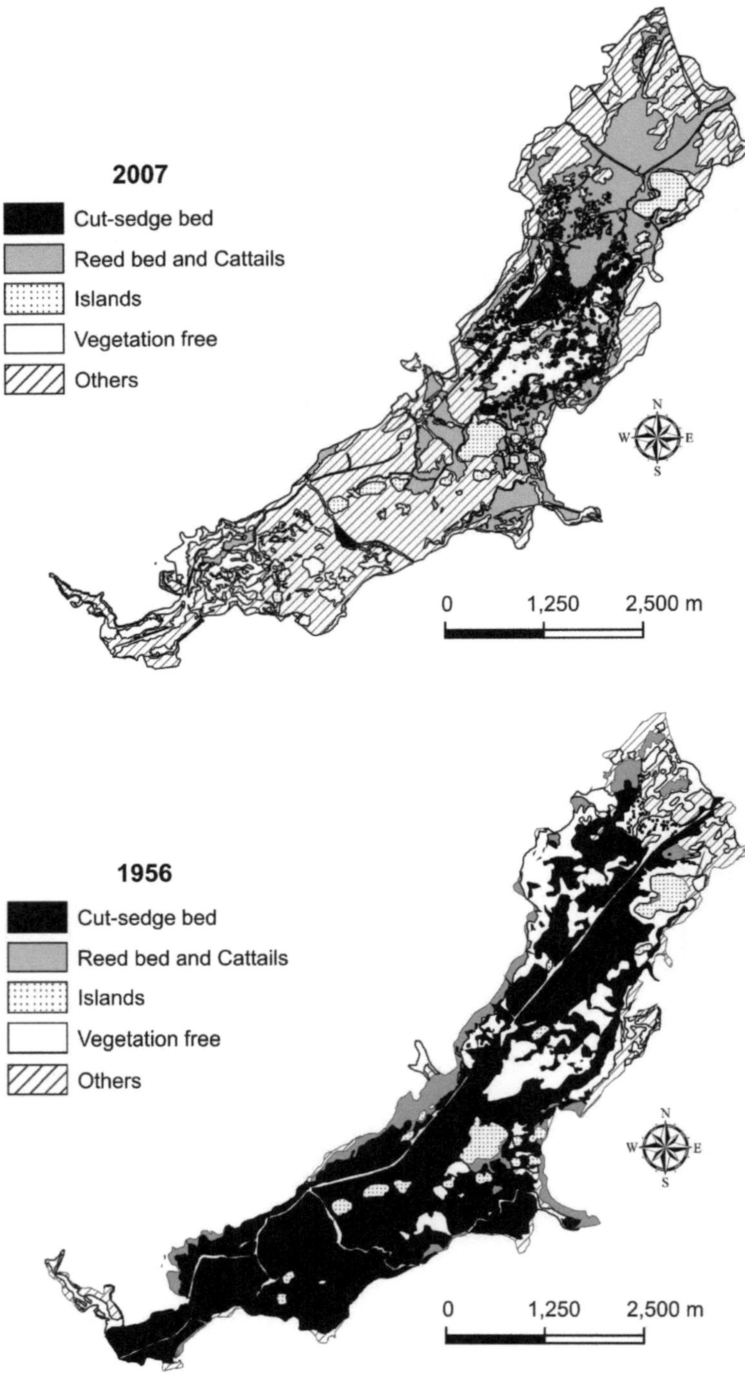

Fig. 7.6 Cover of different vegetation types in 1956, just before wetland degradation, and in 2007

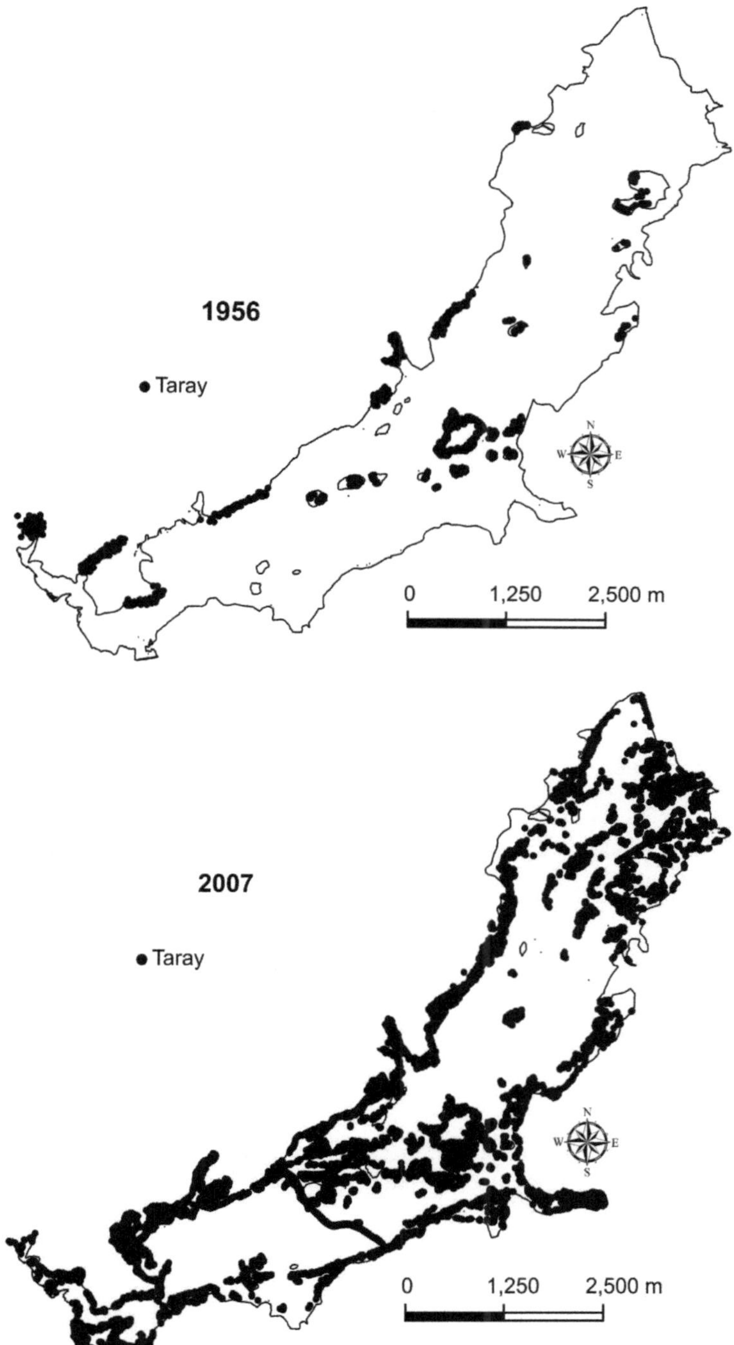

Fig. 7.7 Wetland distribution of woody vegetation of *Tamarix* (*Tamarix canariensis* and *Tamarix gallica*) in 1956 and 2007

Only leaf elongation rates of *Typha* came closer to the highest reported rates (Dykyjová and Kvet 1978). *Cladium* and *Phragmites* never reached the highest reported rates (Conway 1938; Dykyjová and Kvet 1978; Lorenzen et al. 2001). Furthermore, these species presented strong inter-site variability. Therefore, helophyte growth in our studied wetland was suboptimal at the single plant level, as also judged by linear relationships between growth and environmental controls. For shoot elongation, controlling factors were found to differ among species. For *Cladium* it was nitrogen, as in other studies (Urban et al. 1993; Miao and Sklar 1998). Sedimentary N:P ratio was important for *Phragmites*; this has also been reported by Güsewell et al. (2003) for many wetland species, thus providing an explanation of conflicting evidence about abiotic controls of reed growth (Clevering 1998; Kohl et al. 1998). In the case of *Typha*, however, growth rate was negatively related to water N and P, despite positive relationships in other wetlands (Grace 1988; Miao et al. 2000).

Cut-sedge and reed cover were related with average flooding in the preceding year, albeit in an opposite way: as flooding increased, *Cladium* increased and *Phragmites* decreased (Fig. 7.8). These lagged responses to flooding are common in other wetlands (Armentano et al. 2006; Egertson et al. 2004).

Average water level appeared to control *Cladium* biomass up to 20–40 cm, above which water level increase did not enhance biomass increase (Álvarez-Cobelas and

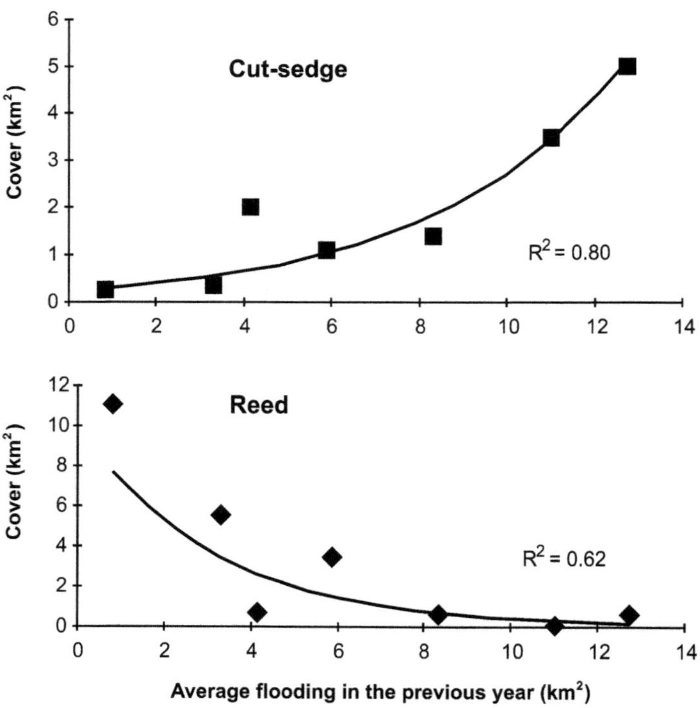

Fig. 7.8 Relationship between helophyte cover and flooding in the previous year at the Las Tablas de Daimiel wetland. Data from 1956 to 2001

Cirujano 2007). We also observed a statistically significant, negative relationship between sedimentary phosphorus and cut-sedge biomass and a 1-year delayed, unimodal response of biomass to average water level, suggesting inhibition of plant growth at deeper sites (Álvarez-Cobelas and Cirujano 2007).

Reed biomass could be controlled by water level variability (positive relationship) as well as by averaged water level of the preceding year (negative relationship, Álvarez-Cobelas and Cirujano 2007). Cattail biomass appeared to be related with sedimentary phosphorus, also showing a positive and nonlinear relationship with average duration of flooding in the preceding year (Álvarez-Cobelas and Cirujano 2007).

7.5 Plant Biomass, Production and Decomposition

Charophyte biomass was promoted by flooding and good water quality. Available data indicated that it could reach more than 3 kg DW m^{-2} (Fig. 7.9), a value that was higher than most reported values for other wetlands (Kufel and Kufel 2002).

Cut-sedge biomass attained the higher values at NE sites, where its populations always grew better. In 2001 and 2002, more biomass was recorded than in 2000 (Álvarez-Cobelas and Cirujano 2007). In 2008, there appeared to be an increase in cut-sedge biomass, but it was the result of the higher variability (Fig. 7.10). Overall, biomass variability ranged between 1 and 11 kg DW m^{-2}. In fact, there were few instances in the wetland where cut-sedge reached its optimal growth. Belowground biomass, only measured in 2002 and 2008, showed strong fluctuations between years, the ratio above: belowground biomass hence showing changing ratios interannually (Fig. 7.10).

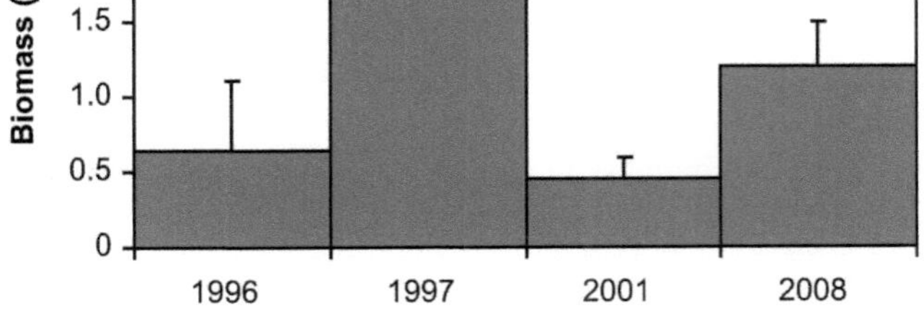

Fig. 7.9 Average values of charophyte biomass (±1 SD) at peak growth in the Las Tablas de Daimiel National Park. Recording sites were chosen to reflect wetland variability. N = 3–5

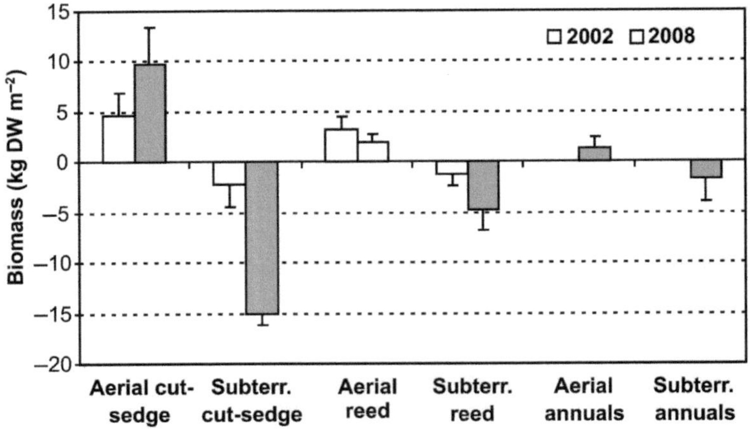

Fig. 7.10 Average biomass of helophytes and terrestrial vegetation (±1 SD) at peak growth at the Las Tablas de Daimiel National Park. Recording sites were chosen to reflect wetland variability. N = 3–10

Reed biomass also appeared to be higher at the NE entrance (the shallower sites of the wetland and river discharge sites), and it was higher in 2000 than in 2001, 2002 and 2008, reaching 4.02 ± 1.76, 2.74 ± 1.18, 3.26 ± 1.50, 2.03 ± 0.80 kg DW m^{-2} in 2000, 2001, 2002 and 2008, respectively (Álvarez-Cobelas and Cirujano 2007; Fig. 7.9). Belowground biomass was higher than aboveground biomass in 2002 and 2008, increasing the ratio from 1.24 to 2.33.

When the hydroperiod lasted throughout the year, *Typha domingensis* thrived and its biomass was similar over time, showing low spatial variability in the wetland (2.63 ± 1.48, 2.14 ± 0.84 and 4.03 ± 2.41 kg DW m^{-2} in 2000, 2001 and 2002, respectively). Belowground biomass was similar to aboveground biomass in 2002 (3.97 ± 1.82 *vs* 4.03 ± 2.41 kg DW m^{-2}; Álvarez-Cobelas and Cirujano 2007). In 2008, *Typha* did not grow in this wetland. In Las Tablas de Daimiel aboveground biomass of these three helophytes was often higher than the values reported in other studies (for instance, Dykyjová and Kvet 1978; Miao and Sklar 1998; Vymazal 1995). The aboveground: belowground biomass ratio of these helophytes, however, was lower in TDNP than in most wetlands (Dykyjová and Kvet 1978; Fiala 1978; Lorenzen et al. 2001).

In this wetland, aerial biomass of annual plant populations was negligible before 2006. As a result of a heavy drought, they thrived later and peaked in 2007 and 2008, when they attained 3.56 ± 3.06 and 1.40 ± 1.21 kg DW m^{-2}, respectively (Fig. 7.9; Rodríguez-Murillo, personal communication). The relationship between above and belowground biomass in annual plants decreased, changing from 7.3 in 2007 to 0.61 in 2008.

In 2008, aerial production of cut-sedge reached 1.67 ± 1.28 kg DW m^{-2} and 1.53 ± 0.3 kg DW m^{-2} for a flooded and a non-flooded site, respectively. In 2008, aerial production of reed has ranged 0.91–3.05 kg DW m^{-2} throughout the wetland (Rodríguez-Murillo, personal communication).

Helophyte decomposition was a slow process for *Cladium* and *Phragmites* in Las Tablas de Daimiel, with 50% of the initial biomass being decomposed after 1.5 years. There were no significant correlations between environmental variables and decomposition rates. An initial phase of leaching with the highest rates ($p < 0.05$) of mass loss and nutrients occurred for both plant species. Throughout the incubation period, no significant differences were observed between both species and experiments, and the average daily rates of mass loss were low: 0.0022 ± 0.001 day^{-1} for *C. mariscus* and 0.0029 ± 0.002 day^{-1} for *P. australis* in the overall decomposition experiment, and 0.0019 ± 0.001 day^{-1} and 0.0016 ± 0.001 day^{-1} due to chemical-microbiological decomposition, suggesting the negligible significance of invertebrates in the plant decomposition process. Decaying mass rates of reed in this wetland were similar to those of Andersen (1978) in Danish wetlands whereas those of *Cladium* were lower than those recorded by Newman et al. (2001) for *C. jamaicensis* in the Florida Everglades. To our knowledge, there are no data on *C. mariscus* decomposition rates. Furthermore, no significant difference was observed for nutrient release, except that *Phragmites* lost more phosphorus than *Cladium,* but differences were not significant ($p < 0.1$). Plant litter decaying rates were roughly an order of magnitude lower than nutrient release rates in both species. Our results suggest that slow decomposition rates of emergent plant biomass increases organic matter sedimentation and reduces water quality at Las Tablas de Daimiel.

There are increasing numbers of long-term studies in wetlands at the landscape level (Leendertsee et al. 1997; Visser et al. 1999; Alvarez et al. 2005). Our study covering the period 1956–2008 showed a trend of reed cover increase and cut-sedge cover decrease over time. In most studies, water level increase and water pollution are the causes for the observed changes. In Las Tablas de Daimiel, water pollution certainly is the most important cause for reed cover dynamics, but water level is the main factor responsible for long-term cover changes of cut-sedge. Notwithstanding, when cover expansion rates are considered, no average flooding effects are observed, and only a weak water quality effect may be suggested for cut-sedge decline at the landscape scale. Saltmarsh et al. (2006) report that, as water stress increases, photosynthetic efficiency of reed is higher than that of cut-sedge. This mechanism may be important in natural habitats, but it only partly explains the patterns observed in TDNP, given that water quality impairment may also affect emergent species in this wetland.

7.6 Spatial Heterogeneity of Helophytes

The size of *Cladium* and *Phragmites* patches, defined as their longest axial dimension, varied from 1956 to 2001 and ranged between 2,300–4,100 m and 1,300–4,500 m for cut-sedge and reed, respectively. The size of cut-sedge patches decreased from 1956 onwards, whereas the trend of reed patch size was inconsistent until 1984. Throughout the study, cut-sedge patches were often bigger than those of reed (Álvarez-Cobelas et al. 2008). Two groups of distinct spatial heterogeneity were recorded for cut-sedge, an earlier natural one (1945–1977) and a later, anthropogenically-driven one, with 1984 as the turning year. Because reed expansion in this wetland occurred in the early 1990s, it was not surprising that there also were two spatially-different groups which then split. The spatial pattern of both helophytes changed more in recent periods, along with anthropogenic change (Álvarez-Cobelas et al. 2008).

Water quality was partly responsible for spatial patterns of *Cladium* and *Phragmites* at the very broad scale (>1,000 ha), while water depth explained variability in plant cover at the very broad and broad scales (500–1,000 ha). The overall variability explained by biotic (the cover of the other species) and abiotic (water depth and water quality) factors never explained more than 39% of overall variability (Álvarez-Cobelas et al. 2008).

These results suggested changing patterns of emergent plant cover in space and time. From the temporal perspective, cut-sedge and reed showed distinct responses between periods dominated by natural variability (1945–1977) and anthropogenic change (1977 onwards). During the period of natural variability, no fragmentation of patches of both emergent plants took place. Man-made change, however, resulted in fragmentation of cut-sedge patches, and the influence of biological traits on its spatial patterns increased. Reed patches showed increasing cover in recent years, but the trend of its spatial pattern was unclear. Patch fragmentation is frequent in species subject to man-made change (Turner 2005).

Based on results obtained from the cold-temperate Delta Marsh (Canada), Seabloom et al. (2001) hypothesized that plant cover is controlled by environmental gradients under stable conditions but biological traits become more important for spatial patterns when the wetland turns to unstable situations. Results from forests support this hypothesis (Camarero et al. 2000; Pu et al. 2005). *Cladium* cover can be controlled by the abiotic environment during stable conditions (*i.e.* in pristine years), whereas species biology partly governs spatial pattern when unstable conditions prevail (Álvarez-Cobelas et al. 2008). By contrast, *Phragmites* did not follow such trends, perhaps because of its faster growth (Ribeiro 2005) which made species cover more fluctuating and unpredictable.

Acknowledgments We are very grateful to Juan Carlos Rodríguez-Murillo for his data on primary production and biomass of helophytes and terrestrial plants in 2008.

References

Alvarez MG, Tron F, Mauchamp A (2005) Sexual versus asexual colonization by *Phragmites australis*: 25-year reed dynamics in a Mediterranean marsh, southern France. Wetlands 25:639–647

Álvarez-Cobelas M, Cirujano S (eds) (1996) Las Tablas de Daimiel: ecología acuática y sociedad. Ministerio de Medio Ambiente, Madrid

Álvarez-Cobelas M, Cirujano S (2007) Multilevel responses of emergent vegetation to environmental factors in a semiarid floodplain. Aquat Bot 87:49–60

Álvarez-Cobelas M, Cirujano S, Sánchez-Carrillo S (2001) Hydrological and botanical man-made changes in the Spanish wetland of Las Tablas de Daimiel. Biol Conserv 97:89–97

Álvarez-Cobelas M, Sánchez-Carrillo S, Cirujano S, Angeler DG (2008) Long-term changes in spatial patterns of emergent vegetation in a Mediterranean floodplain: natural versus anthropogenic constraints. Plant Ecol 194:257–271

Andersen OFO (1978) Effects of nutrient level on the decomposition of *Phragmites communis* Trin. Archiv Hydrobiol 84:42–54

Anderson MG, Idso SB (1987) Surface geometry and stomatal conductance effects on evaporation from aquatic macrophytes. Water Resour Res 23:1037–1042

Armentano TV, Sah JP, Ross MS, Jones DT, Cooley HC, Smith CS (2006) Rapid responses of vegetation to hydrological changes in Taylor Slough, Everglades National Park, Florida, USA. Hydrobiologia 569:293–309

Blindow I (1988) Phosphorus toxicity in *Chara*. Aquat Bot 32:393–395

Blindow I (1992) Decline of charophytes during eutrophication: comparison with angiosperms. Freshw Biol 28:9–14

Camarero JJ, Gutiérrez E, Fortin MJ (2000) Spatial pattern of subalpine forest-alpine grassland ecotones in the Spanish Central Pyrenees. For Ecol Manage 134:1–16

Camargo JA, Cirujano S (1996) Reduction in diversity of aquatic plants in a Spanish wetland: the effect of the size of inundated area. J Freshw Ecol 12:539–543

Casgrain P, Legendre P (2001) The R-Package for multivariate and spatial analysis, version 4.0 d3 – User's manual. Departament de sciences biologiques, Université de Montreal, Montreal

Cirujano S, Casado C, Bernués M, Camargo JA (1996) Ecological study of Las Tablas de Daimiel National Park (Ciudad Real, Central Spain): differences in water physico-chemistry and vegetation between 1974 and 1989. Biol Conserv 75:211–215

Cirujano S, Medina L, Chirino M (2002) Plantas acuáticas de las lagunas y humedales de Castilla-La Mancha. Junta de Comunidades de Castilla-La Mancha – Real Jardín Botánico, Madrid

Cirujano S, Álvarez-Cobelas M, Riolobos P, Ribeiro MD, Sánchez-Carrillo S, Medina L, Moreno M, Angeler DG, Rojo C, Rodrigo MA, Armengol J, Ortega-Mayagoitia E (2003) Seguimiento y recuperación ambiental del P.N. Las Tablas de Daimiel. Informe 1999-2002 para el Organismo Público Parques Nacionales, Madrid

Cirujano S, Camargo JA, Gómez-Cordobés C (2004) Feeding preferences of the red swamp crayfish *Procambarus clarkii* (Girard) on living macrophytes in a Spanish wetland. J Freshw Ecol 19:219–226

Clevering OA (1998) An investigation into the effects of nitrogen on growth and morphology of stable and die-back populations of *Phragmites australis*. Aquat Bot 60:11–25

Clifford P, Richardson S, Hémon D (1989) Assessing the significance of the correlation between two spatial processes. Biometrics 45:123–145

Conover WJ (1999) Practical non-parametric statistics. Wiley, New York

Conway VM (1936) Studies in the autoecology of *Cladium mariscus* R.Br. I. Structure and development. New Phytol 35:177–205

Conway VM (1938) Studies in the autoecology of *Cladium mariscus* R.Br. III. Growth rates of the leaves. New Phytol 37:254–278

Cronk JK, Fennessy MS (2001) Wetland plants. Biology and ecology. Lewis, Boca Ratón, FL

Duarte CM, Planas D, Peñuelas J (1994) Macrophytes, taking control of an ancestral home. In: Margalef R (ed) Limnology now. Elsevier, Amsterdam

Dykyjová D, Kvet J (eds) (1978) Pond littoral ecosystems. Structure and function. Springer-Verlag, Berlin

Egertson CJ, Kopaska JA, Downing JA (2004) A century of change in macrophyte abundance and composition in response to agricultural eutrophication. Hydrobiologia 524:145–156

ESRI (2006) ArcGIS Version 9.2 and the Spatial Analyst Extension, Environmental Systems Research Institute, Redlands, California

Fiala K (1978) Underground organs of *Typha angustifolia* and *Typha latifolia*: their growth, propagation and production. Acta Sci Nat Acad Sci Bohemosl Brno 12:1–43

Grace JB (1988) The effects of nutrient additions on mixtures of *Typha latifolia* L. and *Typha domingensis* Pers. along a water-depth gradient. Aquat Bot 31:83–92

Güsewell S, Koerselman W, Verhoeven JTA (2003) Biomass N:P ratios as indicators of nutrient limitation for plant populations in wetlands. Ecol Appl 13:372–384

Hutchings MJ (1997) The structure of plant populations. Plant litter quality and decomposition: a historical overview. In: Crawley MJ (ed) Plant Ecol. Blackwell, Oxford

Kadlec RH, Knight RL (1996) Treatment wetlands. CRC Lewis, Boca Ratón, FL

Keddy PA (2000) Wetland ecology. Cambridge University Press, Cambridge

Kohl JG, Woitke P, Kühl H, Dewender M, König G (1998) Seasonal changes in dissolved amino acids and sugars in basal culm internodes as physiological indicators of the C/N-balance of *Phragmites australis* at littoral sites of different trophic status. Aquat Bot 60:221–240

Kufel L, Kufel I (2002) *Chara* beds acting as nutrient sinks in shallow lakes-a review. Aquat Bot 72:249–260

Leendertsee PC, Roozen AJM, Rozema J (1997) Long-term changes (1953-1990) in the salt marsh vegetation at the Boschplaat on Terschelling in relation to sedimentation and flooding. Plant Ecol 132:49–58

Legendre P, Fortin MJ (1989) Spatial pattern and ecological analysis. Vegetatio 80:107–138

Legendre P, Legendre L (1998) Numerical ecology, 2nd edn. Elsevier, Amsterdam

Lorenzen B, Brix H, Mendelssohn IA, McKee KL, Miao SL (2001) Growth, biomass allocation and nutrient use efficiency in *Cladium jamaicense* and *Typha domingensis* as affected by phosphorus and oxygen availability. Aquat Bot 70:117–133

Maheu-Giroux M, de Blois S (2005) Mapping the invasive species *Phragmites australis* in linear wetland corridors. Aquat Bot 83:310–320

Mason CF, Bryant RJ (1974) Production, nutrient content and decomposition of *Phragmites communis* Trin. and *Typha angustifolia* L. J Ecol 63:71–95

Miao SL, Sklar FH (1998) Biomass and nutrient allocation of sawgrass and cattail along a nutrient gradient in the Florida Everglades. Wetlands Ecol Manage 5:245–263

Miao SL, Newman S, Sklar FH (2000) Effects of habitat nutrients and seed sources on growth and expansion of *Typha domingensis*. Aquat Bot 68:297–311

Mitsch WJ, Gosselink JG (2001) Wetlands, 3rd edn. Van Nostrand Reinhold, New York

Moran PAP (1950) Notes on continuous stochastic phenomena. Biometrika 37:17–23

Morillo C, González JL (1996) Management of Mediterranean Wetlands 2. Ministerio de Medio Ambiente, Madrid

Newman S, Kumpf H, Laing JA, Kennedy WC (2001) Decomposition responses to phosphorus enrichment in an Everglades (USA) slough. Biogeochemistry 54:229–250

Oden NL (1984) Assessing the significance of a spatial correlogram. Geogr Anal 16:1–16

Ogden JC (2005) Everglades ridge and slough conceptual ecological model. Wetlands 25:810–820

Pascual H (1976) Contribución al estudio ecológico de Las Tablas de Daimiel. I. La vegetación. Anales INIA 2:107–128

Polunin NVC (1982) Processes contributing to the decay of reed (*Phragmites australis*) litter in fresh water. Archiv Hydrobiol 94:182–209

Pu M, Jones RH, Guo D, Lister A (2005) Regeneration strategies, disturbance and plant interactions as organizers of vegetation spatial patterns in a pine forest. Landscape Ecol 20:971–987

Ribeiro MD (2005) Biología de los macrófitos emergentes en Las Tablas de Daimiel. Ph.D. thesis. Universidad Complutense, Madrid

Ribeiro MD, Alvarez-Cobelas M, Riolobos P, Cirujano S (2004) Descomposición de los helófitos en un humedal semiárido hipertrófico. Anal Jar Bot Madrid 61:53–61

Rodewald-Rudescu L (1974) Das Schilfrohr. Die Binnengewässer, 27. E. Schweizerbart'sche Verlagsbuchhandlung, Stuttgart

Rosenberg MS (2002) PASSAGE. Pattern Analysis, Spatial Statistics, and Geographic Exegesis. Department of Biology, Arizona State University, Tempe, Version 1.0

Sáez-Royuela R (1977) Contributión al studio ecológico de las Tablas de Daimiell III, Las aguas (1974–1975). Anal INIA, Serie Recursos Naturales 3:101–149

Saltmarsh A, Mauchamp A, Rambal S (2006) Contrasted effects of water limitation on leaf functions and growth of two emergent co-occurring plant species, *Cladium mariscus* and *Phragmites australis*. Aquat Bot 84:191–198

Sánchez-Carrillo S (2000) Hidrología y sedimentación actual de Las Tablas de Daimiel. Ph.D. Thesis. Universidad Autónoma, Madrid

Sánchez-Carrillo S, Álvarez-Cobelas M (2001) Nutrient dynamics and eutrophication patterns in a semiarid wetland: the effects of fluctuating hydrology. Water Air Soil Pollut 131:97–118

Sánchez-Carrillo S, Angeler DG, Sánchez-Andrés R, Alvarez-Cobelas M, Garatuza-Payán J (2004) Evapotranspiration in semi-arid wetlands: relationships between inundation and the macrophyte cover:open water ratio. Adv Water Resour 27:643–655

Seabloom EW, Moloney KA, van der Valk AG (2001) Constraints on the establishment of plants along a fluctuating water-depth gradient. Ecology 82:2216–2232

Turner MG (2005) Landscape ecology: what is the state of the science? Ann Rev Ecol Evol Syst 36:319–344

Urban NH, Davis SM, Aumen NG (1993) Fluctuations in sawgrass and cattail densities in Everglades Water Conservation Area 2A under varying nutrient, hydrological and fire regimes. Aquat Bot 46:203–223

Visser JM, Sasser CE, Chabreck RH, Linscombe RG (1999) Long-term vegetation change in Louisiana tidal marshes, 1968–1992. Wetlands 19:168–175

Vollenweider RA (1968) Scientific Fundamentals on the Eutrophication of Lakes and Flowing Waters, with particular reference to Nitrogen and Phosphorus as Factors in Eutrophication. OECD Report, Paris

Vymazal J (1995) Algae and element cycling in wetlands. CRC Press, Boca Ratón, FL

Chapter 8
Fish and Avian Communities: A Testimony of Wetland Degradation

M. Álvarez-Cobelas

Abstract Ongoing degradation of this wetland has resulted in biodiversity losses and alterations of fish and bird communities. Fish were represented by 14 species until 1980, five of which were introduced species; nowadays, there are six species left (three of them introduced) in the years when water quality was slightly improved. Habitat destruction, water availability and water pollution are the main processes threatening fish biodiversity in this wetland; toxic algae, cyanide and gill malfunctioning related with eutrophication were responsible for mass mortality of fish. Waterfowl are strongly dependent upon flooding areas, with this wetland acting as an ecological refuge for waterfowl living in nearby, temporary wetlands when these had dried up. Breeding numbers of aquatic birds were related to the spring flooding area. Impaired water quality, with its side effect of diminishing transparency, has resulted in the disappearance of most diving ducks. This chapter will review long-term patterns (1970–2007) of fish and bird communities as related to man-made impacts in the wetland. Some data on passerine species are also provided.

8.1 Introduction

The Tablas de Daimiel National Park was created in 1973 to preserve an important faunal assemblage, whose richest vertebrate groups were fish and birds. 37 years later these groups are impoverished as a result of diminishing quantity and impaired quality of water. This chapter will deal with these changes.

M. Álvarez-Cobelas (✉)
Spanish National Research Council (CSIC), Institute of Natural Resources,
Serrano 115 dpdo, E-28006, Madrid, Spain
e-mail: malvarez@ccma.csic.es

S. Sánchez-Carrillo and D.G. Angeler (eds.), *Ecology of Threatened Semi-Arid Wetlands: Long-Term Research in Las Tablas de Daimiel*, DOI 10.1007/978-90-481-9181-9_8,

8.2 Fish Communities

Previous reports on fish in Las Tablas de Daimiel National Park have been those by Elvira and Barrachina (1996), Elvira (1998) and Ortega-Mayagoitia et al. (2002).

8.2.1 Material and Methods

Fish were collected in July every year, from 2000 to 2005, using the capture per unit of effort approach by means of nets and seines, following procedures described by Ricker (1968). Several sampling sites were selected to cope with wetland heterogeneity. Once caught, fish were measured and weighted. Data on wetland features and flooding were retrieved from the National Park staff.

8.2.2 Results and Discussion

When the National Park was created, fish richness decreased, from 16 species in 1973 to three species in 2007 (Table 8.1). It is likely that most species have disappeared due to water pollution (see Chapter 6), but in recent years with high water inputs, the fish assemblage increased as a result of reinvasions from the

Table 8.1 Fish species richness at Las Tablas de Daimiel National Park in 1973, when the natural reserve was created, and in 2007, after three decades of ongoing environmental degradation (data source: Elvira and Barrachina 1996 and National Park staff)

	1973	2007
Anguilla anguilla	1	0
Barbus comiza	1	0
Barbus microcephalus	1	0
Chondrostoma wilkommii	1	0
Cobitis paludica	1	0
Cyprinus carpio	1	1
Esox lucius	1	0
Gambusia holbrooki	1	1
Lepomis gibossus	1	1
Leuciscus alburnoides	1	0
Leuciscus pyrenaicus	1	0
Micropterus salmoides	1	0
Petromyzon marinus	1	0
Rutilus lemmingii	1	0
Salaria fluviatilis	1	0
Tinca tinca	1	0
Species richness	16	3

catchment, with species such as *Cobitis maroccana, Tropidophoxinellus alburnoides* and *Lepomis gibbosus*. However, the harsh environment, with low oxygen tensions and poor water quality, soon affected these species which again disappeared.

In 2000–2005, we undertook annual surveys to quantify fish populations, using fish nets and trawls to get numbers of individuals on a capture per effort unit (CPUE) basis. The only species that presented sufficient numbers to be tracked were *Cyprinus carpio* (carp), *Lepomis gibbosus* (pumpkinseed) and *Gambusia holbrooki* (mosquitofish). While carp and pumpkinseed populations were higher in the first years, almost disappearing later from the wetland, mosquito-fish showed the opposite trend (Fig. 8.1a). At any rate, spatio-temporal variability was very high, and yearly averages were strongly influenced by their variability. Our results did not suggest preferences for any habitat, possibly because the poor water quality was widespread during the studied period. Average density was inversely related with average wetland water quality for carps and pumpkinseeds (Fig. 8.1b), but *Gambusia* did not show such a statistical relationship. Average flooding in the preceding

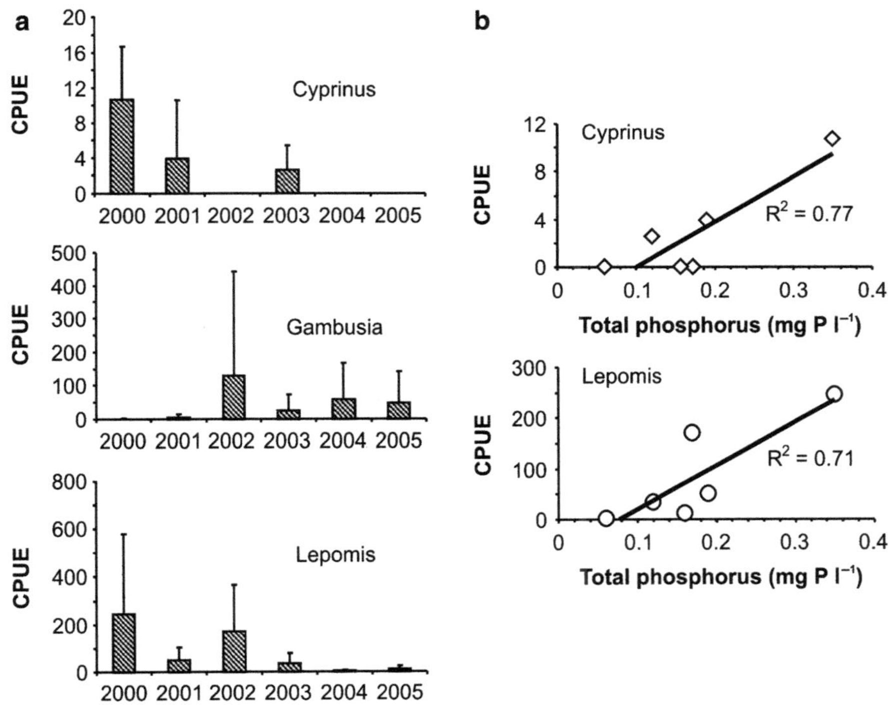

Fig. 8.1 (**a**) Captures per unit effort of the main fish populations gathered during 2000–2005 in the Las Tablas de Daimiel National Park. (**b**) Captures per unit effort of *Cyprinus carpio* and *Lepomis gibbosus vs* an index of water quality. Total phosphorus was averaged for the whole wetland in the 7 months preceding fish surveys. Obviously, such an index is the opposite of water quality, namely, high phosphorus values are suggestive of poor water quality

months was not related to any population size of the three species involved ($p > 0.05$). While effects of carp on internal recycling and its resistance to water pollution are well known (King et al. 1997), thus explaining its prevalence when nutrient enrichment is high in Las Tablas de Daimiel, factors controlling mosquito-fish are more difficult to ascertain. It appears that this tiny fish is strongly dependent on the presence of large crustaceans in this wetland, such as *Daphnia* and omnivorous copepods (Ortega-Mayagoitia et al. 2002), but these zooplankters exhibit a complex time trend in Las Tablas de Daimiel (see Chapter 6) and, hence, it is difficult to explain the patterns observed in their predator, the mosquito-fish. Definitely, there are factors other than food affecting *Gambusia* populations in Las Tablas, but our data limit a further exploration of these relationships.

During the 3 years when enough animal numbers were recorded, two different cohort classes could be observed for carp, one young and other much larger, suggestive of populations from the preceding year (Fig. 8.2). Cohorts of *Lepomis* varied from one year to another: there might be two, or maybe more, in 2000–2001, two in 2003 and 2005 and one in 2003, with 2004 as an anomalous year (Fig. 8.2). Finally, cohorts of mosquito-fish went down to one, except in 2001 and 2003 when there were two (Fig. 8.2).

There are few long-term studies on the responses of fish assemblages to nutrient enrichment, and it is therefore difficult to analyze and compare what has taken place in the fish community of Las Tablas de Daimiel and to predict its future time course. This is unfortunate, because fish are usually affected by trophic cascades, and long-term nutrient enrichment is certainly changing the structural and functional features of fish communities. Nutrient enrichment appears to promote fish productivity and increase older age classes in brown trout (Milbrink et al. 2008), and this population follows a hysteresis process once nutrient enrichment is over, recovering afterwards the pre-enrichment age structure 5 years after the termination of nutrient enrichment. Since nutrient enrichment is still occurring in Las Tablas de Daimiel, further studies are necessary on the responses of fish assemblage.

Massive fish mortalities have also been common in this wetland (Fig. 8.3 upper panel). The number of retrieved corpses ranged between 500 and 40,000. Several causes for these catastrophic events could be suggested, such as cyanide occurrence in water and porewater (see Chapter 5) and sudden sediment re-suspension that clogged fish gills.

8.3 Avian Communities

Las Tablas de Daimiel has always been a waterfowl habitat, used for duck hunting since the fourteenth century (Príncipe Don Juan Manuel 1319–1325; Relaciones topográficas mandadas hacer por el Rey Felipe II 1575; Respuestas Generales al Catastro de Ensenada 1752; Chapman and Buck 1910; Settier 1956). Bird species richness might have been very high, and this was the main reason for including the

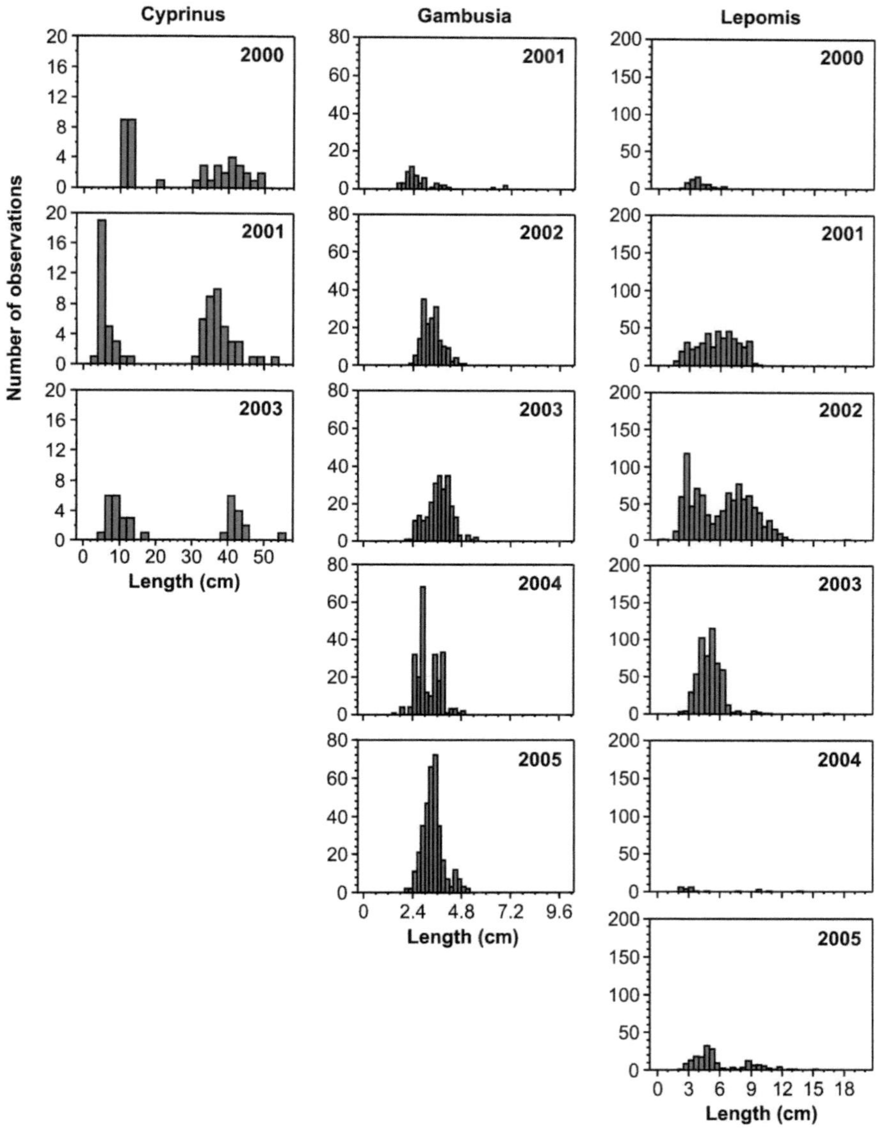

Fig. 8.2 Size classes of *Cyprinus carpio* (*left panels*), *Lepomis gibbosus* (*centre panels*) and *Gambusia holbrooki* (*right panels*) in Las Tablas de Daimiel National Park

area in the Ramsar Convention in 1982 (Bernués 1998). Carrasco and Sánchez Soler (1996), Carrasco and del Moral (1998), Carrasco and Ruiz de la Hermosa (2004), Hernández-García (2004, 2007), Jiménez et al. (1992), del Moral-Fernández del Rincón and del Moral-Molina (2004) and Villanueva (2007) have recently studied avian communities in this wetland.

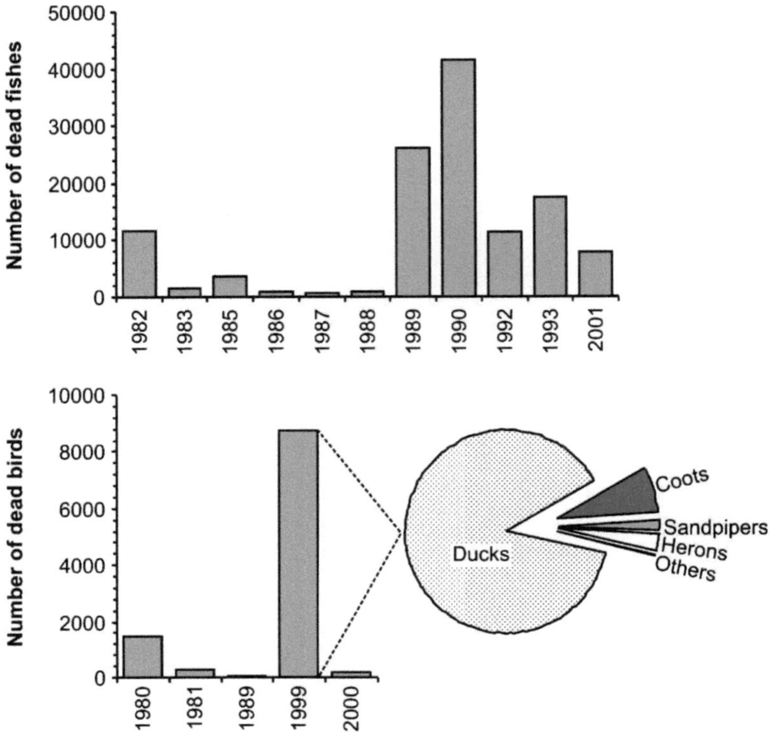

Fig. 8.3 *Lower panel* – numbers of dead birds as a result of botulism in the Tablas de Daimiel wetland and percentage share per bird group in the massive bird kill of 1999 (data source: National Park staff)

8.3.1 Material and Methods

Since 1973, National Park staff has carried out each January an annual waterfowl census of overwintering animals. Breeding pairs have also been recorded since 1983. From 1997 onwards, a net was installed at a SW site called La Quebrada to catch passerine birds. Later, in 2004 and 2007, another net was also installed at a NE site. Data on wetland flooding were supplied by the National Park staff, whereas those of helophyte cover were collected by Santos Cirujano (see Chapter 7).

8.3.2 Results and Discussion

Bird species richness has also experienced a decline, even though this decrease was not as strong as that observed for fish assemblages. In fact, other studies carried out in Spanish protected areas have revealed that those species concentrating in strictly protected natural marshes have tended to decline (Rendón et al. 2008), and this has also held in Las Tablas de Daimiel. Nowadays, the number of breeding waterfowl species amounts to 17, 68% out of the total breeding when degradation of the wetland was not too strong. *Anas strepera*, *Ardea cinerea*, *A. purpurea*, *Ixobrychus minutus*, *Larus ridibundus*, *Podiceps nigricollis*, *Sterna albifrons*, *Tachybaptus ruficollis* and *Vanellus vanellus* no longer breed in Las Tablas de Daimiel (Table 8.2).

The number of wintering ducks was of 45,000 animals when scientific censuses were carried out, starting in 1968 (Fig. 8.4a), the remaining bird populations being much lower. There were 6,000 pairs of breeding ducks in 2004, but these were negligible in some years (Fig. 8.4a). The highest population of wintering ducks is the mallard (*Anas platyrhynchos*), which may reach up to 21,000 animals

Table 8.2 Changes in breeding non-passerine bird species in Las Tablas de Daimiel National Park from 1983 (mild wetland degradation to 2007 (strong wetland degradation). 1: present, 0: absent

		1983	2007
Accipitridae	*Circus aeruginosus*	1	1
Anatidae	*Anas strepera*	1	0
	Anas clypeata	1	1
	Anas platyrhynchos	1	1
	Anas querquedula	1	1
	Aythya ferina	1	1
	Aythya nyroca	1	1
	Netta rufina	1	1
Ardeidae	*Ardea cinerea*	1	0
	Ardea purpurea	1	0
	Botaurus stellaris	1	0
	Bubulcus ibis	0	1
	Egretta garzetta	1	1
	Ixobrychus minutus	1	0
	Nycticorax nycticorax	1	0
Charadriidae	*Vanellus vanellus*	1	0
Ciconidae	*Ciconia ciconia*	1	1
Glareolidae	*Glareola pratincola*	1	1
Laridae	*Larus ridibundus*	1	0
Podicipedidae	*Podiceps cristatus*	1	1
	Podiceps nigricollis	1	0
	Tachybaptus ruficollis	1	0
	Chlidonias hybridus	1	1
Rallidae	*Fulica atra*	1	1
Recurvirostridae	*Himantopus himantopus*	1	1
	Recurvirostra avossetta	0	1
Sternidae	*Sterna albifrons*	1	0
Threskiornitidae	*Platalea leucorodia*	0	1
Species richness		25	17

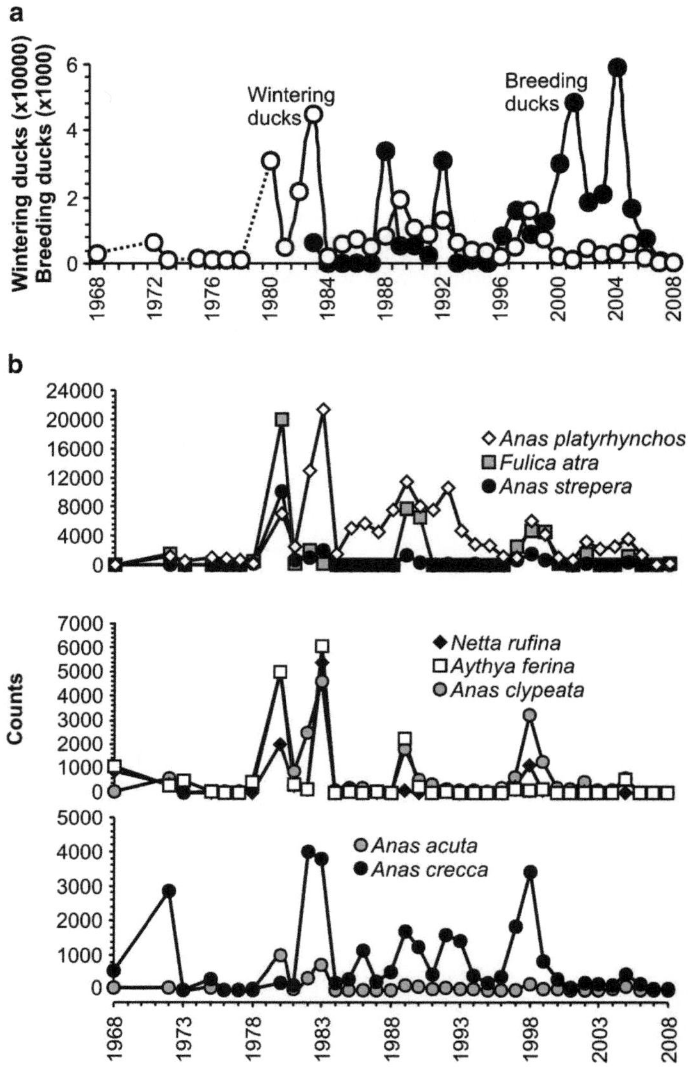

Fig. 8.4 (**a**) Censuses of wintering and breeding ducks in the Las Tablas de Daimiel National Park (data source: National Park staff). (**b**) Time series of overwintering populations of ducks and coots in Tablas de Daimiel (data source: National Park staff)

(Fig. 8.4b). Coots (*Fulica atra*) also reach similar high numbers. Red-crested pochards (*Netta rufina*), shovelers (*Anas clypeata*), teals (*A. crecca*) and gadwalls (*A. strepera*) may be as high as 6,000–7,000 animals. However, all species experience high interannual fluctuations (Fig. 8.4b).

Mallards are the highest breeding species of ducks in this wetland, with up to 3,000 breeding pairs, followed by pochards, red-crested pochards and gadwalls

(Fig. 8.5a). Coots, purple herons, night herons, black-winged stilts, little grebs and great crested grebs may attain up to 1,200 breeding pairs (Fig. 8.5b), albeit with high interannual fluctuations.

Overwintering birds do not appear to be related with autumn flooding ($p > 0.05$). However, most overwintering duck populations, on one hand, and breeding ducks, on the other, co-vary in this wetland ($r = 0.50$–0.92; $p < 0.05$). Breeding ducks as a whole are related to average spring flooding (Fig. 8.6), but the only species with large population that appears to be dependent on flooding is *Anas strepera* ($r = 0.72$; $p < 0.05$). There is no statistical relationship between overwintering and breeding ducks ($p > 0.05$).

Since the early eighties, botulism is a recurrent phenomenon in Las Tablas de Daimiel wetland (Fig. 8.3 lower panel). In 1999, the number of corpses retrieved, most being ducks, was of 9,000 (Fig. 8.3 lower panel).

All these results point to the fact that waterfowl has been very important at Tablas de Daimiel, and that its recovery (e.g. increasing species richness and abundance) is dependent upon recovery of the hydrolevel and improvement of water quality (Carrasco and Ruiz de la Hermosa 2004). Water quality is also a key factor, because many species are diving ducks, feeding on benthic animals and plants, now

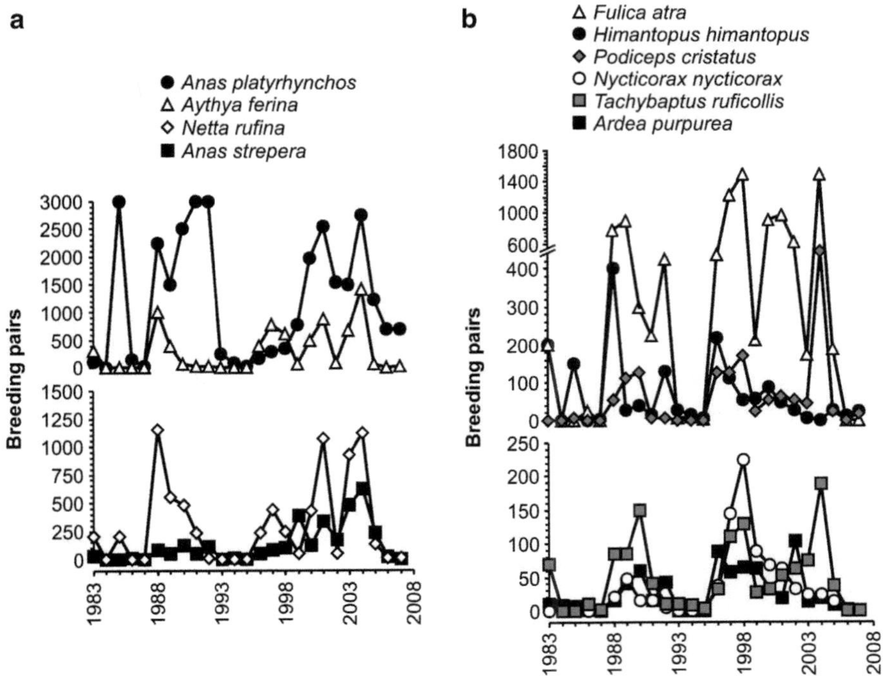

Fig. 8.5 (**a**) Time series of breeding populations of ducks in Tablas de Daimiel (data source: National Park staff. (**b**) Time series of breeding populations of ardeids and grebs in Tablas de Daimiel (data source: National Park staff)

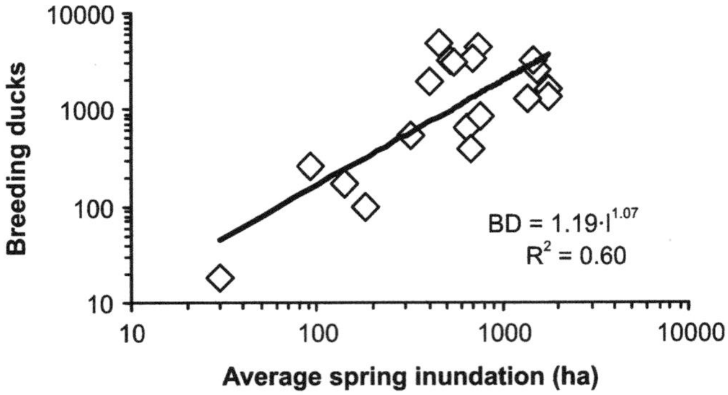

Fig. 8.6 Relationship between average spring flooding (I) and breeding pairs of ducks (BD) in Las Tablas de Daimiel

impoverished as a result of increasing turbidity and accumulation of toxic materials due to water pollution (see Chapter 5).

Passerine birds have been surveyed at a single site since 1997 and at two distinct sites of the wetland in 2004 and 2007. Long-term changes of the most abundant species appear to be very fluctuating (Fig. 8.7), and no control factor of those fluctuations can be envisaged. However, some studies (Desgranges et al. 2006) show that strong hydrolevel fluctuations, such as those that have occurred in Las Tablas de Daimiel, may have deleterious effects on wetland breeding passerines. More species were recorded in 2004 than in 2007, maybe because flooding was higher in the former year. The NE site showed more species richness along with higher flooding, whereas there were more passerine birds at the SW site, with lower flooding (Table 8.3).

Only the kingfisher was related with average flooding ($r = 0.78$; $p < 0.05$). The other dominant passerines (Fig. 8.7) did not show any relationship with yearly-averaged flooding. Another environmental variable that could be important for these birds is reed cover, but no significant relationship was found between their populations and annual reed cover in this wetland. Interestingly, correlations suggest some ecological relationships among dominant passerines (Table 8.4). The reed warbler might be a competing species with nightingale. The house sparrow shows a mutualistic relationship with reed warblers, goldfinches and Cetti's warblers, whereas it appeared to compete with nightingale.

This chapter has summarized the available evidence on fish and avian changes in Las Tablas de Daimiel wetland, as a result of strong degradation of the habitat due to shortages in water availability and impairment of water quality. Floodplains are ecosystems where conservation of vertebrate species is more difficult than in forests (Miller et al. 2004). The resilience of fish and avian communities in this wetland is well-known (Elvira 1998; Carrasco and Ruiz de la Hermosa 2004), and these assemblages may easily be recovered to the richness and abundance levels prior to environmental degradation, providing that water of enough quantity and good quality is supplied to this wetland.

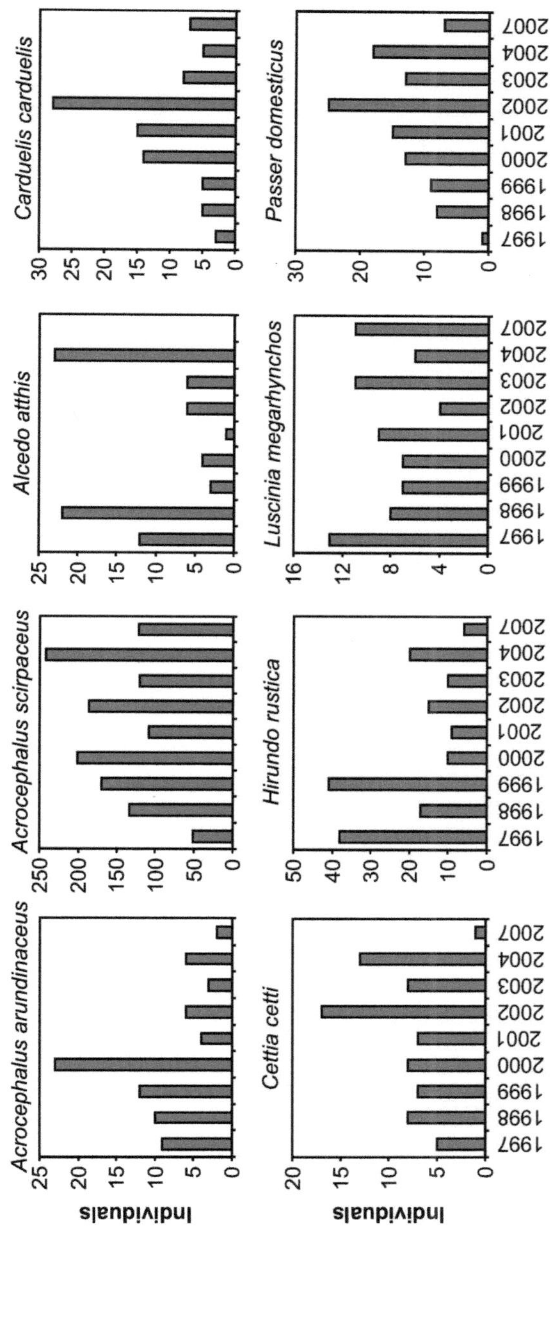

Fig. 8.7 Time series of passerine birds caught with nets for recovering rings at a SW site (La Quebrada) of the Las Tablas de Daimiel wetland (data source: del Moral-Fernández del Rincón and del Moral-Molina (2004) and Hernández-García (2004, 2007)

Table 8.3 Passerine species caught with nets (gray filled) at a NE site (Algeciras) and a SW site (La Quebrada) of Las Tablas de Daimiel in 2004 and 2007 (data source: Hernández-García 2004, 2007)

		NE site (2004)	NE site (2007)	SW site (2004)	SW site (2007)
Alcedinidae	*Alcedo atthis*	■		■	■
Alaudidae	*Galerida cristata*	■		■	■
Columbidae	*Columba palumbus*	■		■	■
Cuculidae	*Clamator glandarius*	■		■	■
Emberizidae	*Emberiza schoeniclus*	■		■	■
	Miliaria calandra	■		■	■
Fringillidae	*Carduelis cannabina*	■	■	■	
	Carduelis carduelis	■	■	■	■
	Carduelis chloris	■	■	■	■
	Fringilla coelebs	■	■	■	■
	Fringilla montefringilla	■		■	■
	Serinus serinus	■		■	■
	Sturnus unicolor	■		■	■
Hirundinidae	*Hirundo daurica*	■		■	■
	Hirundo rustica	■		■	■
	Riparia riparia	■		■	
Laniidae	*Lanius meridionalis*	■	■		
	Lanius senator	■	■	■	■
Meropidae	*Merops apiaster*	■		■	■
Motacillidae	*Anthus pratensis*	■		■	
	Anthus trivialis		■		
	Motacilla alba	■			
Muscicapidae	*Ficedula hypoleuca*	■	■	■	■
	Muscicapa striata	■	■	■	■
Paridae	*Parus caeruleus*	■		■	■
	Parus major	■	■	■	■
Passeridae	*Passer domesticus*	■		■	■
	Passer hispaniolensis	■		■	■
	Passer montanus	■	■	■	■
	Passer sp.	■	■	■	■
Picidae	*Jynx torquilla*	■		■	
	Picus viridis		■	■	
Prunelidae	*Prunella modularis*	■		■	■
Remicidae	*Remiz pendulinus*	■	■	■	■
Scolopacidae	*Actitis hypoleucos*	■		■	■
Sylviidae	*Acrocephalus arundinaceus*	■	■	■	■

(continued)

Table 8.3 (continued)

		NE site (2004)	NE site (2007)	SW site (2004)	SW site (2007)
	Acrocephalus palustris	■		■	
	Acrocephalus schoenobaenus	■	■	■	
	Acrocephalus scirpaceus	■	■	■	■
	Cettia cetti	■	■	■	■
	Cisticola juncidis	■	■	■	■
	Hippolais polyglotta	■	■	■	■
	Locustella luscinioides	■	■	■	
	Locustella naevia	■	■	■	
	Phylloscopus bonelli	■	■	■	
	Phylloscopus collybita	■	■	■	■
	Phylloscopus sibilatrix		■		
	Phylloscopus trochilus	■	■	■	■
	Regulus ignicapillus	■			
	Sylvia atricapilla	■	■	■	■
	Sylvia borin	■	■	■	■
	Sylvia cantillans	■	■		
	Sylvia communis	■	■	■	■
	Sylvia hortensis	■	■		
	Sylvia melanocephala	■	■	■	
	Sylvia undata	■	■	■	■
Timalidae	*Panurus biarmicus*	■		■	
Troglodytidae	*Troglodytes troglodytes*	■		■	■
Turdidae	*Erithacus rubecula*	■	■	■	■
	Luscinia megarhynchos	■	■	■	■
	Luscinia svecica	■	■	■	■
	Phoenicurus ochruros	■		■	
	Phoenicurus phoenicurus	■	■	■	■
	Saxicola rubetra	■	■	■	
	Saxicola torquata	■	■	■	■
	Turdus merula	■	■	■	■
	Turdus philomelos	■	■	■	■
Upupidae	*Upupa epops*	■	■	■	■

Table 8.4 Correlation matrix of the main passerine birds recorded on a long-term basis at La Quebrada, sampling site located in the Southwestern area of Las Tablas de Daimiel. Statistically significant relationships ($p < 0.05$) are shown in bold letters. The extent of reed (*Phragmites australis*) cover and the average flooding of each year were also included in calculations

	Acrocephalus arundinaceus	*Acrocephalus scirpaceus*	*Alcedo atthis*	*Carduelis carduelis*	*Cettia cetti*	*Hirundo rustica*	*Luscinia megarhynchos*	*Passer domesticus*	Average flooding
Acrocephalus scirpaceus	0.32								
Alcedo atthis	0.02	0.25							
Carduelis carduelis	0.01	0.28	–0.38						
Cettia cetti	0.04	0.63	0.36	0.62					
Hirund rustica	0.21	–0.15	0.22	–0.42	–0.02				
Luscinia megarhynchos	–0.28	**–0.85**	–0.16	–0.59	**–0.80**	0.08			
Passer domesticus	–0.12	**0.68**	–0.02	**0.79**	**0.85**	–0.42	**–0.80**		
Average flooding	0.21	–0.18	**0.78**	–0.49	0.07	0.55	0.10	–0.39	
Reed cover	0.44	–0.10	–0.29	0.47	0.26	0.11	–0.20	0.21	0.12

Acknowledgments The author gratefully acknowledge the current and the former directors of Las Tablas de Daimiel National Park (Carlos Ruiz de la Hermosa and Manuel Carrasco-Redondo, respectively) for their advice, help, internal reports and data over the years. Carlos Ruiz de la Hermosa and Manuel López-Sánchez have carefully read a previous draft and made some suggestions. The staff of the National Park has carried out bird surveys and censuses and helped with fish catches.

References

Bernués M (ed) (1998) Humedales españoles inscritos en la Lista del Convenio de Ramsar, 2nd edn. Ministerio de Medio Ambiente, Madrid

Carrasco M, del Moral A (1998) Avifauna. Parque Nacional Las Tablas de Daimiel. Editorial Esfagnos, Talavera de la Reina, pp 203–288

Carrasco M, Ruiz de la Hermosa C (2004) ¿Siguen siendo importantes Las Tablas para las aves? An Ornitol Ciudad Real 2002–2003:239–244

Carrasco M, Sánchez Soler MJ (1996) Avifauna palustre. In: Alvarez-Cobelas M, Cirujano S (eds) Las Tablas de Daimiel, Ecología acuática y Sociedad. ICONA, Madrid, pp 187–202

Chapman A, Buck WJ (1910) Unexplored Spain. Edward Arnold, London

Respuestas generales al catastro de Ensenada, Daimiel (1752) Edition of 1993. Tabapress, Madrid

del Moral-Fernández del Rincón A, del Moral Molina A (2004) La estación de anillamiento "La Quebrada", Parque Nacional de Las Tablas de Daimiel, Programa PASER. An Ornitol Ciudad Real 2002–2003, pp 225–232

Desgranges JL, Ingram J, Drolet B, Morin J, Savage C, Borcard D (2006) Modelling wetland bird response to water level changes in the Lake Ontario-St Lawrence river hydrosystem. Environ Monit Assess 113:329–365

Elvira B (1998) Invertebrados acuáticos y peces. Parque Nacional Las Tablas de Daimiel. Editorial Esfagnos, Talavera de la Reina, pp 151–168

Elvira B, Barrachina P (1996) Peces. In: Alvarez-Cobelas M, Cirujano S (eds) Las Tablas de Daimiel, Ecología acuática y Sociedad. ICONA, Madrid, pp 171–186

Relaciones topográficas mandadas hacer por el rey Felipe II, año. 1575. In: Sarria A (ed) 1986. Ayuntamiento de Daimiel, Daimiel

Hernández-García JM (2004) El anillamiento de aves en el Parque Nacional Tablas de Daimiel, 2004. Report for the National Park, Ciudad Real

Hernández-García JM (2007) El anillamiento de aves en el Parque Nacional Tablas de Daimiel, 2007. Report for the National Park, Ciudad Real

Jiménez J, del Moral A, Morillo C, Sánchez-Soler MJ (1992) Las aves del Parque Nacional de Las Tablas de Daimiel y otros humedales manchegos. Lynx edicions, Barcelona

King AJ, Robertson AI, Healey MR (1997) Experimental manipulations of biomass of introduced carp (*Cyprinus carpio*) in billabongs. I. Impacts on water-column properties. Mar Freshw Res 48:435–443

Milbrink G, Petersson G, Holmgren S (2008) Long-term effects of nutrient enrichment on the condition and size-structure of an alpine brown trout population. Environ Biol Fish 81:157–170

Miller JR, Dixon MD, Turner MG (2004) Response of avian communities in large-river floodplains to environmental variation at multiple scales. Ecol Appl 14:1394–1410

Ortega-Mayagoitia E, Rojo C, Rodrigo MA (2002) Factors masking the trophic cascade in shallow eutrophic wetlands – evidence from a microcosm study. Archiv Hydrobiol 155:43–63

Príncipe Don Juan Manuel (1319–1325) In: Gutiérrez de la Vega J (ed) Libro de la caza, 1879 edn. Imprenta M. Tello, Madrid

Rendón MA, Green AJ, Aguilera E, Almaraz P (2008) Status, distribution and long-term changes in the waterbird community wintering in Doñana, south–west Spain. Biol Conserv 141:1371–1388

Ricker WE (1968) Methods for assessment of fish production in fresh waters. IBP handbook no. 3. Blackwell Scientific Publications, Oxford

Settier J (1956) Caza menor: anécdotas y recuerdos. Imprenta editorial Reus, Madrid

Villanueva C (2007) Un episodio desafortunado: La epidemia de botulismo en el Parque Nacional de Las Tablas de Daimiel durante el verano de 1999. An Ornitol Ciudad Real 2004–2005:87–96

Part IV
Applied Issues

Chapter 9
Biomanipulation: A Useful Tool for Wetland Rehabilitation

D.G. Angeler

Abstract Food web manipulation, or biomanipulation, is a frequently applied lake management tool, aiming to restore water quality and vegetation characteristics through interventions in the fish communities. Despite the strong management appeal of biomanipulations, this tool found so far little application in wetlands. This chapter highlights pros and cons of biomanipulations in wetlands, and suggests that an extension of the current biomanipulation paradigm beyond fish management, to consider interventions in other components of wetland communities, can be useful for rehabilitating degraded wetlands such as Las Tablas de Daimiel.

9.1 Introduction

Las Tablas de Daimiel is one of many examples of large-scale ecosystem degradation as a result of excessive waste water discharge from nearby urban areas, combined with other anthropogenic stressors (e.g., disruption of the natural hydrological regime, introduction of exotic species, increased agricultural use in its catchment (e.g. Chapter 1, Alvarez-Cobelas et al. 2001). These impacts have profoundly altered the ecosystem structure and function, and Las Tablas de Daimiel shows symptoms of degradation that differ little from those observed in temperate shallow lakes affected by anthropogenic eutrophication. Elevated concentrations of nutrients, increased turbidity, shifts in the primary producer community from submerged macrophyte dominance to phytoplankton dominance, decreased biomass of large-bodied zooplankton, increased biomass of planktivorous and/or benthivorous

D.G. Angeler (✉)
Department of Aquatic Sciences and Assessment, Swedish University of Agricultural Sciences, P.O. Box 7050, SE-75 007 Uppsala, Sweden
e-mail: David.Angeler@vatten.slu.se

S. Sánchez-Carrillo and D.G. Angeler (eds.), *Ecology of Threatened Semi-Arid Wetlands: Long-Term Research in Las Tablas de Daimiel*, DOI 10.1007/978-90-481-9181-9_9,

fish and losses of piscivores characterize degraded lakes (Jeppesen 1998; Jeppesen et al. 1998) and wetlands (Whillans 1996; Chow-Fraser et al. 1998; Alvarez-Cobelas et al. 2001). In addition to these symptoms, eutrophied wetlands may show changes in the structure of emergent vegetation, an important biotic compartment in many riparian wetlands (Mitsch and Gosselink 2000; Alvarez-Cobelas et al. 2008).

The similarity of these symptoms suggests that both ecosystem types could benefit from similar rehabilitation strategies. In the practice, however, wetland eutrophication abatement schemes have been curiously different when compared to those of shallow lakes. Most notably, management of shallow lakes has made use of, and contributed itself a great deal to, a large body of scientific evidence that interventions in food webs hold potential to rehabilitate degraded sites (Falk et al. 2006). In addition to interventions in catchments (e.g., control of point and diffuse sources of pollution) and in situ (macrophyte implantation, sediment dredging, nutrient precipitation, water column aeration), lake management has typically focused on top-down manipulations of fish communities as a rehabilitation tool (Jeppesen and Sammalkorpi 2002; Cooke et al. 2005). The enhancement of piscivores or reduction of zooplanktivores and/or benthivores can induce shifts in plankton and benthos community dynamics and lead to improved water quality. Biomanipulation has served as a field test of classic food web models, mainly in north-temperate lakes, and a significant amount of studies has accumulated in the literature that document successes and failures of biomanipulations (Søndergaard et al. 2007; Gulati et al. 2008).

Despite the strong conceptual appeal of biomanipulation for improving the environmental quality of aquatic ecosystems, wetland restoration has frequently ignored or violated ecological theory (Zedler 2000). Based on increased evidence that biotic interactions shape wetland communities and ecosystem processes, some studies suggested that biomanipulations may also be useful for wetland rehabilitation (e.g., Angeler et al. 2003a; Schrage and Downing 2004; Ruggiero et al. 2005; Evelsizer and Turner 2006). Following examples in lakes, studies in wetland biomanipulations have focused chiefly on the top-down manipulation of fish communities (Angeler et al. 2003a; Potthoff et al. 2008).

This chapter will expand on a previous review (Angeler et al. 2003a), and incorporate more recent results to demonstrate potentials and limitations of interventions in the fish communities as a wetland rehabilitation tool. In its original form, biomanipulation was proposed to target different communities and habitats (Shapiro et al. 1975), and the present chapter will also show that biomanipulations targeting communities other than fish can be especially suitable for rehabilitating Las Tablas de Daimiel and perhaps other wetlands. Examples will discuss the potentials and limitations of manipulations of vegetation structure and the populations of an invasive crayfish. A conceptual model will show that the current biomanipulation paradigm could be broadened by extending ecological theory beyond the application of trophic cascades and alternative stable states to consider sedimentation as a threat to wetland disappearance in the long term.

9.2 Biomanipulation: Theory and Rationale

Despite substantial reduction of external nutrient loading, many aquatic ecosystems remain in a degraded state caused by large amounts of phosphorus that have accumulated in sediments (e.g. Havens et al. 2001). Lake sediments can therefore act as nutrient sources to the water column and provide a means to sustain high algal productivity even if allochthonous nutrient loading is effectively controlled. Under such conditions, ecosystem structure is altered in ways that create a self-sustaining feedback loop which works to maintain hysteresis, and the activities of planktivorous and benthivorous fish play an important role in maintaining these internal stabilising mechanisms (Box 9.1).

Making use of the concept of trophic cascades in lake food webs (Carpenter and Kitchell 1993), and alternative stable state theory (Scheffer et al. 1993), biomanipulation has, at least in some cases, proved useful to disrupt the equilibrium condition, favouring the shift of shallow lakes from the degraded, turbid state to the clear-water state. Both concepts are useful for understanding the ecology of wetlands (summarised in Angeler et al. 2007), and several studies suggest that similar state shifts can be induced through fish manipulations (see Section 9.3).

Box 9.1 Trophic cascades and alternative stable states in shallow aquatic ecosystems

In degraded shallow lakes and wetlands, phytoplankton is the dominant primary producer, while zooplanktivores and/or benthivores dominate the fish community. Zooplanktivorous fish contribute to eutrophication, chiefly via food-web mediated effects. The biomass of efficiently grazing zooplankton (the main ones being large-bodied cladocerans) is effectively controlled by zooplanktivores, thereby relieving phytoplankton from top-down control. The resulting high phytoplankton standing crop contributes to high turbidity, which in turn constrains submerged macrophytes. On the other hand, benthivorous fish favours phytoplankton growth by transferring sediment-bound nutrients to the water column during bottom foraging. The foraging activity of benthivores also contributes to sediment re-suspension which causes high non-algal turbidity. Furthermore, benthic feeding inflicts mechanical damage to submerged macrophytes.

Community structure in shallow lakes differs at lower nutrient concentrations, i.e. in the desirable clear water state. Submerged macrophytes are the dominant primary producers, as a result of an improved light climate. Plants consolidate sediments and provide predation refugia for large cladocerans, thus contributing to more control of phytoplankton via zooplankton grazing. Submerged plants also control phytoplankton via competition for light and nutrients and/or by allelopathic activity. In the clear-water state, fish communities tend toward higher piscivore to planktivore ratios, ultimately relieving large daphnids from predation by zooplanktivorous fish.

Ecosystem internal management by means of biomanipulation may be particularly useful in wetlands to reduce eutrophication effects through features that are inherent to wetlands. By definition, wetlands are shallow ecosystems (<2 m). This suggests that the effects of planktivorous and benthivorous fish increase with decreasing water depth in shallow aquatic ecosystems (Jeppesen et al. 1999; Chow-Fraser 1999; Angeler et al. 2002). Such effects are manifest in: (1) high zooplankton production per unit volume, as a result of high primary production; hence, strong top-down control of fish on zooplankton is possible, (2) the limitation of zooplankton refugia in shallow, turbid systems, especially when macrophytes are absent, (3) fish ingestion of sediment with high nutritive value due to low settling times in the water column; thus, populations of obligate and facultative benthivores are sustained, even when invertebrates are scarce, and (4) the more pronounced effects of sediment re-suspension by benthic-dwelling fish in shallower systems, because turbidity may affect the entire water column.

Wetlands can also show considerable variability in flood frequencies and hydroperiods, features that distinguish them well from physically less disturbed shallow lakes (Mitsch and Gosselink 2000). Fish reduction schemes should therefore have potentially positive effects in many deteriorated wetlands when their inundated area is at minimum levels. Catch per unit effort should increase with decreasing water column depth and when fish are hydrologically confined. Thus, it has been suggested that biomanipulation, when appropriately timed as a function of hydrological disturbance regimes of wetlands, may be a low cost–high benefit tool for wetland eutrophication abatement (Angeler et al. 2003a). However, landscape settings, vegetation characteristics and other inherent features of many wetlands may complicate intervention.

9.3 Biomanipulation in Wetlands: Applying a Lake Restoration Tool

Biomanipulation attempts in Las Tablas de Daimiel are limited to an enclosure study which was carried out during a summer draw down in 1999 (Angeler et al. 2002). This study assessed impacts of three exotic fishes that dominate the degraded wetland's fish community (Chapter 8). Because of the seasonally-pronounced changes of inundated area as a result of the climatic conditions, fish concentrate during severe summer droughts, thereby reaching very high biomass levels. Simulating the biomasses of the fish under such hydrological confinements, Angeler et al. (2002) tested for the impacts of common carp (*Cyprinus carpio* L.), pumpkinseed sunfish (*Lepomis gibbosus* L.) and mosquito-fish (*Gambusia holbrooki* Girard) on water quality and zooplankton.

With the addition of carp or pumpkinseed sunfish, chlorophyll *a*, total phosphorus, total nitrogen and turbidity increased. The magnitude of this increase depended on the fish species and was most pronounced in the carp treatment. Mosquitofish did not significantly affect water quality as compared to a fishless control (Table 9.1). Zooplankton biomass was significantly lower in the carp treatment. However, no direct negative effects (predation) of carp were observed.

Table 9.1 Comparison of selected water quality and biotic variables in response to biomanipulation in selected wetlands

Site	Secchi depth	Phytoplankton biomass	Tot-P	Tot-N	Turbidity	Cladocerans	Submerged vegetation	Intervention	Type of study/duration	Source
Prairie wetland (USA)	n.d.	–6.1[a]	–1.5[a]	–2.0[a]	–4.7[a] as NTU	+176.8[a]	n.d.	Planktivore elimination with rotenone	Whole ecosystem study; 4 years	Zimmer et al. (2001)
Cootes Paradise Marsh (Canada)	n.d.	n.s.[b]	–1.6[b]	n.d.	–2[b] as NTU	n.s.[b]	n.d.	Carp exclusion	Enclosure study; 15 days	Lougheed and Chow-Fraser (1998)
Tablas de Daimiel (Spain)	n.s.	–2.6[b,c]	–2.3[b, c]	–1.4[b, c]	–3[b, c] as Tot-SS	+ ≈250[b, c]	n.s.	Carp, sunfish and mosquitofish exclusion	Enclosure study; 6 weeks	Angeler et al. (2002)
Ventura marsh (USA)	+2.9[d]	–11.5[d]	–≈1.2[d]	n.d.	n.s.	+ ≈2.3[d,e]	+(n.c.)	Benthivore elimination with rotenone	Whole ecosystem study; 14 months	Schrage and Downing (2004)
Major Lake (Hungary)	+2.1[d]	–(n.c.)	–1.3[d]	n.d.	n.d.	+(n.c.)	+4.6[d]	Planktivore and benthivore removal, piscivore stocking	Whole ecosystem study; 4 years	Tátrai et al. (2005)
Delta marsh (Canada)	n.d.	n.d.	n.d.	n.d.	n.s.	n.d.	+11.9[b]	Planktivore and benthivore exclusion	Exclosure study; ca. 4 months	Evelsizer and Turner (2006)
Pairie wetlands	n.d.	n.s.	n.s.	n.d.	n.d.	+7.03[a]	n.s.	Piscivore stocking	Replicated whole ecosystem experiment; 3 years	Potthoff et al. (2008)

Values indicate the multiplicative change observed in each variable in response to the manipulation

+/–, Increase or decrease of value, respectively, n.d., no data; n.s., not significant; n.c. no calculations possible based on original study (but trends to increase or decrease after the manipulations are indicated)

NTU, Nephelometric Turbidity Units; Tot-P, total phosphorus; Tot-N, total nitrogen; Tot-SS, total suspended solids

[a]Changes observed between a treatment and a reference wetland

[b]Values calculated by comparing enclosures with highest fish stock and fishless controls

[c]Shown are carp data only, given that its effects were most deleterious for water quality

[d]Comparing periods before and after fish manipulations

[e]Values refer to body length and not biomass

Zooplankton biomass did not differ from the control, either with sunfish or mosquitofish treatments. Nevertheless, both fish species affected zooplankton community composition, contributing to a community of ineffective grazing rotifers and copepods. The cladoceran *Ceriodaphnia reticulata* Jurine developed only in the fishless control, and was able to exert some top-down influence on phytoplankton.

Although beneficial effects of fish exclusion on water quality and plankton communities were evident in this study, consistent with predictions made by the trophic cascade theory, it is currently difficult to evaluate to what extent biomanipulations at larger spatial scales will reflect the findings of the enclosure study in Las Tablas de Daimiel. Although the ecological relevance of enclosure studies is limited, other wetland biomanipulation studies in larger fish exclosures and ecosystem scale interventions also successfully tested the application of the trophic cascade theory. These results are encouraging because studies across different spatial extents and climatic areas have shown that top-down manipulations of the wetland fish communities hold potential to improve the environmental quality of the studied systems (Table 9.1). This suggests that biomanipluation beyond the scale of enclosures could be useful in Las Tablas de Daimiel.

Despite the potential of biomanipulation as a wetland rehabilitation tool, several patterns emerge from the studies summarised in Table 9.1, that highlight potential limitations which may affect restorative interventions in a context-dependant way. Such limitations must be considered if biomanipulation is applied at the ecosystem scale of Las Tablas de Daimiel.

Angeler et al. (2002) and Potthoff et al. (2008) have discussed the importance of fish life history traits in biomanipulations. Potthoff and colleagues found that additions of young-of-the-year (YOY) walleye (*Sander vitreus* Mitchell) was most effective in controlling planktivorous fathead minnows (*Pimephales promelas* Cope) in wetland lakes of the Prairie Pothole Region of North America, resulting in much lower densities of minnows, higher densities of cladocerans and some macroinvertebrates and decreased phytoplankton standing crop relative to sites that were not stocked with piscivores. Few changes in plankton and benthos communities were found when larger size classes of predatory walleye were added. These differences were attributed to the rapid diet shift of YOY walleye and their ability to consume and suppress all life stages of minnows. Consistent with observations made in lakes (Perrow et al. 1997; Hansson et al. 1998), a further limitation of this study was that stocking of piscivores was not efficient in the long term, suggesting that repeated stocking schemes are required for controlling the deleterious effects of minnows.

Angeler et al. (2002) suggested that biomanipulation schemes focusing on mosquito-fish may differ from those of other fish. So far, planktivorous and benthivorous fish removal in temperate lakes has considered species that produce off-spring once a year (e.g., roach (*Rutilus rutilus* L.), bream (*Abramis brama* L.), bleak (*Alburnus alburnus* L.) and perch (*Perca fluviatilis* L.)). Mosquitofish is a highly reproductive species which can have more than three generations per year (Vargas and de Sostoa 1996). Hence, the life history traits of mosquitofish may be of importance in the consideration of large scale fish manipulation designs, if long-lasting effects are desired. Recommendations based on observations from lakes include a reduction of planktivorous fish by at least 75%, if long term effects are to be attained (Perrow et al. 1997; Hansson et al. 1998). In regard particularly to Las Tablas de Daimiel, this number may be well suitable for common carp and pumpkinseed sunfish, but effective biomanipulation in mosquitofish dominated systems can only be attained by a 100% reduction (a seemingly impossible task), otherwise, the pre-manipulation abundance will soon return. However, extermination plans of mosquitofish could even encounter social rejection because these fish are considered to act as a biological control against nuisance mosquitoes.

The re-establishment of submerged vegetation is a desired secondary effect of biomanipulations in shallow aquatic ecosystems. However, several factors, including water colour and wind activity, have been shown to constrain the restoration of submerged macrophytes to extents prior to degradation (Angeler et al. 2003a). The enclosure study in Las Tablas de Daimiel (Angeler et al. 2002) showed that turbidity caused by high water colour remained very high whether or not fish biomass was manipulated. M. Alvarez-Cobelas (2009 unpublished data) found that high humic levels arise from the decomposition of the common reed (*Phragmites australis* Trin. ex Steud.), an emergent plant that has widely replaced sawgrass (*Cladium mariscus* (L.) Pohl), which generates less humic compounds, as a result of hydrological alterations and eutrophication of the wetland (Alvarez-Cobelas et al. 2008). Thus, while Shapiro (1990) suggested that light limitation resulting from high humic levels may enhance biomanipulation potential, because 'low light' refugia reduce fish predation on large-bodied zooplankton, the "facilitation effect" of water colour may work in deep lakes rather than in shallow aquatic ecosystems. Also the effects of wind-induced sediment re-suspension have been frequently considered to be a limiting factor for successful biomanipulations in shallow lakes and wetlands (e.g., Lammens 1988; Lougheed and Chow-Fraser 1998). These examples suggest that additional remedial actions, preferentially in sediments and vegetation, would be required to restore submerged aquatic vegetation, as has been previously suggested (Wilcox and Whillans 1999; Angeler et al. 2003a).

It is important to highlight that the usefulness of fish manipulations as a test of trophic cascade theory and alternative stable state concepts depends on landscape contexts (Angeler et al. 2003a; Reed 2006). The hydrogeomorphic settings and connectivity of wetlands, their flood frequencies and durations, land use characteristics in their catchments and climatic conditions, ultimately regulate wetland communities and ecosystem processes (Mitsch and Gosselink 2000). Thus, interactions among different biotic and abiotic parameters in wetlands add complexity to our understanding of ecosystem ecology, beyond levels known for shallow lakes. Global warming may further increase this complexity and complicate predictions on ecosystem responses to fish manipulations. Jeppesen et al. (2007) recently highlighted climate-related aspects that may limit biomanipulations. Interacting abiotic and biotic characteristics will certainly need thorough consideration in future wetland biomanipulation studies in Las Tablas de Daimiel and other wetlands. The following section will highlight potentials and limitations of an extended biomanipulation paradigm.

9.4 Biomanipulation in Wetlands: Extending the Paradigm

Fish occupy a key position in lake food webs, and it may not be surprising that the biomanipulation paradigm is mainly based on the notion that interventions in fish community structure have cascading effects towards lower trophic levels, ultimately improving water quality. The complexity of wetland food webs and the strong influence of abiotic factors require a focus on different communities and habitats, as initially proposed by Shapiro et al. (1975). In fact, recent studies suggest that the introduction of bivalves (e.g., *Dreissena polymorpha* Pallas) could be used to control cyanobacterial blooms (Gulati et al. 2008); however, the introduction of these species could cause more ecological damage than benefits; therefore, a thorough assessment of impacts is needed. The following examples highlight that targeting multiple communities could extend the usefulness of biomanipulations beyond water quality improvements to control key abiotic wetland processes.

A striking difference between many shallow lakes and riparian wetlands, such as Las Tablas de Daimiel, is the low ratio of open water area to space covered by emergent vegetation in marshes. With the exception of, for example, Lake Neusiedlersee (Austria/Hungary), emergent vegetation in shallow lakes is frequently limited to a littoral fringe. By contrast, vast areas of Las Tablas de Daimiel are covered by large extents of emergent plants (Chapter 7), where they play important roles in many biological (primary production, decomposition, nesting ground for birds, refuge for fish and other wildlife), and abiotic processes (sedimentation patterns (Sánchez-Carrillo et al. 2000, 2001), wetland hydrology (Sanchez-Carrillo et al. 2004)). Thus, emergent vegetation clearly represents a key biotic component in this wetland.

Before large-scale degradation took place during the second half of last century, the emergent macrophyte community of Las Tablas de Daimiel was dominated by the evergreen *Cladium mariscus.* Nowadays, *Cladium* has largely been replaced by annual *Phragmites australis,* which better tolerates the hypertrophic conditions and irregular flooding patterns in the wetland (Alvarez-Cobelas and Cirujano 2007; Alvarez-Cobelas et al. 2008; Chapter 7). Sánchez-Carrillo et al. (2000) demonstrated that internal primary production, mainly through emergent vegetation, accounts for the considerable variability in sedimentation patterns of Las Tablas de Daimiel. The rates determined were substantial (max. 2.88 ± 1.2 cm year^{-1}), suggesting that, if current sedimentation patterns are maintained, the wetland will silt up and convert to a terrestrial ecosystem by the end of this century (Sánchez-Carrillo et al. 2000).

This provides a strong argument in favour of intervention in the vegetation community structure, which could help to decrease the present sedimentation rates and extend the lifespan of Las Tablas de Daimiel. Specifically designed plant harvesting schemes should focus chiefly on controlling *Phragmites*, which contributes with a high detritus biomass after plant senescence. These plant biomanipulation designs must be balanced, to continue providing “vegetation services” to the wildlife of Las Tablas de Daimiel, and to counteract the negative impacts which arise as a result of large amounts of biomass generation.

Biomanipulations of components of the benthic food web of lakes and wetlands have often focused on common carp. Las Tablas de Daimiel has become infested with the invasive American red swamp crayfish (*Procambarus clarkii* Girard) which had a profound effect on benthic ecology. A large body of scientific evidence has accumulated about the negative impacts of *P. clarkii* in invaded ecosystem (Geiger et al. 2005; Rodríguez et al. 2005). These negative impacts include the eradication of submerged aquatic vegetation and the alteration of sediment and water quality which have been also experimentally demonstrated in Las Tablas de Daimiel (Angeler et al. 2001, 2003b). The broad tolerance to a wide range of ecological conditions, the ability to survive droughts by burrowing into the sediment, and its capacity to profoundly alter abiotic and biotic wetland characteristics, confer this species the status of an ecosystem engineer (Jones et al. 1994).

Biomanipulation of crayfish populations could hold potential to lessen the ecological damage caused to the wetland. However, it is currently hard to evaluate which management strategy can be most efficient in controlling crayfish populations in Las Tablas de Daimiel. Control of crayfish biomass emphasizes that extending the biomanipulation paradigm undoubtedly increases the spectrum of potential limitations. Several examples highlight these limitations.

While some studies suggest that piscivorous fish readily feed on *P. clarkii* (Elvira et al. 1996), it is unlikely that management based on piscivore introductions is fruitful in the long term. The environmental harshness in the wetland (frequent hypoxia, poor conditions for reproduction) may inflict high mortality to piscivorous fish, requiring costly and periodic restocking. Furthermore, the ability to seek refuge in sediment burrows could provide a means for *P. clarkii* to, at least temporarily, escape piscivore predation, thereby reducing efficiency of fish stocking. More importantly, the history of Las Tablas de Daimiel, and the Iberian Peninsula in general, shows a lack of large fish predators, suggesting that negative impacts of piscivore introductions on other components of the wetland food web could outweigh beneficial effects.

The efficiency of periodic and selective harvesting of crayfish as a management tool is also difficult to evaluate from a cost-benefit perspective. Crayfish commerce is an important socioeconomic factor in many areas (Avault 1992), and, therefore, it may be feasible to assess whether commercialization of crayfish can compensate the costs resulting from periodic harvesting. A best-case scenario would be profitable crayfish commerce, raising crayfish harvests to levels able to lessen the ecological damage of this species.

Hydroperiod is an important abiotic constraint of *Procambarus clarkii* populations (Gutiérrez-Yurrita and Montes 1999), suggesting that reducing hydroperiods could have management potential. Water levels can be artificially regulated through the terminal dam at Las Tablas de Daimiel (Chapter 1), helping to inflict drought-induced mortality to crayfish and fish. However, artificial management of hydroperiods as a tool to manage *P. clarkii* populations must be balanced in ways that avoid negative impacts on other biotic components of the wetland and its whole ecological integrity.

Procambarus clarkii has a wide tolerance to a range of environmental conditions, and this should be highlighted, as it may currently impede its eradication from the wetland. However, future advances in pest species science and management could provide insight to effective control/extirpation of this species from the wetland, through the development of highly taxon-specific biocontrol agents.

9.5 Biomanipulation: A Model for Las Tablas de Daimiel

As has been noted by Angeler et al. (2003a), the complex nature of abiotic and biotic parameters complicates the predictability of ecosystem responses to biomanipulations in wetland ecosystems. Therefore, the usefulness of biomanipulation as a wetland rehabilitation tool is complicated by this uncertainty. It is apparent, however, that biomanipulation has been a useful tool to improve wetland quality in certain circumstances (overview in Table 9.1), particularly when focusing on benthivorous carp (Lougheed and Chow-Fraser 1998) or YOY walleye (Potthoff et al. 2008). Likewise, targeting emergent vegetation biomass seems promising to counteract excessive sedimentation rates and to avoid the conversion of Las Tablas de Daimiel into a terrestrial ecosystem during this century.

Further exploration on the usefulness of biomanipulations as a wetland rehabilitation tool seems warranted. It may be that the integral restoration of Las Tablas de Daimiel is impossible, due to the inability of restoring the natural disturbance regime and wetland geomorphology to conditions existing before large-scale degradation. New constraints arising from climate change will further complicate restoration at local and catchment scale. Under such limitations, the conceptual model presented in Fig. 9.1 could serve managers as a useful guide for improving the environmental quality of Las Tablas de Daimiel.

The model will focus on primary and secondary effects, extensively studied in lakes as a result of fish manipulations. Primary effects are associated to cascading effects in lake food webs that may lead to improved water quality. In the context of Las Tablas de Daimiel, fish manipulations should focus primarily on benthivorous carp which has been shown to have the most deleterious effects on water quality (Angeler et al. 2002). It will be necessary to evaluate whether the 75% biomass reduction threshold suggested for lakes will be also suitable to achieve the desired long-term improvements in this wetland (Perrow et al. 1997; Hansson et al. 1998). Effective management of crayfish populations will, presumably, also result in improved water quality, in terms of reduced sediment re-suspension and nutrient recycling from sediments to the water column, and relief submerged vegetation from herbivorous pressure (Angeler et al. 2001). Artificial and prolonged reduction of hydroperiods could increase the effectiveness of fish and crayfish reduction schemes.

According to the alternative state theory, secondary effects are based on the notion that fish manipulations disrupt equilibrium conditions in shallow lakes, ultimately allowing re-colonisation of submerged macrophytes. It is reasonable to assume that submerged macrophyte re-colonization will be facilitated at Las Tablas

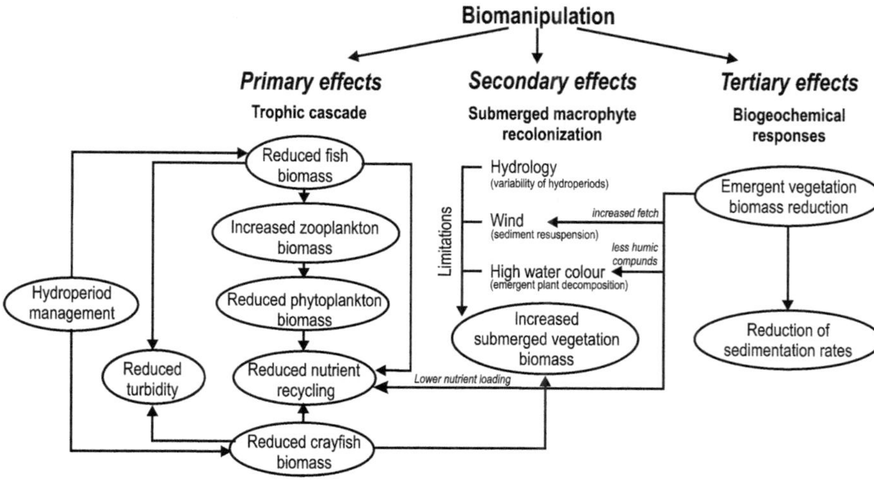

Fig. 9.1 Conceptual model showing potential effects of biomanipulations in the abiotic and biotic environment of Las Tablas de Daimiel (for details see text)

de Daimiel by manipulating benthic species (carp, crayfish), favored by the storage effects (re-growth from seed banks). However, several other factors may limit the establishment and persistence of submerged plants. Jeppesen (1998) suggested that significant changes in the biological community structure cannot be expected to occur until phosphorus concentrations in lakes are reduced to levels below 0.1–0.15 mg L^{-1}. If these thresholds are also applicable to wetlands, then, theoretically, stability in submerged macrophyte reestablishment is not to be expected in the hypereutrophic Las Tablas de Daimiel (Angeler et al. 2002). Furthermore, the high water colour, arising chiefly from emergent plant decomposition, may constrain macrophyte re-colonisation by deteriorating the aquatic light field, even if phytoplankton biomass and turbidity are significantly reduced. A further point to consider is the wetland hydrology. Based on the high seasonal variability of hydroperiods, large-scale submerged macrophyte re-colonization and persistence may be constrained and restricted to species adapted to disturbance (e.g., charophytes, Wade 1990).

The model in Fig. 9.1 will include a novel category representing ecosystem processes that are especially relevant in riparian wetlands. Effects that may arise from manipulation of other wetland communities (e.g., emergent vegetation) are considered tertiary. On one hand, tertiary effects negatively feedback factors that maintain degraded conditions (poor water quality, low submerged macrophyte biomass), and may therefore reinforce effects of manipulations on fish communities. For example, by retrieving high biomass of *Phragmites*, the decomposition of plant litter may generate fewer amounts of humic substances, leading to clearer waters, which could favour submerged plant re-colonisation. Retrieval of emergent plant litter also reduces the autochthonous nutrient load to the system, and may counteract eutrophication phenomena related to nutrient recycling from the sediments to the water column. Conversely, decreasing the area of emergent vegetation cover increases fetch and may favour wind-induced sediment re-suspension. Tertiary effects also contemplate processes that are uncoupled from changes in communities and water quality. Geomorphological and other biogeochemical processes affected by sedimentation fall within this category.

9.6 Conclusions and Perspectives

There exists increasing evidence that the application of theories related to trophic cascading interaction and alternative stable states holds potential to rehabilitate degraded wetlands. Even more than in most shallow lakes, riparian wetland are characterised by a high food web complexity in which many communities play a key role (emergent vegetation, crayfish). This suggests that interventions in other wetland communities, in addition to those of fish, could improve ecological conditions beyond water quality in wetlands, thereby broadening the biomanipulation paradigm.

Biological processes in wetlands are tightly coupled with, and mediated by, abiotic factors such as the hydrological disturbance regime. Although the complexity of interacting abiotic and biotic parameters complicates predictions on ecosystem responses to biomanipulations, ultimately limiting its potential usefulness for wetland rehabilitation, interactions between biotic and abiotic influences need thorough consideration in future wetland studies. Future research in Las Tablas de Daimiel could address this need for information, to the benefit of wetland science and future rehabilitation strategies.

References

Alvarez-Cobelas M, Cirujano S (2007) Multilevel responses of emergent vegetation to environmental factors in a semiarid floodplain. Aquat Bot 87:49–60

Alvarez-Cobelas M, Cirujano S, Sánchez-Carrillo S (2001) Hydrological and botanical man-made changes in the Spanish wetland of Las Tablas de Daimiel. Biol Conserv 97:89–98

Alvarez-Cobelas M, Sánchez-Carrillo S, Cirujano S, Angeler DG (2008) Long-term changes in spatial patterns of emergent vegetation in a Mediterranean floodplain: natural versus anthropogenic constraints. Plant Ecol 194:257–271

Angeler DG, Sánchez-Carrillo S, García G, Alvarez-Cobelas M (2001) The influence of *Procambarus clarkii* (Cambaridae, Decapoda) on water quality and sediment characteristics in a semiarid floodplain wetland. Hydrobiologia 464:89–98

Angeler DG, Alvarez M, Sánchez S, Rodrigo MA (2002) Assessment of exotic fish impacts on water quality and zooplankton in a degraded semi-arid floodplain wetland. Aquat Sci 64:76–86

Angeler DG, Chow-Fraser P, Hanson MA, Sánchez-Carrillo S, Zimmer KD (2003a) Biomanipulation: a useful tool for freshwater wetland mitigation? Freshw Biol 48:2203–2213

Angeler DG, Sánchez-Carrillo S, Alvarez-Cobelas M, Cirujano S, Medina L (2003b) Exotic crayfish activity and its effects on water quality: preliminary implications for the alternative stable equilibria in Mediterranean wetlands. J Mediterr Ecol 4:13–21

Angeler DG, Boulton AJ, Jenkins KM, Sánchez B, Alvarez M, Sánchez S (2007) Alternative states and temporary wetlands: research opportunities for understanding effects of anthropogenic stress and natural disturbance. In: Clarkson PA (ed) Environmental research advances. Nova, New York, pp 5–17

Avault JW (1992) A review of world crustacean aquaculture II. Aquacult Mag 18:83–92

Carpenter SR, Kitchell JF (1993) The Trophic Cascade in lakes. Cambridge University Press, Cambridge

Chow-Fraser P (1999) Seasonal, interannual, and spatial variability in the concentrations of total suspended solids in a degraded coastal wetland of Lake Ontario. J Great Lakes Research 25:799–813

Chow-Fraser P, Lougheed V, Le Thiec V, Crosbie B, Simser L, Lord J (1998) Long-term response of the biotic community to fluctuating water levels and changes in water quality in Cootes Paradise Marsh, a degraded coastal wetland of Lake Ontario. Wetlands Ecol Manage 6:19–42

Cooke GD, Welch EB, Peterson SA, Nichols SA (2005) Restoration and management of lakes and reservoirs, 3rd edn. Taylor & Francis, Boca Raton

Elvira B, Nicola GG, Almodovar A (1996) Pike and red swamp crayfish: a new case on predator-prey relationship between aliens in central Spain. J Fish Biol 48:437–446

Evelsizer VD, Turner AM (2006) Species-specific responses of aquatic macrophytes to fish exclusion in a prairie marsh: a manipulative experiment. Wetlands 26:430–437

Falk DA, Palmer MA, Zedler JB (2006) Foundations of restoration ecology. Island Press, Washington, DC

Geiger W, Alcorló P, Baltanás A, Montes C (2005) Impact of an introduced crustacean on the trophic webs of Mediterranean wetlands. Biol Invasions 7:49–73

Gulati RD, Pires LMD, Van Donk E (2008) Lake restoration studies: failures, bottlenecks and prospects of new ecotechnological measures. Limnologica 38:233–247

Gutiérrez-Yurrita PJ, Montes C (1999) Bioenergetics and phenology of reproduction of the introduced red swamp crayfish, *Procambarus clarkii*, in Doñana National Park, Spain, and implications for species management. Freshw Biol 42:561–574

Hansson L-A, Annadotter H, Bergman E, Hamrin SF, Jeppesen E, Kairesalo T, Luokkanen E, Nilsson P-A, Søndergaard M, Strand J (1998) Biomanipulation as an application of food chain theory: constraints, synthesis, and recommendations for temperate lakes. Ecosystems 1:558–574

Havens KE, Fukushima T, Xie P, Iwakuma T, James RT, Takamura N, Hanazato T, Yamamoto T (2001) Nutrient dynamics and the eutrophication of shallow lakes Kasumigaura (Japan), Donghu (PR China), and Okeechobee (USA). Environ Pollut 111:263–272

Jeppesen E (1998) The ecology of shallow lakes: trophic interactions in the Pelagial. NERI Technical Report No. 247. Denmark: Ministry of Environment and Energy, Silkeborg

Jeppesen E, Sammalkorpi I (2002) Lakes. In Perrow M, Davy T (eds) Handbook of ecological restoration, volumne 2: restoration in practice. Cambridge University Press, Cambridge pp 297–324

Jeppesen E, Søndergaard Ma, Søndergaard Mo, Christoffersen K (eds) (1998) The structuring role of submerged macrophytes in lakes. Ecological studies, vol 131. Springer Verlag, New York

Jeppesen E, Søndergaard M, Kronvang B, Jensen JP, Svedsen LM, Lauridsen TL (1999) Lake and catchment management in Denmark. Hydrobiologia 395/396:419–432

Jeppesen E, Meerhoff M, Jacobsen BA, Hansen RS, Søndergaard M, Jensen JP, Lauridsen TL, Mazzeo N, Branco CWC (2007) Restoration of shallow lakes by nutrient control and biomanipulation – the successful strategy varies with lake size and climate. Hydrobiologia 581:269–285

Jones CG, Lawton JH, Shachak M (1994) Organisms as ecosystem engineers. Oikos 69:373–386

Lammens EHRR (1988) Trophic interactions in the hypertrophic lake Tjeukemeer: top-down and bottom up effects in relation to hydrology, predation and bioturbation during the period 1974–1985. Limnologica 19:81–85

Lougheed VL, Chow-Fraser P (1998) Factors that regulated the zooplankton community structure of a turbid, hypertrophic Great Lakes wetland. Can J Fish Aquat Sci 55:150–161

Mitsch WJ, Gosselink JG (2000) Wetlands, 3rd edn. Wiley, New York
Perrow MR, Meijer M-L, Dawidowicz P, Coops H (1997) Biomanipulation in shallow lakes: state of the art. Hydrobiologia 342/343:355–365
Potthoff AJ, Herwig BR, Hanson MA, Zimmer KD, Butler MG, Reed JR, Parsons BG, Ward MC (2008) Cascading food-web effects of piscivore introductions in shallow lakes. J Appl Ecol 45:1170–1179
Reed JR (2006) Effects of Landscape-Scale Factors on Wetland Biomanipulations. Minnesota Department of Natural Resources Investigational Report 533
Rodríguez CF, Bécares E, Fernández-Aláez M, Fernández-Aláez C (2005) Loss of diversity and degradation of wetlands as a results of introducing exotic crayfish. Biol Invasions 7:75–85
Ruggiero A, Solimini AG, Carchini G (2005) The alternative stable state concept and the management of Apennine mountain ponds. Aquat Conserv Mar Freshw Ecosyst 15:625–634
Sanchez-Carrillo S, Angeler DG, Sánchez-Andres R, Alvarez-Cobelas M, Garatuza J (2004) Evapotranspiration in semi-arid wetlands: relationships between inundation and the macrophyte cover:open water ratio. Adv Water Res 27:643–655
Sánchez-Carrillo S, Alvarez-Cobelas M, Angeler DG, Vizcaíno-Muñoz C (2000) Tasas de accreción y características de los sedimentos actuales del Parque Nacional Las Tablas de Daimiel (Ciudad Real). Estudios Geológicos 56:239–250
Sánchez-Carrillo S, Alvarez-Cobelas M, Angeler DG (2001) Sedimentation in the semiarid freshwater wetland Las Tablas de Daimiel (Central Spain). Wetlands 21:112–124
Scheffer M, Hosper SH, Meijer M-L, Moss B, Jeppesen E (1993) Alternative equilibria in shallow lakes. Trends Ecol Evol 8:275–279
Schrage LJ, Downing JA (2004) Pathways of increased water clarity after fish removal from Ventura mars, a shallow, eutrophic wetland. Hydrobiologia 511:215–231
Shapiro JV (1990) Biomanipulation: the next phase – making it stable. Hydrobiologia 200/201:13–27
Shapiro JV, Lamarra V, Lynch M (1975) Biomanipulation: an ecosystem approach to lake restoration. In: Brezonik PL, Fox JL (eds) Proceedings of a symposium on water quality management through biological control. University of Florida, Gainesville, pp 85–96
Søndergaard M, Jeppesen E, Lauridsen TL, Skov C, Van Nes E, Roijackers R, Lammens E, Portielje R (2007) Lake restoration: successes, failures and long-term effects. J Appl Ecol 44:1095–1105
Tátrai I, Mátyás K, Korponai J, Szabó G, Pomogyi P, Héri J (2005) Response of nutrients, plankton communities and macrophytes to fish manipulations in a small eutrophic wetland lake. Int Rev Hydrobiol 90:511–522
Vargas MJ, de Sostoa A (1996) Life history of *Gambusia holbrooki* (Pisces, Poecillidae) in the Ebro delta (NE Iberian Peninsula). Hydrobiologia 341:215–224
Wade PM (1990) The colonisation of disturbed freshwater habitats by Characeae. Folia Geobot Phytotax, Praha 25:275–278
Whillans TH (1996) Historic and comparative perspectives on rehabilitation of marshes as habitat for fish in the lower Great Lakes basin. Can J Fish Aquat Sci 53(suppl 1):58–66
Wilcox DA, Whillans TH (1999) Techniques for the restoration of disturbed coastal wetlands of the Great Lakes. Wetlands 19:835–857
Zedler JB (2000) Progress in wetland restoration ecology. Trends Ecol Evol 15:402–407
Zimmer KD, Hanson MA, Butler MG (2001) Effects of fathead minnow colonization and removal on a prairie wetland ecosystem. Ecosystems 4:346–357

Chapter 10
Analysis of Applied Environmental Management Strategies for Wetland Conservation During the Last 30 Years: A Local History

S. Cirujano, M. Álvarez-Cobelas, and C. Ruíz de la Hermosa

Abstract This chapter is a popular science essay that examines the environmental history and the effectiveness of management practices applied to Spanish wetlands, with emphasis on Las Tablas de Daimiel (TDNP). We document different management philosophies motivated by social and economic factors that were often detrimental rather than beneficial for wetlands. Sound management and conservation has also been hampered by an all-too-frequent violation of ecological theory. More recent recognition of the environmental values of Spanish wetlands promoted action plans and environmental laws, aiming to restore surface and groundwater water quality and quantity, as well as the establishment of a balanced ecological functioning of aquatic ecosystems. It remains to be evaluated to what extent these plans will achieve their goals, considering the irreversible damage inflicted historically to wetlands over the whole Spanish territory.

10.1 Introduction

There are many wetlands in Spain, with variable area from less than 0.5 up to 28,000 ha in the marshland of Doñana National Park (SE Spain). Their environmental features are also very variable: permanent mountain ponds of low mineralization, permanent karstic lakes, flooding plains, freshwater and marine marshlands, shallow temporary lakes, endorheic hypersaline lakes and so on (Casado and Montes 1995). Nowadays, 63 (281,768 ha) out of the approx 2,500 Spanish of wetlands larger than 0.5 ha are included in the Ramsar Convention, and many have some environmental protection, such as Natural Park or Game Reserve (Bernués 1998).

S. Cirujano (✉)
Spanish National Research Council (CSIC), Royal Botanic Garden, Pza. Murillo 2, E-28014 Madrid, Spain
e-mail: santos@rjb.csic.es

S. Sánchez-Carrillo and D.G. Angeler (eds.), *Ecology of Threatened Semi-Arid Wetlands: Long-Term Research in Las Tablas de Daimiel*, DOI 10.1007/978-90-481-9181-9_10,

Despite this protection, an estimated 60% of wetlands have recently disappeared due to a variety of causes related to human impact. Wetland conservation is further hampered by neglecting or violation ecological theory, which often has undesired consequences. During its recent history, Las Tablas de Daimiel has experienced a series of management intervention that generated more impact than repair. This chapter is a popular science essay that provides an overview of historical (mis) management of Spanish wetlands, using TDNP as a case study. What we document is just a further example of the plethora of information which has accumulated in recent years over the entire globe.

10.2 Wetland Conservation in Spain Until 1970

Until the late 1960s, wetlands were considered as unhealthy and low productivity areas that had to be eliminated and, if possible, devoted to agricultural purposes. The most outstanding case of legal regulation to eradicate wetlands was the Law of July 27, 1918, aimed at desiccating and draining lakes, wetlands and marshlands. This law was issued by Francisco Cambó, the Minister of Public Works. This regulation emerged after some years of economic growth promoted by the First World War, in which Spain was not involved, and brought about State intervention on Spanish national economy in an attempt to develop an agricultural policy to transform unproductive lands. This Law was formulated broad enough to inflict ecological damage to all possible wetland types for the sake of socioeconomic growth. It also contemplated the provision of financial benefit to all parties (individuals, public and private companies) that participated in wetland desiccation works, including the provision of personal ownership rights of converted wetland areas (Ceballos-Moreno 2001).

10.3 Wetland Restoration, Conservation and Management Since 1970: Common Problems Across Spain

It is only after 1970 when wetland management and conservation received more attention from decision makers in Spain, exemplified by the creation of National Parks of the Doñana marshlands in 1969 and Las Tablas de Daimiel in 1973. However, an increased appreciation of wetlands does not mean that these ecosystems were managed in ecologically sound manners. This is highlighted by the first Spanish scientific meeting devoted to wetlands (VV.AA. 1987), where its conclusions and recommendations were not considered in decision making.

Until the late 1970s, nature management was based upon a contemplative idea. Wetlands should be allowed to evolve naturally, without any human intervention. This view never took into account that wetlands have been severely altered by a variety of human action (Table 10.1), and that management intervention is needed to restore ecological conditions (Valladares 2007). At an early stage, wetlands management

Table 10.1 Historical stages on Las Tablas de Daimiel wetland management and its surroundings, including the most important environmental changes (Modified from Cobelas et al. 1996)

Period	Dominant management
1600–1400 BC	First human settlements in the Bronze Age. Culture of "Las Motillas" (small stone villages). Fishing mode of living.
400–300 BC	Iberian culture. Fishing mode of living.
100–200 AC	Romanization of the area. First agricultural settlements. Fishing mode of living.
700 AC	First waterwheels and watermills. Agricultural and fishing modes of living.
1300–1600 AC	First records of the wetland as an important waterfowl hunting area. Intensive functioning of watermills. Oak forest clearing to obtain pasturelands. Agriculture, hunting and fishing.
1700–1800 AC	Endemic malaria. Locust plagues. Elimination of riparian forest. Oak forest clearing. Intensive agriculture and shepherding.
1800–1950 AC	Leasing of game and fish. Founding of game societies. Gambusia (*Gambusia holbrooki*) introduction to control malaria. Property fragmentation. Waterfowl hunting. Intensive agriculture and shepherding.
1950–1970 AC	Desiccation plans start down in 30% of the pristine wetland. Watermill destruction. Creation of the National Game Reserve. Malaria eradication. Exotic species introduction, such as pike (*Esox lucius*) and native crayfish (*Austropotamobius pallipes*). Intensive agriculture and fishing.
1970–1980 AC	Introduction of American crayfish (*Procambarus clarkii*). National Park established in 1973. The area was considered as a Special Bird Protection Zone (ZEPA) in 1979. The National Park area is enlarged in 1980. Intensive agriculture in the surroundings.
1980–2000 AC	The wetland is included in the Ramsar Convention. Desiccation of all springs associated to the Guadiana river. The groundwater aquifer is declared as overexploited (1987). Burning and destruction of one third of the original *Cladium* vegetation (called "masegar"). Implementation of the Hydric Restoration Plan and water diversions from the Tajo catchment (1987). Frequent inputs of strongly polluted waters. Intensive agriculture in the surroundings.
2000–2009 AC	Acquisition of agricultural plots in the surroundings. Frequent inputs of strongly polluted waters. Approval of the Upper Guadiana Special Action Plan (2008). Intensive agriculture in the surroundings.

schemes had a narrow target focus. Many wetlands were managed as a priority habitat for waterfowl, thus maximizing flooding area for rare and endangered bird species, and without taking other wetland functions into account. Management success was evaluated only on the basis of an increase in bird populations and the occurrence of a protected species in censuses. This bird-oriented management scheme promoted unwise actions for the overall protection and preservation of wetlands (change of permanent to temporary hydroperiods), but it enhanced the recovery of some threatened population, such as the white-headed duck, *Oxyura leucocephala*, the marbled teal, *Anas angustirostris* (Ambiental 1992) and the gull-billed tern, *Gelochelidon nilotica* (Matamala et al. 1994; Matamala and Aguilar 2003). Given the limitation of such a narrow-focused management plan, undesired results of degradation (e.g. biodiversity loss) resulting from other forms of impacts (e.g. eutrophication) may not be surprising. The resulting loss of food resources for birds or the increased frequency of massive bird kills through botulism outbreaks even counteracted management goals in the long term. Rather than protecting and enhancing waterfowl populations, the results were depauperate communities and loss of desired species.

Another management focus embracing dynamic and integral ecosystem processes is only recently implemented. Management is carried out based upon global ecosystem studies (water quality and physical, hydrology, fauna, flora, history, etc.). What matters is to restore the wetland physiognomy and to attain the maximal biological diversity, while respecting the environmental characteristics of each wetland. Notwithstanding, there are very few instances where restoration designs take into account, or even know, some basic facts about water balance (water availability, hydrological functioning, storing capacity, evapotranspiration, water renewal, etc.), water quality (ionic features, salinity, nutrient content), plant productivity, and so forth. This makes it difficult to evaluate the success of rehabilitation programs.

While both management models (narrow-focused and dynamic/integral) coexist nowadays, given different administrative targets, large-scale ecological changes at the catchment scale (aquifer overexploitation, pollution runoff, deforestation, exotic species, fragmentation) constrain recovery at the local scale. Thus, more recent management focuses highlight that specific ecosystem and catchment characteristics must be considered simultaneously for an integral management of wetland resources. Despite recognising this need, there is so far no Spanish wetland whose management takes these global requirements specifically into account.

The Water Framework Directive (DOCE 2000) establishes an European setting of action in water policy and its foremost goal is the recovery of surface and subterranean waters and their associated ecosystems. It defines a wide array of concepts related to wetland health, such as the so-called "reference conditions" that involve to those environmental conditions held by wetlands when they were in their pristine condition. But, in most cases, it is not well known how to implement such a recovery and evaluate the return to early conditions. This is in many cases virtually impossible, due to different alterations (diminishing water sources, changes in basin morphology, water pollution) that are practically irreversible.

Nowadays, Spanish wetland management must be focused on either solving or mitigating five common types of impacts: desiccation, eutrophication and pollution, siltation, salinization and colonization by invasive species. In most places, these problems co-occur and contribute to loss of those values that once promoted their environmental protection (Zalidis et al. 2002, Fig. 10.1).

In a region where global change is an undisputable truth (Moreno 2005; Barceló 2008; Sousa et al. 2007; Alcácer 2007), it is unavoidable to include issues of water availability and quality in any project of wetland recovery and conservation. These

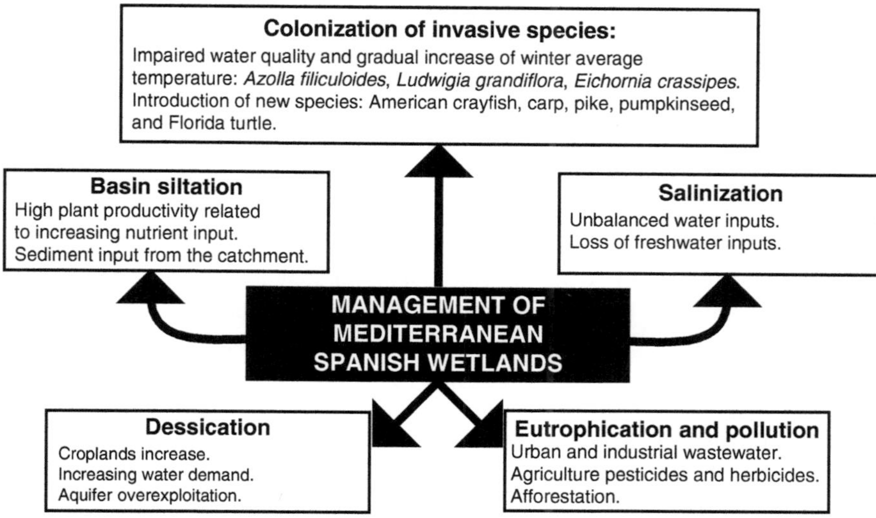

Fig. 10.1 Man-made impacts on the conservation of Spanish wetlands

are the main factors controlling wetland vegetation and animal dynamics. Water availability is key to any wetland recovery project. However, water resources are thought to diminish by some 19% in the long-term (100 years) in Spain (Iglesias et al. 2005). Thus, heavily treated wastewaters to meet required water quantity and quality needs, must be considered as a non conventional resource for wetland conservation (Alcácer 2007; Sánchez-Vila 2008).

It must be understood that wetland restoration is a mid- to long-term process, since any given wetland needs long time periods to set its new structural and functional characteristics. It is crucial that not only water quantity but also suitable water quality is key to sound wetland restoration. Quality and quantity should be two inseparable concepts in the management of wetlands. Spain must learn from other countries that a strong cooperation among environmental managers and scientists is crucial to warrant environmental success of projects. Management models must also be flexible enough to cope with new environmental data as they emerge over time, but such flexibility often collides with the slow pace of administrative processes in Spain.

10.4 Las Tablas de Daimiel: A Case Study

Las Tablas de Daimiel (TDNP) is a wetland where human impacts were moderately low until the nineteenth century. Stronger impacts occurred from the 1950s, mostly derived from man-made actions (unsustainable land use and water resource overexploitation; Chapter 3).

As has been outline in several chapters of this book, TDNP is an example of several forms of impact that have collectively altered ecosystem structure and function (Table 10.1). Most importantly, aquifer overexploitation resulting from excessive agricultural irrigation of crops with high water demands strongly changed the ecosystem hydrology from a semi-permanent wetland (with some 50% of its whole area permanently flooded) to a highly fluctuating wetland with long periods (several years) of almost complete dryness (see Chapter 3). The so-called Hydrologic Regeneration Plan (HRP hereafter), started in the mid-1980s, establishing a critical threshold of 16 hm^3 of water supply for guaranteeing a minimum of an ecologically required hydroperiod for TDNP. This volume enabled a maximal flooding of 1,587 ha. The HRP, however, did not take some essential aspects of wetland restoration into account: water quality, hydrological balance, and, most importantly, a thorough follow-up monitoring to evaluate the success of the HRP.

Without doubt, the main change in the hydrology of TDNP was the disappearance of subterranean water draining through wells and springs from the groundwater aquifer, locally called "ojos" (eyes), the most important of which were the nearby Ojos de Guadiana. Nowadays, flooding sources depend upon external sources, delivered to TDNP through interbasin transfer.

In addition to contemplating interbasin water transfer, a terminal dam was built to assure that the entered water remains in the wetland area. As part of the HRP, also another semi-permeable dam was constructed in the centre of the wetland and wells were constructed to artificially manage hydroperiods during severe drought periods (Fig. 10.2).

The very nature of the semiarid climate, combined with competing demands of water resources of heavily exploited tourist/urban areas at the Mediterranean coast, makes interbasin water diversion for environmental conservation problematic. In addition, diverted water can be highly inefficient, because when they are not optimally timed seasonally, dry channels can contribute to a massive infiltration of water that is designated to flood the wetland. Moreover, the transferred water is often heavily polluted from point sources that are delivered to transfer channels (in the case of TDNP the Gigüela river acts as such a transfer channel) (Fig. 10.3).

What is the outcome of the HRP 23 years after its implementation? Without the 13 water transfers between 1987 and 2007, TDNP would no longer exist, but the price for maintaining this wetland alive is the worst ecosystem health since it was declared a National Park. While the HRP helped to keep TDNP in the landscape, it has not solved its most important threats: water scarcity and impaired water quality. This failure is partially related to the lack of consideration of environmental management at the catchment scale.

The Upper Guadiana Special Plan (Plan Especial del Alto Guadiana, PEAG) is a new environmental plan, approved in 2008 and set in operation for the next 20 years, whose main goal is to develop the basis for a more sustainable and rational use of water in the catchment where TDNP is located. To achieve its goals, croplands and irrigated lands will gradually be abandoned, after their owners have received financial support from the Spanish government. Abandoned land will be reforested using autochthonous species, and more and better wastewater treatment

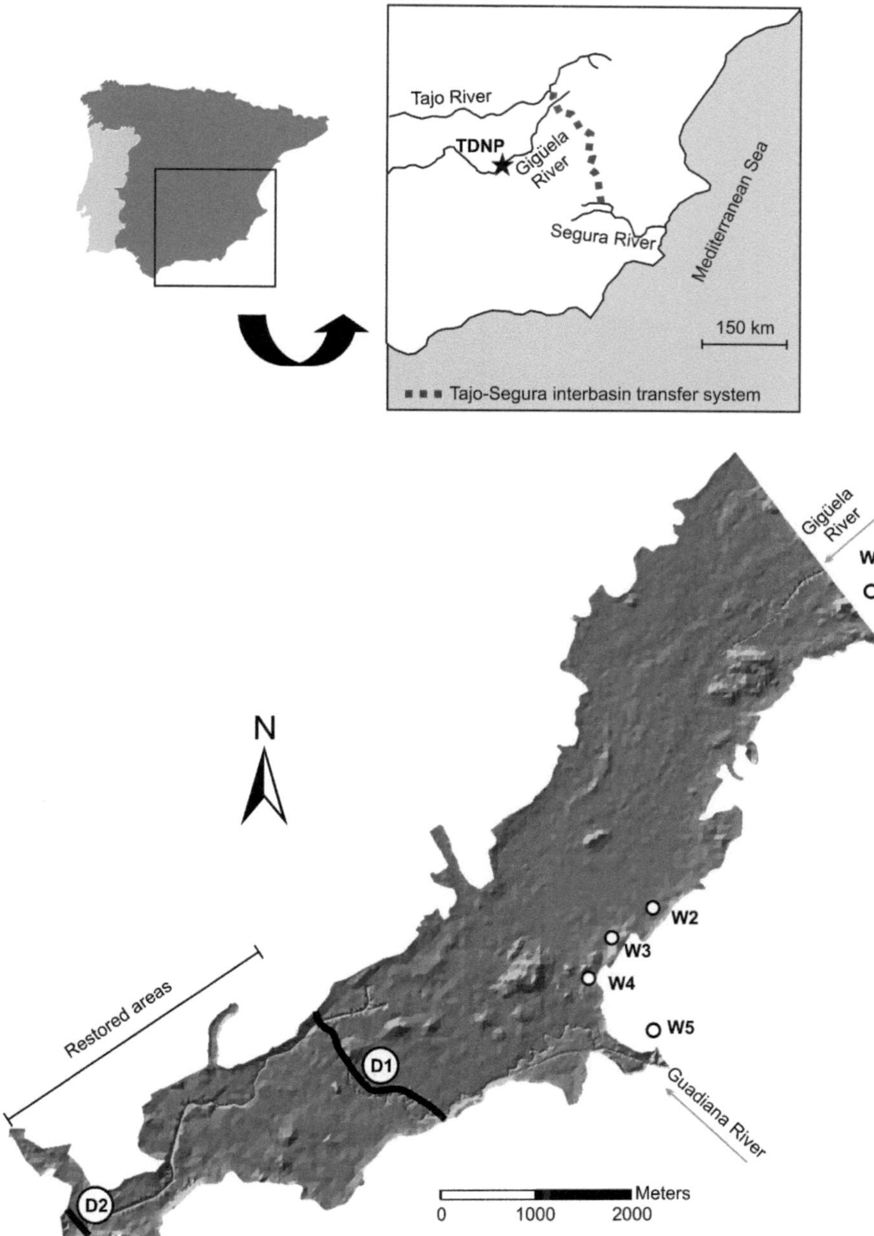

Fig. 10.2 Main actions undertaken in the Hydrologic Restoration Plan, started in middle 1980s for the ecological restoration of Las Tablas de Daimiel (TDNP). The water diversion to the Gigüela river from the Tajo-Segura transfer is shown in the upper panel, whereas actions undertaken at the National Park are shown in the lower panel. D1: Inner hydraulic dam, also called Morenillo dam; D2: Puente Navarro dam; W = wells

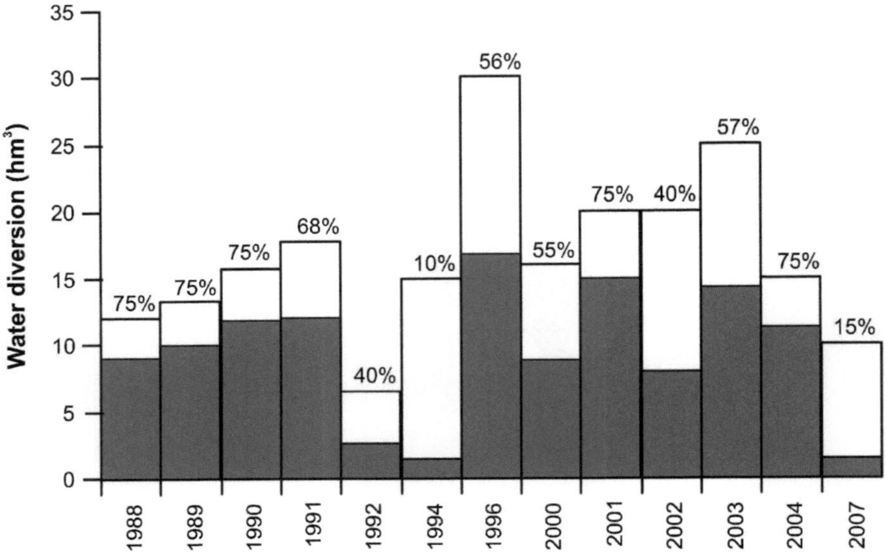

Fig. 10.3 Total volume of water diverted to Las Tablas de Daimiel National Park from the Tajo-Segura inter-basin diversion system (*white bars*) and effective amount of discharged water into the wetland (*grey bars*) from 1988 until the present day. The efficiency of every water diversion is shown as a percentage

plants will be constructed to control and improve water quality. The success of such a plan will be measured by the restoration of freshwater ecosystems lying on the Upper Guadiana catchment, such as the TDNP wetland, which is the single National Park there.

TDNP impairment and changes have been a process lasting 40 years and, hence, given its current environmental situation, its recovery cannot be instantaneous. The PEAG assumes that TDNP problems stem from a lack of agricultural policy planning, which has been based upon an unsustainable demand for groundwater and insufficient wastewater treatment. This means that wetland restoration must be dependent on the adequate and sustainable management of water in the Upper Guadiana catchment.

Current water needs in this catchment and the environmental features of TDNP, however, are quite different from those in the 1950s, when the wetland was in its pristine condition. Thus, complete restoration of TDNP is utopia. The new TDNP will be built on the current water availability, searching for non-conventional water sources (i.e. treated wastewaters whose quality is good enough for ecosystem rehabilitation). This approach and the already extant water sources (water diversion wells and natural inputs from the catchment) will perhaps provide the necessary conditions for maintaining ecosystem performance and a healthy life for its once rich biological communities. However, it remains highly uncertain to what extant the biological recovery of TDNP will take place.

References

Alcácer C (2007) La Gestión de las aguas continentales en el margen septentrional del Mediterráneo. In: Paracuellos M (ed) Ambientes mediterráneos. Funcionamiento, biodiversidad y conservación de los ecosistemas mediterráneos. Instituto de Estudios Almerienses, Almería

Ambiental SL (1992) Plan de recuperación de la cerceta pardilla (Marmaronetta angustiostris) en la Comunidad Valenciana. Consejería de Medi Ambient de la Generalitat Valenciana, Valencia

Barceló D (ed) (2008) Aguas continentales. Gestión de recursos hídricos tratamientos y calidad del agua. Consejo Superior de Investigaciones Científicas, Madrid

Bernués M (1998) Humedales españoles inscritos en la lista del Convenio de Ramsar. Ministerio de Medio Ambiente, Madrid

Casado S, Montes C (1995) Guía de los lagos y humedales de España. Reyero, Madrid

Ceballos-Moreno M (2001) La problemática jurídico-administrativa de las zonas húmedas. Humedales Mediterráneos (SEHUMED) 1:155–162

Cobelas A, Sánchez-Soler MJ, Carrasco M, García-Consuegra B, Escuderos-Hernández J, Álvarez-Cobelas M (1996) Aspectos históricos. In: Alvarez-Cobelas M, Cirujano S (eds) Las Tablas de Daimiel: ecología acuática y sociedad. Ministerio de Medio Ambiente, Madrid

DOCE (2000) Directiva 2000/6/CE del parlamento Europeo y del Consejo por la que se establece un marco comunitario de actuación en el ámbito de la política de aguas

Iglesias A, Estrela T, Gallart F (2005) Impacts on hydric resources. In: Moreno JM (ed) A preliminary assessment of the impacts in Spain due to the effects of climate change. Spanish Office of Climatic Change, Madrid

Matamala JJ, Aguilar FJ (2003) Humedales almerienses. In: Paracuellos M (ed) Ecología, manejo y conservación de los humedales. Instituto de Estudios Almerienses, Almería, Spain

Matamala JJ, Aguilar FJ, Ayala JM, López JM (1994) Distribución actual de la malvasía (*Oxyura leucocephala*) en España. Importancia de los humedales almerienses para la recuperación de una especie amenazada. In: VV.AA, La malvasía común. Especies singulares almerienses. Consejería de Cultura y Medio Ambiente, Junta de Andalucía, Sevilla

Moreno JM (ed) (2005) A Preliminary assessment of the impacts in Spain due to the effects of climate. Spanish Office of Climatic Change, Madrid

Sánchez-Vila X (2008) Gestión de recursos hídricos en zonas áridas y recarga artificial de acuíferos. In: Barceló D (ed) Aguas continentales. Gestión de recursos hídricos tratamientos y calidad del agua. Consejo Superior de Investigaciones Científicas, Madrid

Sousa A, García-Barrón L, Jurado V (eds) (2007) El cambio climático en Andalucía: evolución y consecuencias medioambientales. Consejería de Medio Ambiente de la Junta de Andalucía, Sevilla

Valladares F (2007) Hábitat mediterráneo continental: un sistema humanizado, cambiante y vulnerable. In: Paracuellos M (ed) Ambientes mediterráneos. Funcionamiento, biodiversidad y conservación de los ecosistemas mediterráneos. Instituto de Estudios Almerienses, Almería

Vallès C (ed) (1989) Els Aiguamolls de l'Empordà. Aspectos ecològics, històric i socials del Parc Natural. Quaderns dels Indiketes, 3rd edn. ART, Figueres

VV.AA (1987) Bases Científicas para la protección de los Humedales Españoles. Real Academia de Ciencias Exactas, Físicas y Naturales, Madrid

Wu J, Loucks OL (1995) From balance of nature to hierarchical patch dynamics: a paradigm shift in ecology. Quart Rev Biol 70:439–466

Zalidis GC, Crisman TL, Gerakis PA (eds) (2002) Restoration of mediterranean wetlands. Hellenic Ministry of the Environment, Physical Planning and Public Work, Athens and Greek Biotope/Wetland Centre, Thermi, Greece

Part V
Cultural and Societal Aspects

Chapter 11
The Man and Las Tablas de Daimiel

M. Alvarez-Cobelas, S. Cirujano, and A. Meco

Abstract Human settlement in Las Tablas de Daimiel started 3600 years BP and developed into a rich culture of hunters, fishermen and plant and animal gatherers that survived until the last third of the twentieth century. Very recently, unsustainable economic development in the catchment resulted in strong pollution and depletion of water resources, forcing people to leave, impairing the wetland ecological health and threatening the future of this important European wetland. This chapter pays special attention to human uses of this wetland, and outlines the main historical and ecological relationships between the wetland and human societies.

11.1 Early Settlement

The first known settlements of man in Las Tablas de Daimiel (TDNP, hereafter) can be traced back to 3600 BP. These first settlers built small villages called "*motillas*", which were little artificial mounds 4–10 m high, providing shelter against flooding and wildlife. Some *motillas* were fortified, and were mostly used for living and storing agricultural tools and cattle. Twenty such villages were recorded in the area (Terriza and Clemente 2000), but nowadays there are no remains left of most of them.

One of the best preserved *motillas* is that of Las Cañas (Fig. 11.1), located in the southwestern part of TDNP. It belonged to the Argar Culture, which was dominant in SE Spain in 1600–1400 BC. It is a 6–10 m high and 50 m diameter circular mound, built of stone and made up by a central tower and several concentric walls (see two well preserved *motillas* in Fig. 11.2). Grain was stored within the motilla and there were some ovens too. A small village was attached to the fortress, whose

M. Alvarez-Cobelas (✉)
Spanish National Research Council (CSIC), Institute of Natural Resources,
Serrano 115 dpdo, E-28006 Madrid, Spain
e-mail: malvarez@ccma.csic.es

S. Sánchez-Carrillo and D.G. Angeler (eds.), *Ecology of Threatened Semi-Arid Wetlands: Long-Term Research in Las Tablas de Daimiel*, DOI 10.1007/978-90-481-9181-9_11,

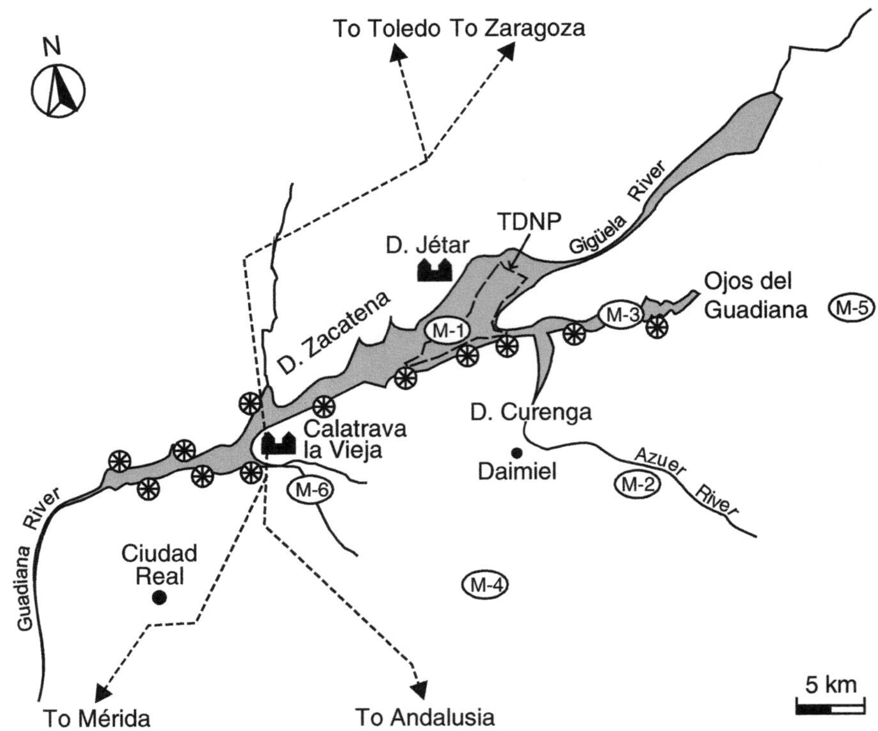

Fig. 11.1 Map of the area, showing the main sites cited in the text. D: dehesa (holm oak pasture), M: motilla (Bronze Age village), M-1: Las Cañas motilla, TDNP: current location of Tablas de Daimiel National Park. Wheels suggest location of watermills

buildings were rectangular or oval huts, built with clay on stone foundations. This *motilla* was closer to a natural road that linked North and South Spain, crossing the Toledo Mountains.

Some ceramic remains were found, the most important of which were the careened pot, belonging to cultural facies of *motillas*. Vessels, cheese-maker remains, stones for carpentry and mining, bone burins and a limestone bracelet were also gathered from the Las Cañas motilla. Surprisingly, almost no metal remains were found, only some burins and a small chisel.

This montilla was later abandoned and recolonized by Iberian people some 400–500 years BC. They earthed the slopes up and built small rooms with stone walls and adobe, surrounded by a fortress wall. Ceramic remains made it possible to connect this culture with that of Iberian culture in High Andalusia (Southern Spain) and, more specifically, with the city of the dead of Baena in Peal del Becerro (Baza, Granada). Some Greek remains were also found. Later this motilla was abandoned, and recolonized once again in the late Middle Age.

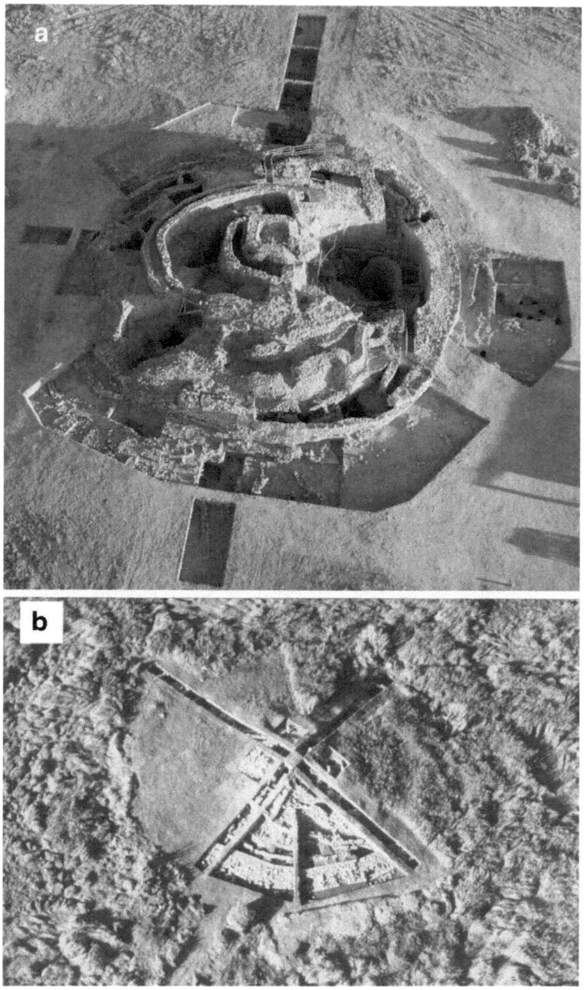

Fig. 11.2 Aerial views of the archaeologically-restored Azuer motilla (**a**) and Las Cañas motilla (**b**). The Azuer motilla is very close to TDNP de Daimiel (M-2 in Fig. 11.1), albeit much more complex than Las Cañas, which is located in the wetland. Photographs belong to the collection of Daimiel Art Museum (Central Spain)

11.2 From Roman Colonization to Lower Middle Age

All the area was already influenced by Romans around 130 BC. Romans created the nearby farm villages in Carrión, Bolaños (southwest of the wetland) and Ruidera (southeast of the same wetland), and a stone road that later crossed the Guadiana river across a ford, called Calatrava la Vieja (Fig. 11.1), though no remains were found there. Some remains of a stone road were also found within the National Park, and it is likely that an oak pasture was named by the Romans (Polistena, later called Zacatena, Fig. 11.1), which suggests that there may have been a Roman settlement there.

No further information is available for TDNP until the arrival of Arab muslims in Spain, in the early eighth century AC. The area was transferred to Berberian tribes that accompanied the invading Arabs, and they colonized all fluvial valleys throughout. To control the territory, they built some fortresses, such as those of Jétar and Calatrava la Vieja (Fig. 11.1), the latter being built in 714 AC. This is also mentioned in Muslim chronicles of the tenth century AC, good for cereal cultivation and cattle breeding. It is also likely that an Arab village was settled in Zacatena.

In 1147 AC, the Christians conquered Calatrava la Vieja castle, and transferred it to the Military Order of Calatrava which established its headquarters there. This fortress owned baths, ovens, watermills and fisheries. However, as Christians conquered more land to the South of Spain, the castle was abandoned, although this has not been proven; it is also likely that malaria forced Muslims to withdraw from this wetland surroundings. Some recent findings revealed that the castle inhabitants ate cyprinid fishes that were likely caught in the nearby wetland. A Muslim castle also rose in the western area of the wetland, close to the Zacatena oak pasture.

At any rate, this area was owned by the Calatrava Order and remained in their hands until the nineteenth century. This Military Order encompassed two economic organizations: the Master Table and the territories owned by the Master, called "Encomiendas". Several watermills in the area have been dated back to late Middle Age (Figs. 11.1 and 11.3). The Zacatena oak pasture belonged to the Master Table.

The Infante Don Juan Manuel (Prince John Emmanuel) was a Spanish aristocrat and writer, the author of "The Game Book" (Infante Don Juan Manuel, 1325), written in the early fourteenth century. In it, he reported that TDNP was a very good game (mostly waterfowl) site in those years.

11.3 Modern Age

In the early fifteenth century, the Encomienda of Daimiel for the current city and the Jétar Encomienda were created. The latter owned two watermills and the right to fish in the Gigüela River. The Master Table owned the fishing rights in the Guadiana River, the Zacatena oak pasture and five watermills (Fig. 11.1). King Philip II (he built the Spanish Armada, defeated by the English fleet and storms in 1588) visited the wetland three times in the second half of sixteenth century, most likely to hunt. In King Phillip's Account on Spanish Wealth, it was reported that TDNP was a good for fishing eels and barbels.

In the sixteenth century, the constant flooding of the wetland was an already known fact.

Fig. 11.3 Three watermills that were in operation from late Middle Ages until the early 1970s (**a**) Griñón watermill, located in the Guadiana River, upstream of current the National Park, (**b**) Molemocho watermill, located at the entrance of the National Park in the Guadiana River, (**c**) Puente Navarro watermill, located at the current outlet of the National Park. Photographs belong to the Image Collection of the Water and Wetland Centre of La Mancha (Daimiel, Central Spain)

There are many records of floodings by the Guadiana River and other streams since the eighteenth century. The first malaria epidemics are dated back to 1745–1749, 1755–1769 and 1785–1787, with many more episodes occurring later, until malaria disappeared in the late 1940s of the twentieth century. Many locust plagues took also place in the area from eighteenth century onwards. To diminish the effects of plagues, watermill dams were opened in summertime, to enable water to run, hence enhancing water renewal.

Another royal account of the Spanish wealth was performed in 1752, where 3 fishermen, 7 watermills, 2 fulling mills and 11 water-powered oil mills were recorded in the area. There were also gypsum saltworks, whose waters were useful to treat malaria.

From mid-nineteenth century onwards, watermills were the center of human activity in TDNP. In the larger watermills, one could eat, meet other people, buy or sell commodities, sleep, drink wine, find sex and so on. Most transportation took place over the water, using flat bottom, 2–4 m long boats, propelled by a big stick that was dragged through the wetland mud. These boats were imported from the Albufera Lake (Valencia, Eastern Spain) in the middle nineteenth century; later boats were built locally. There are no earlier records of any vessel in TDNP, but some legends tell about magnificent vessels commanded by an Arab prince, navigating up and downstream the Guadiana River 1,000 years ago. Because of strong vegetation growth (mostly cut-sedge, *Cladium mariscus*) water tables were soon occupied and closed to navigation. Fishermen had to burn vegetation in some areas to produce water trails ("trochas" as they called them) to enable boat movements.

The Daimiel Encomienda was disentailed in 1831, and the Military Orders were abolished. As a result of this, some wetland zones were purchased by three private owners (Rafael Sevillano, Francisco Martí de Veses and Julián Settier) who transformed them into waterfowl preserves, where hunting was only allowed to members (Settier 1956). Hence, TDNP was a group of waterfowl preserve from the middle nineteenth century until 1973. Since only rich people could afford the expense of belonging to game societies, a lot of resentment grew among the low class folks, and some of them, known as "black guns" started poaching there.

There were some executions in the wetland during the first Spanish Civil war (the first Carlist War in 1833–1840) (Asensio-Rubio 1987). European crayfish (*Austropotamobius pallipes*) was introduced in the wetland in the mid-nineteenth century and soon developed large populations.

11.4 The Twentieth Century

The knowledge on human settlement in TDNP from the late nineteenth century until the third Spanish Civil War (1936–1939) is very scant. No fighting occurred in TDNP, but there were three agricultural collectivities in Daimiel, run by anarchists, communists and socialists (Alía 1994). The anarchist collectivity was that of the Zacatena property, which was shortly before owned by a duchess. After the war, game preserves were resumed and three hunting parties were organized every year.

TDNP were soon colonized by people devoted to hunting and crop gathering, in a manner closely resembling that of the Arab people in the Tigris and Eufrates marshland (Thesiger 1964; see also Alwan 2003 film), and African people in the Okavango delta (Kgathi et al. 2006). Crayfish was caught in enormous numbers, enabling 200–300 families to make a living (Pardo 1941), using net-like devices made up of willow wood or rush, called "garlitos" (Fig. 11.4). There were also cyprinid fishermen who used several fishing nets that they soaked and tarred themselves. Fishing used to be done by night, using a trident, called "rijaca" (Fig. 11.4) and a lamp fueled by calcium carbide (Fig. 11.4). Some occupations emerged to take

Fig. 11.4 (**a**) fishing trident, (**b**) carbide light, (**c**) crayfish keel, (**d**) tools for making fishnet weights (photographs by Miguel Alvarez-Cobelas and Salvador Sánchez-Carrillo)

advantage of the natural wealth of TDNP. Leeches were gathered by river people to use their anticoagulant compounds. Reed (*Phragmites australis*) was collected to build roofs (Figs. 11.5 and 11.6). Cut-sedge (*Cladium mariscus*) was also cut and used as fuel for heating ovens. Cattail (*Typha* spp.) was collected for furniture carpentry (Fig. 11.6). Cattle-raisers bred cows and mules on the wetland banks.

All in all, river people earned three times more than peasants and this was the cause of frequent quarreling and hatred between these social groups. The river people used to live in small stone huts, called "casillas" covered by reed roofs (Fig. 11.5). They grew up and lived more freely than their agricultural counterparts of Daimiel, but they were more subject to water-borne diseases, such as malaria, that was erradicated in the late 1940s. Previously, in the 1920s, mosquito-fish had been

Fig. 11.5 Outdoor pictures of a fisherman's family and their hut (locally called "casilla" in the 1940s. Notice roofs made up by reed stems (photographs belong to the Image Collection of the Water and Wetland Centre of La Mancha (Daimiel, Central Spain))

Fig. 11.6 Inside view of a traditional fisherman's hut, showing the main room (a mixture of living-room and cooking area). The photograph belongs to the Image Collection of the Water and La Mancha Wetland Centre (Daimiel, Central Spain). 1: cattail (Typha) carpet, 2: cattail seat, 3: rush crayfish keel, 4: water pumpkins, 5: reed roof, 6: bunch of aromatic herbs, 7: framework of poplar beams, 8: gypsum covered reed at the chimney, 9: Esparto (Stipa) bag, 10: cattail bench cloth, 11: fishing net with cork floaters, 12: rush broom, 13: olive wood seat, 14: holm oak wood

introduced to fight this disease. To provide an education for river children, a rich woman funded a school in the area, where they were taught primary education and the Catholic religion. Most of these cultural developments have now disappeared, an otherwise common problem in Mediterranean wetlands (Papayannis 2008).

Some parts of the wetland at the Guadiana entrance to TDNP were devoted to rice crops from 1940 to late 1960.

Austropotamobius was decimated and finally disappeared from the wetland as a result of an infection by the fungus *Aphanomyces astaci*, around the mid-1970s. This put an end to fishing, but an attempt to reintroduce it, using the American crayfish *Procambarus clarkii*, resulted in massive growth of this resistant, omnivorous species and proved deleterious for many other species in the wetland, such as charophytes. The biological activity of this introduced crayfish species also impaired water quality because it re-suspended particles and nutrients, thus encouraging phytoplankton increase and affecting submerged plant growth due to increasing light scarcity.

Another introduced species was pike, which entered TDNP from a nearby reservoir in the 1960s and threatened other fish populations (mostly barbels) until it disappeared as a result of cannibalism.

In 1956, a government plan for the Alto Guadiana catchment, where TDNP is located, promoted desiccation of many wetlands, with two purposes: getting new land plots for agriculture and fighting malaria, which had been mostly eradicated in the earlier decade. Locally, desiccation was mainly promoted by two landowners: Teresa Aldea and the marquis of Perinat. Watermills along the wetland were destroyed, but attempts to grow anything in the reclaimed wetland soils proved ineffective, due to high soil salinity. More than 100 km^2 were desiccated in the Alto Guadiana catchment in 1973 (the true value is hard to ascertain) (Alvarez-Cobelas et al. 2009). In 1973, in order to preserve a small spot of what the wetlands had been, the Spanish government declared a National Park, with the name of TDNP de Daimiel and an extension of some 1,000 Ha. This was later increased to 1,928 Ha in 1981. An area formerly desiccated, Las Cañas, was recovered, albeit not restored to its pristine plant vegetation in its whole extension. All the river people were gradually removed.

11.5 Ecosystem Health and the Future of Man in TDNP de Daimiel

The main fact impinging on TDNP performance is that it is located in a large catchment (Alto Guadiana, 215,000 km^2), TDNP being a very small ecosystem (some 20 km^2). Human activities in that catchment have greatly increased from 1970 onwards. Irrigation agriculture evolved dramatically in the 1980s, well beyond sustainability of the water resource, and strongly impairing the quality of surface water entering TDNP. Al terminal dam was also built in TDNP in 1987, increasing water and pollutant retention in the wetland, part of which arose from strong plant production promoted by nutrient-enriched water. Thus, TDNP became a hypertrophic environment that, due to the aquifer overexploitation, often lacked water. 1986 was the last year of groundwater out-welling. Some years later, the Spanish government promoted a water transfer plan to get some flooding water from a nearby catchment, but the efficiency of those transfers was greatly variable as a result of illegal water abstractions and high soil porosity in some areas over which the transferred water travelled, and, in some years, there was not enough water to be transferred.

The wetland was nutrient enriched for decades, but its closing by a dam in 1986 and the increase of wastewater inputs from 1975 onwards greatly impaired its water quality. Earlier it was common to have wetland water in the Guadiana area of the wetland where salinity was not high (Fig. 11.7).

The wetland suffered an intentional fire in 1986 that eliminated more than 100 ha of cut-sedge vegetation and its underlying peat. In 2009, a self-combusted, natural fire has been triggered, burning peat in other wetland areas, but chances to extinguish it are pretty low, unless a strong water input is supplied, which, according to Spanish government plans, might occur early in 2010.

Unfortunately, these processes of ecosystem damage are not new. They all belong to what Garrett Hardin (1968) termed "the tragedy of the commons" and has occurred and is still taking place in many sites of the world. In short, what Hardin has said is that common goods and commodities are no one's own, and this, over

Fig. 11.7 Fisherman seeping water from TDNP through the stem of a water lilly. This procedure was common in the Guadiana area of the wetland (a fresher water area) when water quality was not as impaired as it is nowadays. Pencil drawing by Marta Chirino-Argenta

time, has resulted in appropriation and degradation of the ecosystem in question, particularly from the twentieth century onwards. The ongoing process of ecological degradation of TDNP started with the abstraction of public groundwater by irrigation farmers far away from the wetland, i.e. a public resource was depleted for private use and benefit.

The wetland has inspired some local painters, such as Juan D'Opazo and Ignacio Meco (Figs. 11.8 and 11.9).

It is hard to tell what the future of TDNP might be. There are too many threats to overcome, and local society is not interested in the future of this wetland. The Spanish government is reluctant to implement a sustainable plan of ecological

Fig. 11.8 Paintings by local artists, inspired by TDNP de Daimiel wetland. Juan D'Opazo paintings are exhibited in the permanent collection of the Daimiel Art Museum (Central Spain). Paint (**a**) is entitled "El río Guadiana" (Guadiana River), and is a 33 × 46 cm pastel painting on paper, whereas picture (**b**) is called "Flor Ribera" (Riverside flower, which is the name of a wetland site), and is a 16 × 24 cm water ink painting on paper. The former was painted in 1991, but no date is available for the latter

restoration for TDNP. Over the years, this wetland has shown great resistance to disappear. We do not know what its fate may be, but it is likely that we are to witness the end of a rare European wetland, if no urgent and sustained measures are implemented very soon.

Further information on humans in TDNP can be found in the Chapter 10, Cobelas et al. (1996) and Escuderos-Córdoba (1996).

Fig. 11.9 Paintings by local artists, inspired by TDNP de Daimiel wetland. Ignacio Meco paintings belong to Ana Meco, daughter of the artist. Picture (**a**) is entitled "Molino de Molemocho" (Molemocho watermill, 1997), and is a 20 × 34 cm oil on cardboard painting. Picture (**b**) is entitled "TDNP con sierra y bando volando" (TDNP and its nearby mountain range with waterfowl flock, ca. 2000–2001), and is a 35.5 × 49 cm India ink painting on paper

Acknowledgments The Water and Wetland Centre of La Mancha and the Daimiel Art Museum (Central Spain) are gratefully acknowledged for their loan of images of TDNP de Daimiel photographs and D'Opazo paintings. More specifically, we warmly thank Alicia López and Diego Clemente for their prompt and insightful help with these images.

References

Alía F (1994) La Guerra Civil en retaguardia, Ciudad Real (1936–1939). Diputación de Ciudad Real, Ciudad Real

Alvarez-Cobelas M, Montero E, Cirujano S, Moreno-Pérez M (2009) El origen del Guadiana, desvelado tras 2000 años de discusiones. Consorcio del Plan Especial del Alto Guadiana, Alcázar de San Juan

Alwan A (2003) Zaman, l'homme des roseaux. An Iraqui/French film

Asensio-Rubio M (1987) El carlismo en la provincia de Ciudad Real, 1833–1876. Diputación de Ciudad Real, Ciudad Real

Cobelas A, Sánchez-Soler MJ, Carrasco M, García-Consuegra B, Escuderos-Hernández J, Alvarez-Cobelas M (1996) Aspectos históricos. In: Alvarez-Cobelas M, Cirujano S (eds) TDNP de Daimiel, Ecología acuática y sociedad. Organismo Autónomo Parques Nacionales., Madrid, pp 219–234

Escuderos-Córdoba J (1996) El último pescador. In: Alvarez-Cobelas M, Cirujano S (eds) TDNP de Daimiel, Ecología acuática y sociedad. Organismo Autónomo Parques Nacionales, Madrid, pp 235–254

Hardin G (1968) The tragedy of the commons. Science 162:1243–1248

Infante Don Juan Manuel (1325) Libro de la Caza. Edition of J.M. Fradejas in 1990 for Editorial Casariego. Madrid

Kgathi DL, Kniveton D, Ringrose S, Turton AR, Vanderpost CHM, Lundqvist J, Seely M (2006) The Okavango; a river supporting its people, environment and economic development. J Hydrol 331:3–17

Papayannis T (2008) Action for culture in Mediterranean wetlands. Med-INA, Athens, Greece

Pardo L (1941) Astacicultura elemental. Ministerio de Agricultura, Madrid

Settier J (1956) Caza menor. Anécdotas y recuerdos, 2ªth edn. Instituto Editorial Reus, Madrid

Terriza CA, Clemente D (2000) El enigma de las motillas. La motilla del Azuer. Concejalía de Turismo, Daimiel

Thesiger W (1964) Marsh arabs. Longmans, London

Part VI
Concluding Chapter

Chapter 12
Synthesis: The Past, Present and Future of Las Tablas de Daimiel

David G. Angeler and Salvador Sánchez-Carrillo

Abstract Great strides have been made in our understanding of the ecology of a semiarid floodplain wetland, Las Tablas de Daimiel National Park (TDNP). Continuous interdisciplinary and collaborative research efforts during the last decades helped increase our understanding of basic and applied aspects of the wetland's ecology and biogeochemistry. The main message of this book is that human activity irreversibly altered a unique ecosystem in a few decades which evolved over thousands of years. Hydrological disruptions due to aquifer overexploitation resulting from excessive agricultural irrigation, contamination and a series of management interventions that caused more damage than repair form the core of the problem. However, our knowledge is far from complete, especially concerning means to manage the wetland in a way that guarantees both sustainability and development in times of over-exploitation by humans. In addition to summarizing the contents of this book, we will advocate research approaches that could fill remaining information gaps. Critical to the survival of this wetland will not only be scientific progress. An integration of scientific, cultural and historical knowledge in the interaction cycles between ecological, social, political and economic systems should be the ultimate goal. Without this panarchic approach to understanding ecosystems and their management, sustainable development will remain an eternal oxymoron.

"All human actions have one or more of these seven causes: chance, nature, compulsion, habit, reason, passion, and desire." Aristotle

D.G. Angeler (✉)
Department of Aquatic Sciences and Assessment, Swedish University of Agricultural Sciences, P.O. Box 7050, SE 750 07 Uppsala, Sweden
e-mail: David.Angeler@vatten.slu.se

S. Sánchez-Carrillo and D.G. Angeler (eds.), *Ecology of Threatened Semi-Arid Wetlands: Long-Term Research in Las Tablas de Daimiel*, DOI 10.1007/978-90-481-9181-9_12,

12.1 Background and Current State of Las Tablas de Daimel

Wetland science is a rather recent discipline within aquatic sciences (Mitsch and Gosselink 2000). While historically regarded as useless land, recent appreciation of the provision of the many ecosystem services of wetlands (e.g. biodiversity, flood prevention, nutrient and pollution sinks, and resource supply; Constanza et al. 1997) has motivated conservation and restoration efforts worldwide. In addition to the well-known Ramsar List created for wetland conservation purposes, the environmental, social and cultural value of wetland ecosystem around the world has also received the international recognition of the UNESCO through the Man and the Biosphere (MAB) Programme. MAB recognizes the importance of 120 wetland sites as biosphere reserve ecosystems, which is ca 22% of the total MAB sites, and the number continues to rise each year. In fact, Spain is along with Mexico the country with most wetland sites included in the MAB (i.e. seven wetlands). With regard to Europe, since 1992, the European Union LIFE Programme has co-funded around 120 wetlands-related projects across the nature, environment and third countries' thematic components (http://ec.europa.eu/environment/life/publications/lifepublications/lifefocus/documents/wetlands.pdf). The European appreciation of wetlands is also reflected in the recent establishment of the European section of the Society of Wetland Scientists (SWS) and by the activities of the European Pond Conservation Network, among other, often volunteer-based, organizations.

Given this rather recent appreciation, and the fact that Mediterranean limnology is generally not as well advanced as the limnology of temperate aquatic ecosystems (Alvarez-Cobelas et al. 2005), the great strides which have been made in our understanding of the ecology of the unique semiarid floodplain wetland, Las Tablas de Daimiel National Park (TDNP), are therefore all the more remarkable. This book summarizes a continuous, interdisciplinary and collaborative research effort, initiated in the 1980s by S. Cirujano and M. Alvarez-Cobelas and joined later by all contributors of this book and others. It touches on a broad spectrum of topics, including hydrology, sedimentology, paleolimnology, planktology, nutrient dynamics, vegetation science, food web theory, pollution research, and landscape ecology. It allows blending basic and applied research of a wetland at different spatial and temporal scales in a way that has been seldom documented before in the Mediterranean area of Europe and North Africa. We hope that this book serves to stimulate similar research efforts in other Mediterranean areas. Undoubtedly, such efforts will increase our understanding of the uniqueness of Mediterranean aquatic ecosystems, thereby broadening our current limnological paradigm which is biased mainly towards temperate lakes and rivers (Alvarez-Cobelas et al. 2005).

While the wetland has undergone several transitional ecosystem states over geological history, mainly as a result of natural causes (climatic variability) (Chapter 2), this book highlights that the devastating imprints of human action on this wetland have occurred in very recent history; so recent that even the elderly persons in local populations do not recognize the ecosystem they knew from their childhood! Hydrological disruptions due to aquifer overexploitation resulting from excessive agricultural irrigation, contamination and several management interventions that caused more damage than repair form the core of the problem (for an overview of historical impacts see Table 10.1). Figure 12.1 and Box 12.1 summarize the many direct and indirect impacts on the ecosystem resulting from human activity. Most importantly, Fig. 12.1 shows a sustainable human use of wetland resources (vegetation, fish, crayfish) in the 1960s relative to a recent period where no direct use of these resources is made. By contrast, the major human impacts that contributed to large-scale wetland degradation, mainly in form of hydrogeochemical alterations, are highlighted.

It is clear from people's collective memory, and the research documented in this book, that humans changed a wetland in decades which evolved in hundreds of thousands of years. Las Tablas de Daimiel moved inexorably to an alternative, irreversible stable ecosystem state (Angeler et al. 2007), one that is of much concern, however, because the ecological health of TDNP compares with a coma patient whose vital functions are maintained by artificial life support. Without this life support, mainly in form of artificial hydrological management (interbasin water diversion, local groundwater pumping), TDNP would have passed away some time ago, sharing the fate of so many other wetlands, ultimately leading to a loss of cultural, evolutionary and natural heritage from our memories through phenomena related to transgenerational amnesia.

Las Tablas de Daimiel is one example were stakeholders, decision makers and local societies had no clues on the far-reaching and complex consequences arising from the causes of their action. Even the layman has the chance to appreciate that something is fundamentally wrong, when contemplating the massive irrigation during the hottest hours of summer days when traveling across central Spain. However, the situation seems not hopeless. As pointed out in the book (Chapter 10), political good will is giving rise to an increased allocation of taxpayer's money and other resources to move away from unsustainable agricultural practices and conserve wetlands, including Las Tablas de Daimiel. But the situation is not without burden. Despite the great advances being made in our understanding of TDNP several knowledge gaps remain. Many of these are of ecological nature, which we highlight in the following section, and which have relevance for rehabilitation purposes. In addition, consideration of the complex interplay between social, political and economic factors will be essential to sound environmental conservation (Gunderson and Holling 2002); otherwise the increased knowledge gain will be merely of scientific value with little application for solving the core of the problem.

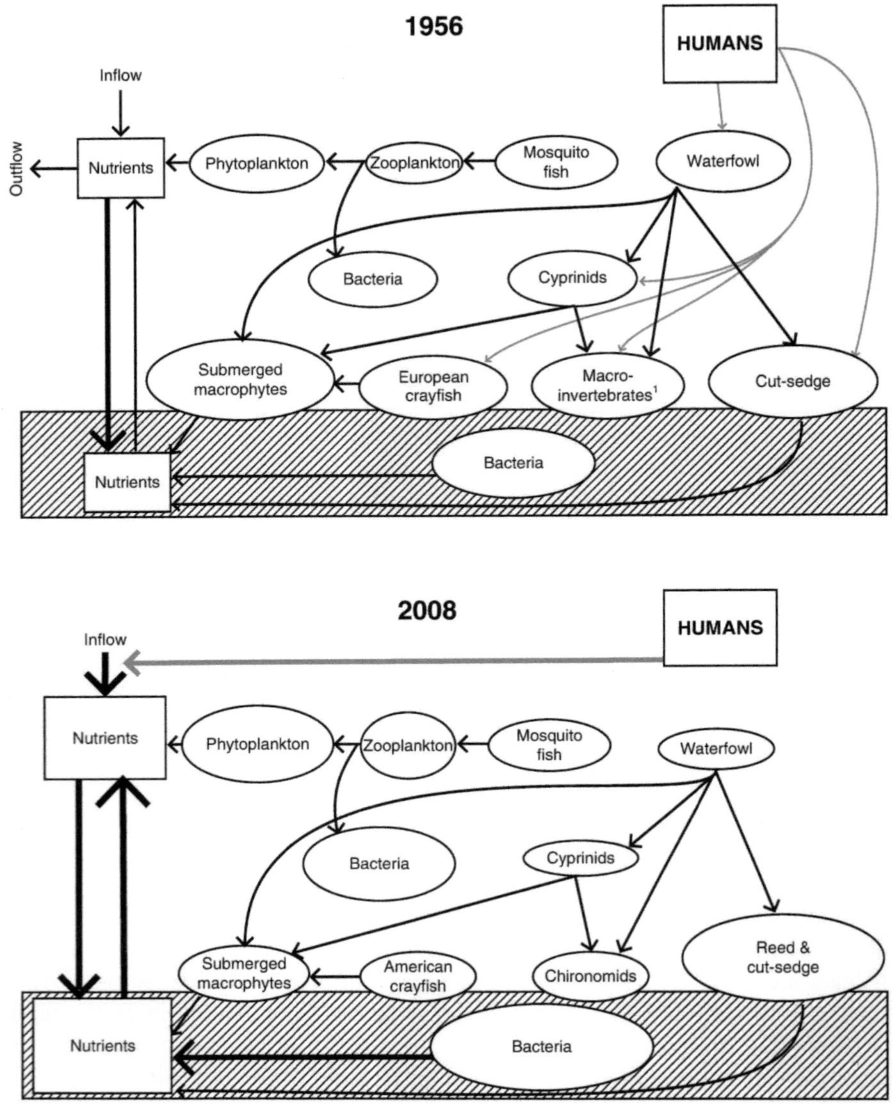

Fig. 12.1 Models comparing simplified ecosystem structure of Las Tablas de Daimiel between periods before (1956) and after (2008) degradation. Boxes and ellipsoids represent qualitatively the abundance of compartments and biological communities. Although bacteria are the sink of all organic matter and waste substances in biota, the arrows have been suppressed to simplify the graph. Note than human influence has been represented differently in each period: in 1956, although humans used the natural resources provided by the wetland, their influence on the ecosystem is almost negligible (*grey fine lines*); contrarily, the influence of human activity on the ecosystem is nowadays negative. Macroinvertebrates from 1956 was composed by mussels, shrimps, insects, crustaceans and sponges

Box 12.1 A comparison of current and past (first half of the twentieth century) conditions of Las Tablas de Daimiel, with emphasis on key hydrological, biogeochemical, ecological and catchment features

Features	Past	Present
Catchment		
Agricultural production	Low	High, unsustainable
Irrigation	Low	Massive
Aquifers	Intact	Severely overexploited
Hydrology		
Conditions	Riverine	Impounded
Surface inflow (Gigüela)	Seasonal	Irregular
Surface inflow (Guadiana)	Regular	Discontinued
Groundwater discharge	Yes	No
Hydroperiod	Regular	Very irregular
Flood frequency	Annual	Decadal
Interbasin transfer	No	Yes
Geomorphology		
Water mills	Functional	Not functional
Dams	Absent	Present
Peat fires	Infrequent	Frequent
Terrain subsidence	No	Yes
Water quality and nutrient cycling		
Quality of water source	Fresh	Oligohaline
Waste water discharge	No	Yes
Eutrophication	Eutrophic-hypereutrtophic[a]	Hypereutrophic
Carbon budget	Higher sink	Lower sink
Nitrogen budget	Transformer	Sink
Phosphorus budget	Transformer	Sink
Bird kills	Rare	Frequent
Fish kills	Rare	Frequent
Turbidity	Low	High
Invertebrates	Diverse, rich	Species poor
Vegetation		
Submerged plants	Diverse, rich	Sporadic, species poor
Emergent plants	Cut-sedge	Reed dominance
Continuity in plant cover	Low	High
Animals		
Fish	Species rich	Species poor
Fish type	Native	Exotic dominance
Crayfish	European[b]	American[c]
Bird abundance	High	Low

[a] The ancient wetland was eutrophic although the clear water state was maintained by nutrient outflows which currently occur only during extreme floods

[b] *Austropotamobius pallipes*, which was also introduced during the seventeenth century

[c] *Procambarus clarkii*, introduced during the 1950s

12.2 Missing Research Links: The Ecological Challenge

This book clearly demonstrates the past of TDNP before agricultural expansion in central Spain and its current ecological status as a result of ecologically unsound approaches taken to water resource management (Angeler et al. 2007). The reader may have noted that restoration has been invoked, albeit timidly, several times in the book to guide future development of TDNP. This timidity is intentional! The major state shift produced in TDNP (e.g. dam construction) created new self-reinforcing feedback loops, which act in concert with other stressors (contamination and exotic species). Although in the very unlikely case that a "self-sustaining" hydrology (e.g. aquifer restoration, approximation of natural hydroperiod and flood frequency regimes through artificial water resource management; see Chapter 3) is re-established in TDNP, restoration of its pristine conditions will still be impossible. Climate change will further increase the uncertainty regarding rehabilitation outcomes (Harris et al. 2006), particularly considering the increased aridity predicted for the Mediterranean region (Gao and Giorgi 2008).

Wetland scientists have traditionally put much emphasis in restoring abiotic disturbance regimes, hoping that nature does the job of bringing the biota back. However, such a restoration strategy hardly achieved desired wetland rehabilitation goals (see e.g. the Field of Dreams hypothesis by Palmer et al. (1997)). Zedler (2000) highlighted that wetland science is burdened with examples of ignoring or violating ecological theory. While it is perhaps true that hydrological disturbance regime is the single most important driver of ecological processes in wetlands (Mitsch and Gosselink 2000), biotic components must not be ignored in wetland restoration. However, it is not just the species or sets of species that should receive primary focus, but critical functional groups that may be able to enhance resilience to future disturbance (Bellwood et al. 2004). Restoration ecology as a science can only advance through research at the interface between landscape ecology and ecosystem ecology, linking abiotic and biotic processes, and this in turn will permit solving lingering ecological questions (Falk et al. 2006). It is clear that wetland restoration and conservation success depends on the advance on ecological knowledge. We believe that expanding the research spectrum in TDNP can contribute to this advance.

In the following we will highlight several research approaches that could be useful for this task. We acknowledge that our list is not complete and the outlined research approaches are speculative and intentionally described superficially because the maturation of these ideas will require further interdisciplinary collaboration in an extended research network to cope with particular topics that fall outside our scientific expertise. Furthermore, converting these ideas into practice will ultimately depend on many factors that the authors and editors of this book cannot control (see Section 12.3). However, we believe that these points represent core issues that could provide relevant information for a sound adaptive management in the future. We highlight with an asterisk topics which we think should receive priority in future research agendas.

12.2.1 Research at the Local Scale

12.2.1.1 Wetland Hydrology*

Las Tablas de Daimiel, as most wetland ecosystems, is not a homogeneous system. Some questions about the water cycle of Las Tablas de Daimiel remain unsolved (see Chapter 3). Water infiltration through the wetland is the least well known component of the water cycle of this wetland. Whilst wetland infiltration rates are always assumed to be spatially uniform and almost constant on monthly scales, water budget computations show that it operates with great spatial and temporal variability as a function of soil physical properties. Despite the recognized importance of infiltration in the water cycle, it still needs to be thoroughly assessed in TDNP. A detailed spatial model of wetland inundation is also lacking currently, but will be needed to understand the main flooding patterns during wetting and drying cycles. Therefore, the lack of this information currently limits optimal management designs regarding water supply to the wetland, and complicates predictions of climate change impacts on wetland hydrology. The following aspects highlight hydrological research needs for the future:

- In situ infiltration essays, using isotopic tracers, need to be carried out with high spatial and temporal resolution, taking the heterogeneity in soil physical properties into account.
- Modeling hydrological processes at adequate spatial and temporal scales.

12.2.1.2 Wetland Biogeochemistry and Climate Change

Numerous uncertainties about wetland biogeochemistry need to be explored in the future. For example, while nutrient cycling in the water column is well known, very little research has been devoted to sediment biogeochemistry. Because of their implications for global warming as well as for ecosystem functioning, decomposition processes need to be examined in wetland soils and microbial processes should be studied under anaerobic conditions during flooding. Stable isotope enrichment experiments could help to trace most biogeochemical processes in wetland sediments. The immediate effects of climate change (atmospheric CO_2 enrichment and increase of air temperature) on macrophtye and invasive plant growth and nutrient transformations will also need to be explored with detail. The following points provide some suggestions for the research agenda in the next years:

- Main nutrient transformations at the water–sediment interface.
- Assessment of carbon, nitrogen and phosphorus transformations in wetland soils, particularly those occurring under anaerobic conditions during flooding events. Denitrification process and rates in the wetland under contrasted hydrological conditions.
- Nutrient releases from wetland sediments and wind-induced resuspension.

- The effects of drying/rewetting cycles on wetland nutrient dynamics.
- Net carbon exchange and CO_2/CH_4 emission rates under contrasted hydrological conditions at the ecosystem level.

12.2.1.3 Plant Ecology

Although macrophytes are probably the most studied wetland community (see Chapter 7), there are still some scientific gaps that need to be evaluated in future research:

- Genetics of cut-sedge and population dynamics
- Genetics and population dynamics of charophytes
- Macrophyte dynamics and nutrient cycling under combined effects of elevated atmospheric CO_2 and air temperature using a FACE (Free-air CO_2 enrichment) approach

12.2.1.4 Macroinvertebrate Ecology

Despite macroinvertebrates forming a key group of organisms in wetlands (Batzer et al. 1999), there is not much information available for TDNP. It is reasonable to assume that the structure of the invertebrate community and the functions they fulfill in the wetland have been dramatically altered as a result of the compounded impacts (contamination, exotic species invasions) inflicted by humans. Preliminary studies using benthic macroinvertebrates to evaluate water quality according to the BMWP' assessment scheme (an adaptation of the British Monitoring Working Party Programme for quality assessment of Spanish waters; Alba-Tercedor and Sánchez-Ortega 1988) supports this conjecture. Judged from the calculated scores, the wetland can be classified as strongly degraded (M. Alvarez-Cobelas, 2007 unpublished data).

The wetland has also been invaded by an exotic crayfish, the American red swamp crayfish (*Procambarus clarkii* Girard), which caused an extinction of the European crayfish *Austropotamobius pallipes* (also introduced during seventeenth century) through transmission of fungal parasites (Aphanomycosis). *Procambarus clarkii* is a highly fecund, voracious predator and efficiently feeds on submerged vegetation. It also shows an extensive burrowing behaviour, causing impact in ecosystems through structural habitat alteration and biological interference with other species (Geiger et al. 2005). Given its ability to alter biological and biogeochemical processes in wetlands, this species certainly qualifies as an ecosystem engineer (Jones et al. 1997). In TDNP only some preliminary impact assessment studies have been carried out. Angeler et al. (2001, 2003b) have studied the impact of *P. clarkii* on water and sediment quality and submerged vegetation using experimental enclosures. The results showed that the cover of submerged macrophytes in the mesocosms was soon eradicated in the presence of the crayfish. Upon loss of the vegetation cover, this species increased water turbidity and contributed to a nutrient transfer from the sediments to the water column. This highlights that the effects of *P. clarkii* work to maintain TDNP in its degraded state. The following points highlight research approaches to be pursued in the future:

- Management plans to control population densities of *P. clarkii* (see discussions in Chapter 9).
- Understanding the trophic ecology of *P. clarkii*, including its impacts on food web structure and function, using correlative and manipulative studies. Complementary approaches based on gut content and stable isotope analyses could be useful for this task.
- Routine monitoring of invertebrate community structure at adequate spatial and temporal scales. From the applied perspective this could make an important contribution for complying with the European Water Framework Directive, and serve the administration as a tool for habitat quality evaluation and management decision making. From the basic ecological side, the monitoring data could help address a variety of research questions related to environmental drivers of structural and functional dynamics of macroinvertebrate communities.

12.2.1.5 Microbial Ecology

Microbial processes have not been well studied in TDNP. Our information is so far restricted to the dynamics of heterotrophic bacteria and autotrophic picoplankton in the water column (Chapter 6 and references therein). The recent application of molecular methods started to reveal the diversity of microbial communities in the wetland (MA Rodrigo et al., 2009 unpublished data). Addressing the following points could help increase our understanding of the ecological role of microbial communities in TDNP:

- Microbial communities as a food web base that fuels secondary production through the microbial loop
- Quantifying the relative importance of autotrophic versus heterotrophic processes in ecosystem metabolism
- Microbial communities and processes in wetland soils involved in carbon dioxide, methane or nitrogen emissions
- Microbial processes in wetland soils as a function of recurring wetting/drying events
- Changes of microbial processes and community compositions under elevated atmospheric CO_2 and temperature
- Assessing the role of virus ecology in microbial processes
- Sanitary issues: microbial communities and processes as drivers of disease outbreak (botulism)

12.2.1.6 Vertebrate Ecology

Fish and bird ecology have not been enough studied in the wetland. Most existing data provided information relative to abundance and species richness; however, many basic aspects on their community ecology remain unknown. Besides, the ecology of these groups should be approached from larger spatial scales (regional or watershed). Although vertebrates could be considered marginal on ecosystem functioning, our ecological knowledge of the wetland ecosystem would be incomplete if these communities are neglected. Some aspects that could be addressed in the future:

- Fish population dynamics
- Effects on water quality
- Waterfowl and passerine population dynamics
- Habitat distribution along with man-made change in the catchment

12.2.1.7 Food Web Ecology

Chapter 9 presents a conceptual model that summarizes how manipulation of specific target communities in the wetland could help improve the environmental quality of TDNP. Despite the management appeal of the model, its reductionist nature should be emphasized because it was largely inspired by theory and empirical evidence coming from only small-scale manipulative experiments. Its value for guiding management, through manipulation of key communities (vegetation, fish, crayfish), should be tested in the future with a focus on:

- Allochthonous and autochthonous organic matter pathways through food webs using stable isotopes
- C:N:P stoichiometry in the food webs
- Food web lengths and connectivity in manipulated and non-manipulated food webs
- Trophic position and energy supply to food web components in manipulated and non-manipulated food webs evaluated through stable isotopes
- Food web stability as a function of natural and anthropogenic disturbance regimes

12.2.1.8 Evolutionary Ecology

Many wetlands harbor an impressive reservoir of dormant propagule stages in their sediments that can be useful for evolutionary research. These propagule banks integrate the environmental history of a site and its terrestrial catchment (Brendonck and De Meester 2003), and provide different research opportunities:

- Assessing the ecological integrity and/or the impact of anthropogenic stress in wetlands through comparative studies of the structure of emerged communities (Angeler and García 2005).
- Using a resurrection ecology approach to assess ecological and evolutionary change over time (Kerfoot et al. 1999; Kerfoot and Weider 2004). Resurrection ecology has so far made use of resting eggs from water flea (*Daphnia*) that have undergone different dormancy periods (years to decades) within a lake. Resurrected organisms can be studied comparatively using evolutionary/genetic and experimental approaches. Both approaches combined can provide insights

into how the dimensions of species' ecological niches have shifted over time and could help reveal whether direct effects of, for example, climate change (increased temperatures and atmospheric CO2 concentrations and hydrological alterations) or other anthropogenic stressors (e.g., contamination, landuse change, exotic species) have caused microevolution (Angeler 2007). A resurrection ecology approach seems particularly suitable for disentangling the effects of the multiple anthropogenic stressors that have affected TDNP in recent history, and it will be interesting to compare patterns between organisms that occupy different trophic positions in the food webs (i.e. *Daphnia* vs algae vs protozoa).

12.2.1.9 Genetic Diversity

Molecular techniques are increasingly finding their place in ecological research (Hughes et al. 2008). Recent studies demonstrate that diversity at the genetic level of populations enhances resilience to disturbance events (Altermatt and Ebert 2008; Reber et al. 2008). Genetic data can also provide insight into extinction risks of sexually reproducing species resulting from loss of genetic diversity (Stockwell et al. 2003). Research in TDNP is almost exclusively based on "morphological" biodiversity. Determining the genetic structure and variability in populations of native and exotic species could address the following points:

- Biogeographic patterns, dispersal ecology, and conservation status of emblematic and rare wetland taxa at local and regional scales
- Impact of exotic species and other stressors on the genetic structure of populations of native species
- Community and ecosystem resilience based on the genetic patterns of populations
- The impact of genetic diversity in food web configurations and functions

12.2.2 Research Beyond the Wetland Boundary

Several of the points raised in the last section highlight that research approaches must be extended from the local wetland scale to regional scales to fully understand the ecology of TDNP. Given the higher ratio between catchment size and habitat size in Mediterranean areas relative to temperate regions (Alvarez-Cobelas et al. 2005), catchment-scale processes could have a disproportionate effect on local wetland biota and biogeochemical processes. Although the catchment of TDNP and other catchments are heavily influenced by agricultural practices (Chapter 4), local scale phenomena (pollution) seem to swamp the negative effects arising from land use in the surroundings of TDNP (Chapter 5). However, scales and hierarchies are critical to evaluating landscape-level impacts in local wetland ecology. For example, Angeler et al. (2008) have shown that the negative impacts of land use on the populations of two emblematic branchiopods are visible at very broad spatial scales that extend the catchment boundaries. Their study suggests that atmospheric flux of pollutants can contribute to degrade distant ecosystems. This example highlights an extreme challenge to wetland conservation. In this section we will deal with landscape ecological approaches that could help increase our understanding of the interaction of ecological processes in TDNP with those in its surrounding catchment(s).

12.2.2.1 Landscape Ecology and Biogeochemistry*

Here we refer to a landscape approach for highlighting the relevance of landscape structure and processes and landuse patterns across different spatial and temporal scales on local wetland integrity (including biogeochemical and ecological processes). While spatial hierarchies of landuse patterns can provide insight to contemporary landscape-level pressures on TDNP, a historical component should not be ignored. In their seminal paper, Harding et al. (1998) invoked "the ghost of land use past" to explain present-day diversity of stream invertebrates and fish in watersheds with different land-use history. Whole watershed land use in the 1950s was the best predictor of present-day diversity, whereas riparian land use and watershed land use in the 1990s were comparatively poor indicators. Their findings indicate that past land-use activity, particularly agriculture, may result in long-term modifications to and reductions in aquatic diversity, regardless of reforestation of riparian zones. The following research approaches could be useful for determining landscape effects on the ecology of TDNP.

- Accurate assessment and delineation of the historical extent (before 1940s) of wetlands at "La Mancha Humeda" in order to evaluate main landscape factors and processes (geomorphology, geology, hydrology, land uses, etc.) contributing to wetland occurrence and persistence. This could reveal wetland degradation patterns and the main pathways of biodiversity loss.
- Discerning between historical and current landuse effects on wetland integrity. This could help reveal whether current biodiversity elements in the wetland are still the result of the ghost of land use past before large-scale conversion into agricultural lands or whether agricultural use of the catchments has already contributed to diversity loss after accounting for the effects of local contamination. If current diversity is still explained by historical factors, when and at which temporal lags will the negative effects of landuse change be manifest in TDNP? What will be the magnitude of impact, in term of species extinctions, arising from a potential lagged landuse impact? To what extent do other landscape-level phenomena (e.g. habitat fragmentation) interact with landscape structure and landuse patterns to mediate in the strength of impact?
- A logical point following from the previous is that restoration and conservation efforts adopting a Forbes' (1887) view of aquatic ecosystems as isolated microcosms in a terrestrial matrix are erroneous. In fact, Chapter 8 highlights that despite local conservation efforts, the number and diversity of waterfowl is decreasing in the wetland. This suggests that diversity is impoverishing at regional scales (loss of γ diversity) leading to a reduction of spatial turnover (decreased β diversity), and finally local (α) diversity loss. Research approaches

outlined in the next point could be helpful for studying such phenomena with more detail.

- So far, the hydrology of the Upper Guadiana catchment has been assessed only very superficially, causing high uncertainty regarding potential consequences of climate change on catchment hydrology. More gauge stations are needed to develop accurately distributed hydrological models at the watershed scale.
- Nutrient cycling throughout watershed, including surface and below-ground processes, must be assessed for understanding nutrient export dynamics in the catchment during both dry seasons and storm events. In addition, in-stream biogeochemical processes need to be evaluated, because processes related to nutrient transformation in, and supply to, the hyporheic zone are not well understood.
- Diffuse pollution assessment is still lacking which is important given the great weight of agriculture in the catchment. Nonpoint sources of nutrients trigger wetland eutrophication processes during flood pulse events. Alternative improved land use management practices in uplands must be addressed in order to reduce the overall load to TDNP.

12.2.2.2 Metacommunity Ecology*

Metacommunity ecology emphasizes the connection of local communities within defined spatial units through dispersal (Holyoak et al. 2005). While metacommunity ecology is still based largely on theory, a growing body of studies tests its assumptions empirically. Research in TDNP holds potential to inform metacommunity ecology from a hierarchical point of view:

- Local-scale metacommunity processes: Floodplain wetlands may serve as excellent model ecosystems for studying community structure across spatial and temporal scales. The constituent communities of mosaic-like floodplains are arrayed along disturbance gradients and are connected by flood pulsing (Middleton 1999). Preliminary research suggests that the extreme spatial and temporal variability of flood regimes in Mediterranean areas provides unique opportunities to study metacommunity dynamics during periods of high connectivity (flooding situation) and/or fragmentation of sites (drought effects) within the wetland, respectively (Angeler et al. 2010). Further research could provide insight into mechanisms regulating hierarchical metacommunity dynamics.
- Regional-scale metacommunity processes: TDNP forms part of a previous extensive network of aquatic ecosystems within "La Mancha húmeda" wetlands area (Chapter 1). Since the onset of agricultural overexploitation, the network has become dramatically altered as a result of wetland loss and increased fragmentation between sites. A regional metacommunity approach holds potential to determine the connection of TDNP with other wetlands in the region under current landscape settings. This could help determine several aspects, including TDNP as a source or sink of populations, species and consequently gene flow between TDNP and other aquatic ecosystems, and the community specificity of such processes.

12.2.2.3 Water Resource Management**

Several calls have been made in the book for the need of an effective water resource management at the catchment scale. Aquifer restoration must be only considered as a strategy that may work in the long-term after substantial changes in current productivity regimes. However, ultimate goal of water management at the catchment scale must be the restoration of aquifer to levels existing in the 1970s when groundwater discharge was a very significant source for wetland inundation. This is the most problematic issue because aquifer recovery depends on the balance between water inputs (mainly rainfall) and outputs (mainly irrigation) and it is unclear that groundwater recharge will be enough to achieve aquifer replenishment. Groundwater recharge will be impaired by ongoing rainfall decrease resulting from climate change in Central Spain. In the meantime, other water sources such as water diversions or the use of treated waste waters could provide an option to artificial management of wetland hydroperiod. Currently, interbasin water diversions appear the most feasible alternative to sustain the wetland hydroperiod in the short-term. However, this alternative must carefully consider the efficiency of water transfer to the wetland because much of the diverted water is currently lost either through evaporation, infiltration or illegal extractions. Furthermore, it carries risk of exotic species introductions and deprives other places of water. Several recommendations are given in Chapter 10 to increase the success of this approach.

The environmentally sound application of treated waste water is currently illusive given the limitations to reduce nutrient contents and contamination loads in treated waste waters to levels acceptable for release to natural systems. Unfortunately, there exists precedence where discharge of treated waste water caused severe eutrophication problems and a fundamental change of natural temporary to permanent hydroperiods. If the use of treated waste water is to become an option the following points must be considered:

- Creation of subsurface flow wetlands to guarantee further treatment of waste waters before discharge to TDNP. This technology should be designed to reduce as much as possible the evaporative and transpiration loss of water, while decreasing the nutrient content and contamination loads below critical levels.
- This technology should be used to simulate as much as possible the fluctuating hydroperiod and flood frequency regime present in Mediterranean-type aquatic ecosystems. Excess water accumulating during dry periods could be used for recharging exhausted aquifers and/or moderate irrigation purposes.
- Landuse practices should move towards command and control schemes, whereby the types of crops, allocated water for, and the timing of irrigation are more strictly controlled. Such schemes should also envision tracking down, and destroying, illegal wells and imposing severe punishments to those who continue to construct them.

12.3 The Future: The Socioeconomic Challenge

Even if we succeed in advancing our ecological knowledge of TDNP, will this knowledge gain be enough for sound management of the ecosystem and its catchment? It is increasingly recognized that ecological systems depend a great deal on, and interact with social and economic factors (Folke et al. 2005; Levin 2006), which give rise to a complex system known as a panarchy (Gunderson and Holling 2002). Changes in the interaction cycles between the components in these panarchies is definitely the main reason for the environmental deterioration of TDNP, with the changes in the social and economic components (increasing productivity and wealth) being the main drivers of change in the ecological component (degradation). The major effects of socioeconomic factors, which ultimately resulted in the deterioration of TDNP, support the conjecture that in present-day societies, social parameters might be the control of the greater panarchical system (Leuteritz and Ekiba 2008). The capacity to adapt to environmental change, i.e. resilience, can therefore be explained as the collective variable that captures and demonstrates the overall behavior of the system resulting from ecological, social and economic factors. As such, if an area loses environmental resilience, social assets and institutional capacity, in may become "… an accident waiting to happen" (Holling 2001, p. 396).

Throughout human history, we can observe how "large-scale" socioeconomic and policy forces have led to low-resilience land use systems that eventually collapsed. For example, Fraser (2003) shows how a combination of economic policy, population growth and industrialization caused land use patterns in Ireland to shift from being relatively diversified (ca. 1820), towards being dominated by the potato. This system collapsed resulting in famine in the 1840s. Similarly, laissez-faire approaches to trade and the British thirst for tea in the Victorian period led to almost deforestation of Sri Lanka (Schweinfurth 1982) and created a system that collapsed in the 1870s when El Niño-induced droughts claimed tens of thousands of victims (Davis 2001). The importance of socioeconomic and policy drivers in creating vulnerable rural economy and land use systems is such that some development economists argue that these collapses are not so much environmental tragedies as human-engineered problems (Sen 1981). Nevertheless there is an environmental component to many such crises and as we look towards the future it is likely that the frequency and/or duration of environmental shocks increase, in Spain (Moreno 2005) and elsewhere (IPCC (International Panel on Climate Change) 2007).

It seems counterintuitive that humans deliberately continue to destroy their environment in times when their unsustainable actions are well recognized. However, these attributes may in essence reflect the inability of taking action because past socioeconomic drivers may have created "rigidity traps" that reduced the margin of taking effective action, i.e. the possibility to adapt and maintain resilience. Even though the examples from Ireland and Sri Lanka provide proof for the existence of such rigidity traps, it is by far not the aim of this essay to speculate about the existence of these in the case of the "TDNP panarchy". We have neither the necessary means, nor is it the scope of this book, to empirically evaluate how narrow such rigidity traps could be in Spain, nor do we have the necessary training in economy or sociology to evaluate their contributions to shaping future panarchies. In fact, the whole panarchy seems to be currently (still) resilient! However, as the focus of this book is the ecology of TDNP, we are interested in how social or economic factors can influence the ecological component of this particular panarchy in the future. On the basis of some simple indicators, we suggest that socioeconomic factors can provide obstacles to efficient ecological management in general, and especially in the case of the TDNP wetland, in the future.

The first obstacle is of economic and institutional nature. During the last decade, Spain has seen an economic growth to miraculous extents, according to media and general public opinion. However, this economic growth, as in other countries such as Ireland, was based mainly on a dramatic expansion of the construction business while relatively few efforts have been made to invest in alternative sources to sustain economic growth and create a functional redundancy in economy (i.e. creating conditions that would allow mitigation of the impacts if the construction business collapses). This situation especially affects rural areas where the rapid economic growth has not led to changes in patterns of natural resource exploitations which already gave signs of exhaustion decades ago. The created rigidity in Spain's economy soon became notable upon the onset of the global economic crisis, induced by the crash of the loan system in the USA, and which brought the Spanish construction business close to collapse. The consequences of this collapse are far-reaching, and are particularly notable in sectors that do not generate money in the short term. Although governmental plans considered an increase of the budget for R&D with the aim to create mechanisms that could better mitigate similar crisis in the future, these intentions are currently frustrated by the high unemployment rates derived from the economic crisis. Money is needed for social purposes rather than for R&D. As a result, the stipulated budget for R&D has been reduced by the Ministry of Science and Innovation and the Ministry of Environment. If according to the hypothesis of Leuteritz and Ekiba (2008) this budgetary reduction reflects a socioeconomic or institutional parameter that helps explain Spain's sustainability, it is reasonable to assume that a revolution of the ecological component in Spain's panarchy will be unlikely. In other words environmental conservation will likely not be straightforward.

In addition to this institutionally mediated limitation there is a social component which may ultimately feed back to reinforce institutional processes. Social movements develop through several stages (Moyer et al. 2001), from initial rising of concern, through to gaining popular attention and political support, to a point where the concerns and values of the movement are widely accepted in the community and the moral authority of powerholders and decision makers resisting values of the movement collapses. In many rural parts of Spain the movement is still in the very early steps, if non-inexistent, with overexploitation of water resources or loss of biological capital not being seen as an environmental threat by a grand part of the population. Paradoxically, dedicating water resources to environmental management is often interpreted as a waste of water that could be used for agriculture, tourism or other forms of human consumption. A recent example clearly highlights the overwhelming influence of the sociological component in the overall functioning of Spain's society and environmental perceptions.

The Spanish media offer generally a biased vision of environmental problems through sensationalist news where only extreme phenomena such as prolonged droughts or, very recently, peat fires in TDNP are highlighted. By contrast, the factors that caused large-scale degradation of the wetland in the long term receive hardly any coverage by the media. Because environmental degradation is not a main concern of the population, politicians are not under pressure to solve these problems. As a result, mitigation measures are taken in response to medial sensationalism to show the public operational capacity rather than trying to implement large-scale rehabilitation measures to solve the core of the problem. After the recent explosive news that TDNP's peat is on fire, the regional and national governmental authorities decided to show its political will to change, once and for all, the severe environmental situation of the wetland. Hence, they declared to divert urgently 20 hm^3 of water to TDNP, using an unfinished pipe designed to transport exclusively drinking water to the cities Ciudad Real and Puertollano from the Tagus basin (El País newspaper, October 14, 2009; http://www.altoguadiana.es/ES/AREA_COMUNICACION/DOSSIER_PRENSA/2009/DOC/PDF/2009_OCTUBRE_14.pdf). This measurement like most others adopted in the historical management of the wetland (see Chapter 10) will extinguish the peat fires, moving the attention of the media and the public to other topics, and let the wetland in a continued struggle of survival.

After the blind overexploitation of own resources the claim of water from other regions is becoming a solidarity issue in Spain. Castilla-La Mancha diverts water from the Tagus basin to Murcia area to maintain its productive orchards after this region exhausted its water resources. We are thus far away from inducing a socially mediated transformation of the "irrigation catchments" to a "post-irrigation society". While farmers of Murcia rise in masse to seek water from the Tagus basin, which is situated over 500 km away, the people of La Mancha region make no claims for their own water resources, especially regarding its use for mitigating environmental problems. In fact, environmental problems are not perceived by the population of Castilla-La Mancha as one of the top ten concerns, which include unemployment and economic crisis (54%). Shortage of water for agricultural activities (12%) occupies the fourth place (survey conducted by the regional government in 2008; http://www.soitu.es/soitu/2008/07/15/info/1216144867_524007.html). Without fundamental changes in basic perceptions within societies, there will also be no resistance of powerholders and decision makers to make change. On the contrary, even if an environmental consciousness exists at institutional levels it may be the fear to lose votes, and an election, if changes are made that the broad public does not understand or likely not accept.

On the basis of the socioeconomic and ecological characteristics outline above we provide a conceptual construct for TDNP (Fig. 12.2), which could be useful for evaluating and understanding complexly interacting factors in this particular socio-economic-ecological system. Although the model is clearly of a reductionist nature, the highlighted interactions will definitely play a major role in determining whether

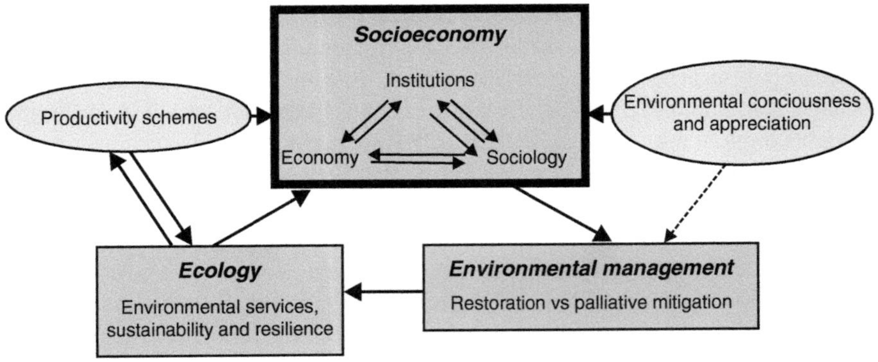

Fig. 12.2 Model of interaction cycles between ecological and socioeconomic components in the panarchy of Las Tablas de Daimiel. The strength of the arrows reflects the strength of potential relationships. For more details see text

environmental management of the wetland can adopt integral rehabilitation schemes to achieve a certain self-maintenance of the wetland, or continue to be of purely mitigative nature to palliate the adverse effects of human impact, i.e. maintaining the current artificial life support machinery. The model shows complex feedback mechanisms between socioeconomic factors and ecological components that are mediated by productivity schemes (i.e. factors that currently contribute to economic sustainability (i.e. agricultural productivity) at the cost of ecological sustainability in the TDNP area) and social perceptions of the environment (ellipsoids in Fig. 12.2). The model emphasizes especially the mutual dependence between the current productivity scheme and the ecological component in this complex interaction cycle. It reflects the adverse impacts of current agricultural practices on the ecological component which in turn will feed back negatively to the productivity regime once critical thresholds are exceeded. It seems logical if current productivity regimes based on agriculture continue to degrade the environment in the long-term, ecological resilience may be lost, and the whole system, including the socio-economic component in this area, may become sensitive to large-scale environmental perturbations. Once the whole system loses the capability to adapt to disturbances, and thus its ability to sustain itself, it will require interventions by national or international agencies to mitigate the impact from these perturbations. It is obvious in our human-centered world that under such circumstances the major focus of rescue will be on securing the quality of life for humans rather than aiming at securing the basis which supports all life on earth, i.e. a healthy environment.

The current economic crisis is very instructive how the "health" of the socioeconomic component itself feeds back to the ecological component through decreased devotions of part of the governmental budgets to R&D. The emphasis of the box comprising the socioeconomic component in Fig. 12.2 highlights the key role which social and economic factors play in the overall functioning of a panarchic system (Leuteritz and Ekiba 2008). It is obvious that socioeconomic constraints, reflected for example through decreased funding opportunities, will limit on one hand the advance of science, and on the other hand the implementation of sound environmental management schemes. In this chapter we have highlighted several research gaps that currently limit our understanding of the ecological functioning, and consequently restoration and conservation of TDNP. If these gaps are not closed through research, it is unlikely that a straightforward adaptive management of this wetland will be able. It is clear from the model that a change in the productivity regime will be required to reduce the competition for limited water resources between economic productivity and environmental conservation. It is unclear, however, to what degree climate change will affect the whole panarchical systems in a future which is predicted to be substantially drier, and when new scenarios for water resource competition will arise. Spain is in serious need of a national water plan that satisfies a minimum of ecological criteria. Crucial to this task will be the involvement of communities and the establishment of a new, nationally consistent water entitlement and trading system that provides security to both water users and the environment (e.g. Wenthworth Group 2003).

12.4 Conclusion

Las Tablas de Daimiel is becoming one of the most well-studied Mediterranean wetland ecosystems in Europe. It serves as a case study that demonstrates the adverse impacts of human activities on wetlands ecological integrity in Mediterranean Europe where one resource is fundamentally limiting: water. Within the "TDNP panarchy", water availability was a driver of the interaction cycles between its ecological and socioeconomic components in the past, and will with great certainty continue to affect these interactions in Spain's warmer future. Climate change adds uncertainty making their outcomes hard to envisage. However, one fact is clear: without revolutions in the socioeconomic components, mainly in form of changed productivity systems, the ecological part of these interaction cycles faces a bleak future. Thus, although scientists continue to enquire about the ecological understanding about our environment, the risk is currently high that social or economic constraints limit the implementation of this knowledge to a sound management of resilience.

Acknowledgments The authors are most grateful to Andrew Boulton and Miguel Alvarez-Cobelas for helpful comments on previous manuscript drafts.

References

Alba-Tercedor J, Sánchez-Ortega A (1988) Un método rápido y simple para evaluar la calidad de las aguas corrientes basado en el de Hellawell (1978). Limnetica 4:51–56

Altermatt F, Ebert D (2008) Genetic diversity of *Daphnia magna* populations enhances resistance to parasites. Ecol Lett 11:918–928

Alvarez-Cobelas M, Rojo C, Angeler DG (2005) Mediterranean limnology: current state, gaps and the future. J Limnol 64:13–29

Angeler DG (2007) Resurrection ecology and global climate change research in freshwater ecosystems. J N Am Benthol Soc 26:12–22

Angeler DG, García G (2005) Using emergence from soil propagule banks as indicators of ecological integrity in wetlands: advantages and limitations. J N Am Benthol Soc 24:740–752

Angeler DG, Sánchez-Carrillo S, García G, Alvarez-Cobelas M (2001) The influence of *Procambarus clarkii* (Cambaridae, Decapoda) on water quality and sediment characteristics in a semiarid floodplain wetland. Hydrobiologia 464:89–98

Angeler DG, Sánchez-Carrillo S, Alvarez-Cobelas M, Cirujano S, Medina L (2003) Exotic crayfish activity and its effects on water quality: preliminary implications for the alternative stable equilibria in Mediterranean wetlands. J Mediterr Ecol 4:13–21

Angeler DG, Boulton AJ, Jenkins KM, Sánchez B, Alvarez M, Sánchez S (2007) Alternative states and temporary wetlands: research opportunities for understanding effects of anthropogenic stress and natural disturbance. In: Clarkson PA (ed) Environmental research advances. Nova, New York, pp 5–17

Angeler DG, Viedma O, Sánchez-Carrillo S, Alvarez-Cobelas M (2008) Conservation issues of temporary wetland Branchiopoda (Anostraca, Notostraca: Crustacea) in a semiarid agricultural landscape: What spatial scales are relevant? Biol Conserv 141:1224–1234

Angeler DG, Alvarez-Cobelas M, Rojo C, Sanchez-Carrillo S (2010) Phytoplankton community similarity in a semiarid floodplain during contrasting hydrological connectivity regimes. Ecol Res (DOI: 10.1007/s11284-009-0681-7)

Batzer DP, Rader R, Wissinger SA (1999) Invertebrates in freshwater wetlands of North America: ecology and management. Wiley, New York

Bellwood DR, Hughes TP, Folke C et al (2004) Confronting the coral reef crisis. Nature 429:827–833

Brendonck L, De Meester L (2003) Egg banks in freshwater zooplankton: evolutionary and ecological archives in the sediment. Hydrobiologia 491:65–84

Constanza R, Arge R, de Groot R, Farber S, Grasso M, Hannon B, Limburg K, Naeem S, O'Neill RV, Paruelo J, Raskin RG, Sutton P, van den Belt M (1997) The value of the world's ecosystem services and natural capital. Nature 387:253–260

Davis M (2001) Late Victorain Holocausts: El Niño Famines and the Making of the Third World. Verso, London

Falk DA, Palmer MA, Zedler JB (2006) Foundations of restoration ecology. Island Press, Washington, DC

Folke C, Hahn T, Olsson P, Norberg J (2005) Adaptive governance of social-ecological systems. Ann Rev Environ Resour 30:441–473

Forbes ST (1887) The lake as a microcosm. Bulletin of the Peoria (Illinois) Scientific Association Reprinted in Bulletin Ill Nat Hist Surv 15:537–550 (1925)

Fraser E (2003) Social vulnerability and ecological fragility: building bridges between social and nautral sciences using the Irish potato famine as a case study. Conserv Ecol 7 (online)

Gao X, Giorgi F (2008) Increased aridity in the Mediterranean region under greenhouse gas forcing estimated from high resolution simulations with a regional climate model. Global Planet Change 62:195–209

Geiger W, Alcorlo P, Baltanas A et al (2005) Impact of an introduced Crustacean on the trophic webs of Mediterranean wetlands. Biol Invasion 7:49–73

Gunderson LH, Holling CS (2002) Panarchy: understanding transformations in human and natural systems. Island Press, Washington, DC

Harding JS, Benfield EF, Bolstad PV, Helfman GS, Jones EBD (1998) Stream biodiversity: the ghost of land use past. Proc Natl Acad Sci USA 95:14843–14847

Harris JA, Hobbs RJ, Higgs E, Aronson J (2006) Ecological restoration and global climate change. Restoration Ecol 14:170–176

Holling CS (2001) Understanding the complexity of economic, ecological, and social systems. Ecosystems 4:390–405

Holyoak M, Leibold MA, Holt RD (2005) Metacommunities: spatial dynamics and ecological communities. University of Chicago Press, Chicago
Hughes AR, Inouye BD, Johnson MTJ, Underwood N, Vellend M (2008) Ecological consequences of genetic diversity. Ecol Lett 11:609–623
IPCC (International Panel on Climate Change) (2007) Claimte Change 2007: the physical science basis. IPCC, The Hague
Jones CG, Lawton JH, Shachak M (1997) Positive and negative effects of organisms as physical ecosystem engineers. Ecology 78:1946–1957
Kerfoot WC, Weider LJ (2004) Experimental paleoecology (resurrection ecology): chasing Van Valen's Red Queen hypothesis. Limnol Oceanogr 49:1300–1316
Kerfoot WC, Robbins JA, Weider LJ (1999) A new approach to historical reconstruction: combining descriptive and experimental paleolimnology. Limnol Oceanogr 44:1232–1247
Leuteritz TEJ, Ekiba HR (2008) Not all roads lead to resilience: a complex systems approach to the comparative analysis of tortoises in arid ecosystems. Ecol Soc 13:1. Online at: www.ecologyandsociety.org/vol13/iss1/art1/
Levin SA (2006) Learning to live in a global commons: socioeconomic challenges for a sustainable environment. Ecol Res 21:328–333
Middleton B (1999) Wetland restoration: flood pulsing and disturbance dynamics. Wiley, New York
Mitsch WJ, Gosselink JG (2000) Wetlands, 3rd edn. Wiley, New York
Moreno JM (2005) Evaluación preliminar de los impactos en España por efecto del cambio climático. Ministerio de Medio Ambiente, Madrid, Spain
Moyer B, McAllister A, Finley ML, Soifer S (2001) Doing democracy: the MAP model for organizing social movements. New Society Publishers, Gabriola Island, BC, Canada
Palmer MA, Ambrose RF, Poff NL (1997) Ecological theory and community restoration ecology. Restoration Ecol 5:291–300
Reber A, Castella G, Christe P, Chapuisat M (2008) Experimentally increased group diversity improves disease resistance in an ant species. Ecol Lett 11:682–689
Schweinfurth U (1982) Landscape change and geomedical consequences in the highlands of Sri Lanka (Cylon). Mountain Res Devel 2:195–199
Sen A (1981) Poverty and famines. Clarendon, Oxford
Stockwell CA, Hendry AP, Kinnison MT (2003) Contemporary evolution meets conservation biology. Trends Ecol Evol 18:94–101
Wenthworth Group (2003) Blueprint for a national water plan. The Wentworth Group, WWF Australia
Zedler JB (2000) Progress in wetland restoration ecology. Trends Ecol Evol 15:402–407

Index

A

Aboveground biomass, 177, 190
Abramis brama, 220
Acanthocyclops robustus, 155, 160
Accretion, 128, 129, 131
Acidobacteria, 141
Acreman, M., 12, 46, 70
Additions of young-of-the-year (YOY), 220
Adrian, R., 138
Age of Bronze, 33, 231, 242
Agricultural development, 104, 230
Agricultural lands, 95, 101, 268
Agricultural uses, 100, 101, 215, 268, 272
Agriculture
  employment, 15, 99
  sector, 95, 99, 104
Air temperature, 47–50, 72, 73, 263, 264
Alba-Tercedor, J., 264
Albufera Lake, 246
Alburnus alburnus, 220
Alcácer, C., 232, 233
Alcazar de San Juan, 115
Alcorlo, P., 223, 264
Alía, F., 246
Allan, J.D., 86
Allanson, B.R., 4
Allelopathic activity, 217
Allen, J.R.M., 24
Allen, T.F.H., 138
Alley, W.M., 62
Allochthonous, 130, 217, 266
Almarza, C., 62
Alnus, 35
Alonso Zarza, A.M., 24, 25, 28
Altermatt, F., 267
Alternative stable states, 216, 217, 221, 226
Alvarez-Cobelas, M., 17, 140, 215, 216, 221, 222, 241–253, 258, 264, 267
Alvarez, M.G., 191
Alwan, A., 247
Amaranthaceae, 33, 35, 38
Ambiental, S.L., 231
Ambrose, R.F., 262
American red swamp crayfish. *See* *Procambarus clarkii*
Amezaga, J.M., 64
Ammonia, 119, 121, 124, 125, 140, 168
Anabaena sphaerica, 143, 144
Anaerolineae, 141
Anarchist collectivity, 246
Anas angustirostris, 231
Anas clypeata, 203, 204
Anas crecca, 8, 204
Anas platyrhynchos, 8, 203
Anas strepera, 203–205
Andersen, O.F.O., 191
Anderson, M.G., 68, 175
Anderson, R.V., 71
Angeler, D.G., 3–18, 86, 109–131, 138, 168–170, 215–226, 257–275
Anthropogenic activity, 25
Anthropogenic pressure, 89
Anthropogenic stressors, 85–104, 215, 266, 267
Anticoagulant compounds, 248
APHA, 111
Aphanomycosis, 264
APP. *See* Autotrophic picoplankton
23/04.04 Aquifer. *See* Llanura Manchega Occidental Aquifer
Aquifer overexploitation, water overexploitation, 12–14, 17, 46, 51, 65, 89, 112, 139, 176, 234, 250
24 or 04.06 Aquifer *See* Campo de Montiel Aquifer
Aquifer storage coefficient, 9
Aquifer transmissivity, 9
Arabs, muslims, 244
Arctodiaptometum saline, 12

Ardea cinerea, 203
Ardea purpurea, 203
Argar Culture, 241
Arge, R., 258
Aridity, 4, 24, 29, 31, 33, 51, 262
Armengol, J., 89
Armentano, T.V., 188
Artemisia, 31, 33, 35, 36, 38
Asensio-Rubio, M., 246
Asteraceae, 28, 29, 31, 33, 35
Aster squamatus, 184
Atmospheric deposition, 115, 123
Austropotamobius pallipes, 231, 246, 261, 264
Autochthonous, 130, 225, 234, 266
Autogenic trajectory, 170
Autotrophic, 142, 165, 265
Autotrophic picoplankton (APP), 138, 140–144, 163–169, 265
Azuer inflows, 60, 63

B

Bachmann, R.W., 111
Bacillarophyceae, 144, 147, 152, 154, 156, 157
Bacterioplankton, 138, 140–142, 163–168
Bacteroidetes, 141
Baena, 242
Baik, K., 141
Bailarón, L., 24
Barbier, D., 25
Barceló, D., 232
Barone, R., 61
Barrachina, P., 198
Basu, B.K., 141
Batzer, D.P., 264
Bayley, P.B., 71
Bazemore, D.E., 130
Bdelloid rotifers, 155
Beaver, J.R., 155
Beckerman, A., 138
Bedford, B.L., 98
Behre, K.E., 25
Bell, T., 166
Bellwood, D.R., 262
Belowground biomass, 178, 189, 190
Benfield, E.F., 268
Bengtsson, L., 128
Benthic-dwelling fish, 218
Benthivorous, 215, 217, 218, 224
Berberian tribes, 244
Bergkamp, G., 50
Berglund, B.E., 25
Berka, C., 86
Bernier, J.C., 97
Bernués, M., 201, 229
Betaproteobacteria, 141
Betula, 29, 35, 37
Betulaceae, 28
Bhaduri, B., 104
Bicarbonate, 113
Biogeochemical processes, 46, 86, 226, 263, 264, 267–269
Biogeochemistry, 263–264, 268–269
Biogeographic patterns, 267
Biomanipulation, 216–224
Biomass, 111, 140, 143, 151, 153, 155–157, 160–163, 165–170, 177–178, 188–191, 215, 217–225
Biovolume, 140, 150, 152, 153, 166, 169
Birks, H.H., 24
Birks, H.J.B., 24
Black-winged stilts, 205
Blanco catchment, 88, 91–93, 95, 97
Bleak. *See Alburnus alburnus*
Blindow, I., 183
Bloesch, J., 111
Bolaños, 243
Bond, G., 23
Bond, N.R., 62, 128
Botulism, 202, 205, 232, 265
Boulton, A.J., 217, 259, 262
Bouraoui, F., 64, 78
Bovo-Scomparin, V.M., 168, 169
Bradley, C., 4, 46, 51, 53, 59
Bradley, P., 23
Brassicaceae, 33
Bream. *See Abramis brama*
Breeding, 202–206, 244
Breitburg, D.L., 110
Brendonck, L., 266
Brieé, C., 141
Brient, L., 123, 130
Brinson, M.M., 89, 97, 104
Broecker, W.S., 24
Bronstert, A., 55, 72
Brooks, R.P., 129
Brown, A.G., 46
Bryant, R.J., 176
Buck, W.J., 200
Buffer area, 17
Building employment, 15, 95, 100
Building sector, 15, 90, 95, 101, 102, 104
Burba, G.G., 67, 68
Burjachs, F., 24
Burke, S., 78, 79, 87
Butler, D.R., 131

C
Calatrava catchment, 87
Calatrava la Vieja, 243, 244
Calatrava Order, Military Order, 244
Calcium carbide, 247
Calcium-rich water, 113
Calder, I.R., 86
Calluna, 33
Camarero, J.J., 192
Camargo, J.A., 176
Cambón, G., 27
Campo de Montiel aquifer, 9, 91
Canfield, D.E., 111
Cannibalism, 249
Cano-Soler, D., 95
Capture (catch) per unit of effort (CPUE), 198, 199, 218
Carex riparia, 181, 182
Carignan, R., 4
Carp. *See Ciprinus carpio*
Carpenter, S.R., 217
Carrasco, M., 201, 205, 206
Carrión de Calatrava, 31
Carrión, J.S., 24, 25
Casado, S., 3, 229
Casgrain, P., 178
Casillas, 248
Castanea, 29, 35, 37
Castaño-Castaño, S., 59, 72
Castilla-La Mancha, 6, 14, 15, 89, 273
Castillo de Calatrava, 31–33
Cattail. *See Typha domingensis*
CC-17 paleorecord, 25, 27, 31–35
Ceballos-Moreno, M., 230
Central Spain, 4, 6, 79, 243, 245, 248, 249, 252, 253, 259, 262, 270
Ceramic remains, 242
Ceratophyllum submersum, 181, 183
Ceriodaphnia dubia, 155, 160
Ceriodaphnia reticulata, 160, 220
Cetti's warblers, 206
Chapman, A., 200
Chara spp., 12
  C. aspera, 180, 181
  C. canescens, 180, 181
  C. connivens, 181, 183
  C. hispida, 140, 180–183
  C. hispida var. major, 180, 183
Charophyceae, 33, 37
Charophyte, 75, 116, 177, 180, 181, 183, 184, 189, 225, 249, 264
Charophyte meadows, 116, 183, 184
Chase, J.M., 138
Chenopodiaceae, 29, 31, 33, 35, 38
CHG, 14–16
Chloroflexi, 141
Chlorophyceae, 144, 147, 156, 157
Chlorophyll-*a,* 111, 118, 120, 122, 124–126, 140, 153, 156, 157, 163, 165, 166, 218
Chow-Fraser, P., 216, 218, 219, 221, 224
Christian reconquest, 33
Christians, 33, 38, 244
Chromatiaceae, 141
Chrysophyceae, 144, 146, 156, 157
Ciliates, 138, 140, 161–163
Cirujano, S., 3–18, 90, 91, 109–131, 175–192, 202, 222, 229–236, 241–253, 258
Cladium jamaicensis, 191
Cladocerans, 155, 161, 163, 170, 217, 219, 220
Claps, M.C., 170
Class A tank, 59
Clear Lake, 183
Clear-water state, 217
Clemente, D., 241
Clevering, O.A., 188
Clifford, P., 179
Climate, climatology, 47–50
Climatic change, 24, 38, 61
Climatic oscillation, climate fluctuations, 24
Cobelas, A., 3–18, 45–79, 90, 91, 109–131, 138–140, 175–192, 197–210, 215, 216, 221, 222, 229–236, 241–253, 258, 264, 267
Cobitis maroccana, 199
Cohorts, 200
Collatz, G.J., 66
Conan, C., 63, 64, 78
Conductivity, 89, 90, 98, 99, 103, 111–114, 124–126, 140, 165, 168, 169
Conflicts (wetlands management), 13, 38, 46, 123, 188, 230
Conforti, V., 140
Conifers, 89
Conly, F.M., 62
Connectivity, 111, 118, 128, 138, 221, 266, 269
Conover, W.J., 177
Constanza, R., 177
Consumption, 14, 16, 17, 63, 64, 69, 71, 78, 79, 183, 272
Conte, M., 62
Continental sequences of Mediterranean region, 24
Controversial water management, 46
Conway, V.M., 178, 188
Cooke, G.D., 216
Coots. *See Fulica atra*
Copepods, 155, 160, 161, 169, 170, 200, 220
CORINE, 89

Corn, 78, 89, 181
Coronado, R., 8
Corylus, 29, 35, 37
Costa Tenorio, M., 25
Coûteaux, M., 27
CPUE. *See* Capture (catch) per unit of effort
Crayfish, 183, 216, 223–226, 231, 246, 247, 249, 259, 261, 264, 266
Cronk, J.K., 176, 178
Crops with high water requirements, 78
Cruces, J., 9, 12, 13, 66, 78
Crustaceans, 12, 140, 155, 169, 200, 260
Cryptophyceae, 144, 146, 154, 156, 157
Cuadrat, J.M., 62
Cubasch, U., 78
Cumulative effects, 6, 46
Cupressaceae, 28, 29, 31, 35, 36
Currie, D.J., 163
Cut-sedge. *See Cladium mariscus*
Cut-sedge cover, 8, 74, 179, 184, 191
Cyanide, 110, 113, 114, 200
Cyanobacteria, 12, 143, 144, 152–154, 166, 169, 222
Cyanobacterial blooms, 222
Cyperaceae, 33, 35, 37, 38
Cyprinus carpio, 183, 198, 199, 201, 218

D
Daimiel, 4–8, 10, 11, 17, 23–38, 46, 48–52, 54–56, 58, 63, 64, 72, 87, 88, 91–95, 97, 110, 115, 116, 118, 123, 128, 130, 138, 139, 141, 144, 158, 166–169, 176, 179, 181–183, 188–191, 197–208, 210, 211, 215–216, 218–226, 230–236, 241–253, 257–275
Daimiel Art Museum, 243, 252, 253
Daimiel catchment, 51, 115, 215, 224, 231, 250, 261
Dale, V.H., 86
Dansgaard, W., 23, 32
Daphnia, 155, 160, 200, 266, 267
Daphnia curvirostris, 155, 160
Dark Ages, 33, 36, 37
D'Auria, G., 140, 151
Davis, M., 271
Davis, S.M., 123, 131
Day, J.W. Jr., 116
de Blois, S., 177
Decay, decaying, 18, 176, 177, 180, 191
Deciduous forests, 37, 89
Decomposition. *See* Decay decaying
DeFries, R., 86
del Giorgio, P.A., 163
del Moral, A., 201
del Moral-Fernández del Rincón, A., 201, 207
del Moral Molina, A., 201, 207
Delta Marsh, 192, 219
Deltaproteobacteria, 141
DEM. *See* Digital elevation model
de Meester, L., 266
Denitrification, 263
Desgranges, J.L., 206
Desiccation, 4, 16, 28, 37, 75, 87, 91, 98, 99, 181, 230–231, 249
de Sostoa, A., 220
Desprat, S., 33, 35
Desulfobacteraceae, 141
Detritus, 110, 131, 222
DGCN, 3, 4
Díaz-Delgado, C., 46
Digital elevation model (DEM), 89
Dinophyceae, 144, 145, 153, 156, 157, 169
Dissolved organic carbon (DOC), 111, 118, 124, 125, 165
Disturbances, 33, 36, 38, 46, 68, 138, 140, 150, 155, 169, 170, 218, 224–226, 262, 266, 267, 269, 274
Diving ducks, 205
DOCE, 232
Doñana National Park, 229
Dorado Valiño, M., 25, 28, 31
Downing, J.A., 216, 219
Drained, 46, 96, 116
Draw down, 51, 168, 218
Dreissena polymorpha, 222
Drought, 12–14, 29, 51, 53, 55, 60–64, 69, 71, 89, 96, 113, 128, 138, 149, 170, 183, 190, 218, 223, 234, 269, 271, 273
Drought frequency, 62
Dry conditions, arid periods, 29, 37
Dry croplands, 89, 92–95, 99–102
Dry-irrigated croplands, 89, 92, 94, 100, 101
DTD paleorecord, 27
Duarte, C.M., 176
Duplessy, J.C., 23
Dupré, M., 24, 25
Dykyjová, D., 176, 188, 190

E
Ebert, D., 267
Eckblad, J.W., 129
Ecological challenge, 262–270
Ecosystem
  scale interventions, 220

services, 258
stability, 50
EEA. *See* European Environmental Agency
Eisler, R., 114
Ekholm, P., 123
Ekiba, H.R., 271, 272, 274
Elmers, M.C., 113
El Niño, 61, 271
Elvira, B., 198, 206, 223
Emergent macrophytes, 8, 47, 66, 67, 69, 70, 111, 130, 131, 222
Enclosure studies, 218–221
Encomienda of Daimiel, 244, 246
Encomiendas, 244
Energy budget, 47, 68
Environmental policy, 104
Eolian sequences, 25, 29, 30
Ephedraceae, 29
EPTISA, 17
Equitability, 140, 151
Eriksson, H., 115
Erygeron bonariensis, 184
Erygeron canadensis, 184
Escuderos-Córdoba, J., 252
Eshleman, N.K., 86
ESRI, 90, 177
Esteller, M.V., 46
Euclidean distance, 178, 179
Euglenophyceae, 144, 145, 156, 157
EU regulation (Directive 91/271), 116, 117
European Agro-Environmental Program (UE 2078/92 and 1257/1999), 17, 94
European crayfish. *See Austropotamobius pallipes*
European Environmental Agency (EEA), 89
EUROSTAT, 15
Eutrophic, 98, 99, 141, 166, 170, 261
Eutrophication, 4, 17, 35, 38, 89, 103, 109–131, 161, 182, 215–218, 221, 225, 232, 261, 269, 270
Evapotranspiration, 9, 45, 50, 53, 58–60, 66–70, 75, 76, 78, 79, 232
Evelsizer, V.D., 216, 219
Everglades, 68, 123, 130, 131, 176, 191
Evolutionary ecology, 266–267
Exotic fishes, 218, 264

F

FACE. *See* Free-air CO_2 enrichment
Faegri, K., 24, 27
Fagus, 29
Falk, D.A., 216, 262
FAO, 9
Fathead minnows. *See Pimephales promelas*
Felipe II, 8, 200
Fennessy, M.S., 176, 178
Fiala, K., 178, 190
Findlay, C.S., 86
Finlayson, C.M., 4
Fire, 17, 87, 176, 180, 182, 250, 261, 273
Fish communities, 198–200, 216–218, 220, 222, 225
Fishermen, 246, 247
Fish mortality, 110, 114, 223
Fish productivity, 200
Flat bottom boats, 246
Flooded area, wetland inundation. *See* Inundation
Floodplain (surface and groundwater), 89, 92
Floodplain (surface water), 89, 92
Floodplain wetland, 3–6, 9, 46, 51, 53, 55, 89, 91, 98, 129, 258, 269
Flood pulse concept, 71
Florchütz, A., 31, 33
Florín, M., 11, 12
Folke, C., 271
Food web stability, 266
Forbes, S.T., 268
Fornés, J., 64
Fortin, M.J., 178
Francisco Cambó, 230
Fraser, E., 271
Fraxinus, 35
Free-air CO_2 enrichment (FACE), 264
Frenzel, B., 24
Freshwater, 4, 8, 10, 28, 37, 111, 114, 128, 130, 141, 166, 179, 182, 229, 236
Fulica atra, 203, 204
Fulling mills, 246
Fungal parasites. *See* Aphanomycosis
Furniture carpentry, 248

G

Gadwalls. *See Anas strepera*
Galop, D., 25
Gambusia holbrooki, 198, 199, 201, 218, 231
Gao, X., 4, 262
García Antón, M., 25, 31–33
García, G., 266
García-Hidalgo, J.F., 31
García-Rodríguez, M., 59
GDP. *See* Gross domestic product
Geiger, W., 223, 264
Gelochelidon nilotica, 231
Genetics, 264, 266, 267

Geology, 9, 268
Gergel, S.E., 104
Gigüela inflows, 53, 56, 60, 63, 71, 261
Gigüela profile 4.2 paleorecord, 25, 27
Gigüela river, 6, 8, 9, 33, 53–56, 64, 71, 91, 112–114, 116–118, 139, 179, 180, 182, 234, 235, 244
Gil-García, M.J., 23–38
Giller, P.S., 86
Gilman, K., 46
Giorgi, F., 4, 262
Giralt, S., 24
Girard, M., 25, 27
Gislason, A., 163, 169
Gómez-Orellana, L., 24
González, J.L., 180
González-Sampériz, P., 24, 25
Gosselink, J.G., 46, 50, 55, 70, 109, 110, 131, 176, 216, 218, 221, 258, 262
Grace, J.B., 188
Gradual Restoration of Las Tablas, 17
Grand-Lieu, 130
Grasslands, 38, 89
Great crested grebs. *See Podiceps cristatus*
Greek remains, 242
Grimm, E.C., 27
Gross domestic product (GDP), 15
Groundwater
  pumping, 12–14, 16, 17, 46, 64–66, 71, 259
  recharge, 10, 12, 13, 46, 64, 65, 78, 79, 270
  spring, 53, 234
Groundwater/aquifer discharge, inflow or input, 63
Groundwater-fed wetlands, 98
Groundwater overexploitation. *See* Aquifer overexploitation
Groundwater table, groundwater level, 13, 53, 78
Growth, 14, 16, 87, 95, 96, 99, 101, 104, 130, 143, 167, 169, 177, 183–190, 192, 217, 230, 246, 249, 263, 271, 272
Grubaugh, J.W., 71
Guadiana inflows, 53, 63, 261
Guadiana river, 5, 6, 13, 28, 31, 47, 53, 78, 85–104, 114–116, 180–182, 231, 243–246, 252
Guadiana Water Authority (Confederación Hidrográfica del Guadiana), 9
Gulati, R.D., 216, 222
Gull-billed tern. *See Gelochelidon nilotica*
Gunatilaka, A., 110
Gunderson, L.H., 259, 271
Güsewell, S., 188
Gutierrez-Elorza, M., 33, 35
Gutiérrez-Yurrita, P.J., 223

H

Habitat quality, 66, 85–104, 265
Hamilton, S.K., 4
Hanson, P.J., 62
Hansson, L-A., 220, 221, 224
Hardin, G., 250
Harding, J.S., 268
Harris, J.A., 262
Harris, T.T., 123
Harvesting of crayfish, 223
Havens, K.E., 217
Heating ovens, 248
Heinrich events, 29, 32
Heinrich, H., 29
Hellström, T., 128
Helophytic vegetation, 180
Hensel, P.F., 128
Heras, G., 4, 91
Hernández-García, J.M., 201, 207, 208
Heterotrophic, 141, 265
Hierarchical levels, 138
High-value crops, 87
Hildrew, A.G., 86
Historical extent, 268
Historical land use, 86
Hobbs, R.J., 262
Hofmann, H., 61
Holling, C.S., 259, 271
Hollis, G.E., 86
Holm oaks, 15, 89, 242, 249
Holocene, 24, 25, 29, 31, 36–38, 46
Holocene Climatic Optimum, 37
Holyoak, M., 269
Hopkinson, C.S. Jr, 115
Horn, H., 165, 167
Horn, W., 165, 167
Horra de la, J.L., 9
Houlahan, J.E., 86
House sparrow, 206
Howard-Williams, C., 71, 110, 123
HRP. *See* Hydrologic Remediation (regeneration) Plan
Hughes, A.R., 267
Hughes, T.P., 262
Human consumption, urban supply, 16
Human settlement, 231, 246
Humic compounds, 221
Humid conditions, 29, 35, 36
Hunters, 241,
Huntley, B.A., 23

Hupp, C.R., 130
Hutchings, M.J., 177
Hydrocharis morsus-ranae, 180, 181
Hydrodynamic wetland types, 87, 91–92, 98
Hydrogenophilaceae, 141
Hydrogeomorphic, 89, 221
Hydrological continuum, 46
Hydrological disturbance, 46, 150, 170, 218, 226, 262
Hydrological model, 47, 64, 70–78, 269
Hydrological periods, 49
Hydrologic Remediation (regeneration) Plan (HRP), 17, 234
Hydroperiod, 47, 50–55, 57, 59–61, 182, 190, 218, 223–225, 231, 234, 261, 262, 270
Hydroperiod restoration, 47, 70–78
Hypereutrophic, 98, 99, 110, 167, 225, 261
Hysteresis, 200, 217

I
Iberian culture in High Andalusia, 242
Iberian Peninsula, 4, 24–26, 31, 33, 37, 38, 62, 223
Idso, S.B., 68, 175
Iglesias, A., 46, 233
IGME. *See* Instituto Geológico y Minero de España; Spanish Geological Survey
Ignacio Meco, 251, 253
Increases on atmospheric CO_2, 78
Industrialization, 101, 102, 104, 271
Industry employment, 15, 95, 100
Industry sector, 15, 90, 101, 104
INE, 90, 95
Infante Don Juan Manuel, 8, 244
Infiltration, 51, 56, 59, 60, 72, 76, 123, 234, 263, 270
Inland wetlands, 4, 11, 50
Instituto Geológico y Minero de España (IGME), 6, 10–14, 47, 64, 89
Internal loading, 110, 111, 115, 123
Inundation, flooding, 6, 71
INYPSA, 111, 113
IPCC, 271
Irrigated croplands, 89, 92, 94, 100
Irrigation, 4, 12–14, 16, 64–66, 78, 79, 86, 87, 92, 94, 95, 99, 101, 115, 234, 250, 251, 259, 261, 270, 273
Isla de Algeciras, 6
Isla del Pan, 6, 8
Islands, 6, 7
Ixobrychus minutus, 203

J
Janus, L.L., 130
Jeppesen, E., 143, 163, 216, 218, 221, 225
Jétar Encomienda, 244
Jiménez, J., 201
John, D.M., 4
Johnsen, S., 23
Johnson, W.C., 66
Johnston, C.A., 110, 123, 128, 129
Jones, C.G., 223, 264
Jones, K.B., 104
Juan D'Opazo, 251, 252
Julià, R., 24, 35
Juniperus, 29
Junk, W.K., 4, 71

K
Kadlec, R.H., 176
Kalff, J., 166
Karl, T.R., 62
Keddy, P.A., 168, 175, 179
Kerfoot, W.C., 266
Kgathi, D.L., 247
King, A.J., 200
Kitchell, J.F., 217
Knight, R.L., 176
Knuuttila, S., 111, 123
Koch, M.S., 68
Köhler, J., 4
Kohl, J.G., 188
Krause, S., 55, 72
Kufel, L., 176, 183, 189
Kvet, J., 176, 188, 190

L
Laguna Cueva Morenilla, 95
Laguna de Alcahozo, 98
Lagunas de Ruidera, 5
LAI. *See* Leaf area index
Lake Erie, 128
Lake Neusiedlersee, 222
Lake Pavin, 141
Lake, P.S., 62
Lake Superior, 128–130
La Mancha Húmeda Biosphere Reserve, 4, 5, 8–12
Lambin, E.F., 86
Lammens, E.H.R.R., 221
Landscape ecology, 258, 262, 268–269
Land use-and cover changes, 86, 92–95
Larus ridibundus, 203
Las Animas catchment, 87, 96, 97

Las Cañas area, 17
Last Glacial Maximum, 29, 36, 37
Last Glaciation, 46
Late Iron Cold Period, 33
Law 25/1980 (BOE 110 of May 7, 1980), Legal Classification of National Park, 17
Law of July 17th, 16
Leaching, 111, 123, 177, 191
Lead, 13, 70, 114, 216, 224
Leaf area index (LAI), 47, 67, 68
Leaf conductance, 47
Leaf elongation rates, 188
Lecane closterocerca, 155, 158
Legal regulation, 230
Legendre, P., 178
Le Houérou, H.N., 61, 62
Lehours, A.C., 141
Lemna gibba, 181
Lemna minor, 181, 182
Lepadella patella, 155, 159
Lepomis gibbosus, 199, 201, 218
Leuteritz, T.E.J., 271, 272, 274
Levin, S.A., 271
Levins, R., 138
Lewis, W.M., 4
Libro de caza, 8
Life stages, 220
Liguliflorae, 29, 33
Limonietalia, 11
Linacre, E.T., 69
Linacre's theory, 69, 70
Lionard, M., 149, 151, 152
Little grebs, 205
Little Ice Age, 35, 36, 38
Liu, H., 104
Llamas, M.R., 10, 46, 51, 86
Llanura Manchega Occidental Aquifer, 6
Loess-like type facies, 29
López-Camacho, B., 58, 59
López Sáez, J.A., 25
Lorenzen, B., 188, 190
Lougheed, V.L., 219, 221, 224
LT paleorecord, 27
Lund, J.W.G., 140
Luzón, A., 24

M

Macklin, M.G., 37
Macroinvertebrates, 141, 220, 260, 264–265
Macrophyte
  cover, 58, 66, 68–70, 72, 74, 75, 111, 131
  implantation, 216
  stands, 68, 69
Magurran, A.E., 140
Maheu-Giroux, M., 177
Main economic activity, 90, 95
Malanson, G.P., 131
Malaria, 4, 231, 244, 246, 248, 249
Mallard. *See Anas platyrhynchos*
Maltby, E., 109
Malvárez, A.I., 97, 104
Management models, 232, 233
Man and the Biosphere (MAB) Programme (UNESCO), 10, 258
Man-made changes, 192, 266
Manny, B.A., 123
Marbled teal. *See Anas angustirostris*
Marion, L., 123, 130
Marker, A.F.H., 111
Marquet, P.A., 138
Marquis of Perinat, 250
Martín Arroyo, T., 24
Martín de Santa Olalla Mañas, F., 78
Martínez-Cortina, L., 7, 14–16
Martínez-Santos, P., 10
Masegones, 180
Mason, C.F., 176
Master Table, 244
Matamala, J.J., 231
McAllister, A., 272
McCauley, E., 140
McIntyre, A., 23
McKee, T.B.N., 47
Medina, L., 90
Mediterranean basin, 4
Mediterranean limnology, 258
Meeker, J.E., 128–130
Menéndez Amor, J., 33
Mesophylous. *See Salix*
Metacommunity ecology, 269
Meteorology, weather, 48, 62
Miao, S.L., 188, 190
Microbial ecology, 265
Microbial loop, 265
Microcoleus chtonoplastes, 12
Micrograzers, 169
Microzooplankters, 167
Middle Age, 242–245
Middleton, B.A., 70, 71, 269
Milbrink, G., 200
Miller, J.R., 144, 206
Mineral precipitation, 131
Ministry of Environment, 272
Miracle, M.R., 90
Mississippi River delta plain, 97
Mitsch, W.J., 46, 50, 55, 70, 109, 110, 128, 131, 176, 216, 218, 221, 258, 262

Modern Age, 244–246
MO eolian sequence, 25
Molemocho outflows, 60
Molemocho water mill, 6
Molina, F., 33
Molinillo, M., 87
Molles, M.C. Jr., 46
Monolete spores, 35
Montes, C., 3, 11, 12, 90, 223, 229
Montserrat, J., 24
Moore, P.D., 24, 25, 27
MOPU, 90
Moran, P.A.P., 178
Morellón, M., 24
Morenillo dam, 7, 17, 139, 235
Moreno, A., 23
Moreno, J.M., 23, 232, 271
Moreno-Ostos, E., 169, 170
Morillo, C., 180
Morphometric features, 6, 7
Mosquitofish. *See Gambusia holbrooki*
Moss, B., 104
Motilla de las Cañas, 16, 243
Motilla del Azuer, 243
Motillas, 16, 241, 242
Moustafa, M.Z., 110, 123
Moyer, B., 272
Mulligan, M., 78, 79, 87
Murcia, 273
Muslim-Christian conflicts, 38
Muslim dominance, 33
Muslim pasture exploitations, 38
Mutualistic relationship, 206
Muylaert, K., 170

N
Nano and microphytoplankton, 138, 166
NAO. *See* North Atlantic Oscillation
Naselli-Flores, L., 61
National Land Use/Land Cover map, 89
Natural vegetation, 89, 92–95, 99–102, 104
Navarro, A., 13
Neiff, J.J., 4
Net carbon exchange, 264
Netta rufina, 8, 203, 204
Niehoff, D., 86
Night herons, 205
Nightingale, 206
Nitrate, 119, 121, 124, 125, 140, 165, 168
Nitrification, 119, 263
Noges, P., 128
Non-point (diffuse) pollution, 89, 103
North America, 97, 220
North Atlantic Oscillation (NAO), 62
Nuphar luteum, 180, 181
Nutrient budged or balance, 123
Nutrient dynamics, 110, 116–124, 258, 264
Nutrient fluxes, 110
Nutrient import, 116
Nutrient precipitation, 216
Nutrient removal efficiency, 123
Nutrient retention, 123, 250
Nutrient sink or source, 110, 123, 258, 261
Nymphaceae, 35
Nymphaea alba, 180, 181

O
Oasis effect/clothes-line effect, 68
Ochromonas sp., 143, 147
Oden, N.L., 178
OECD (OCED) standards, 14, 118
Ogden, J.C., 176
OIS-5, 37
OIS-7, 28
Ojos del Guadiana, 6, 13, 53–55, 59
Ojos, ojillos, 6, 234
Okavango delta, 4, 247
Olea, 31
Oligohaline, 111, 179, 261
Olives, 15, 16, 89, 249
Olmedo, A., 14, 15
Omnivory, 168
Oñate, J.J., 62
Ondok, J.P., 66, 68
Open water evaporation, 47, 68–71, 73
Ordinance 1874/1973 (BOE 181 of June 30, 1973), National Park declaration, 16
Organic loading, 116
Organic matter, 4, 115, 119, 130–131, 143, 165, 176, 191, 260, 266
Orlando, B., 50
Ortega-Mayagoitia, E., 138, 140, 144, 155, 160, 161, 163, 167, 168, 170, 198, 200
Ovas, 180
Oxidation, 113
Oxyura leucocephala, 231

P
Paleoclimatic reconstruction, 23, 25
Paleoecological reconstructions, 24–25
Paleorecords, 24–29, 31–35
Palmer, M.A., 262
Palutikov, J.P., 62
Panarchy, panarchical systems, 271–275
Papayannis, T., 249

Pardo, L., 247
Parkhurst, R.S., 68
Pascual, H., 176
Passerine birds, 202, 206–210, 266
Pata Gallina, 7, 111
PEAG. *See* Upper Guadiana Special Plan
Peñalba, M.C., 24
Peña Monné, J., 33, 35
Perca fluviatilis, 220
Per capita income, 15
Perch. *See Perca fluviatilis*
Pérez-González, A., 9
Pérez-González, M.E., 9, 47
Peters, R.H., 163
Phillips, J.D., 110
Phosphorus enrichment, 98
Phycocyanin-containing coccoid cyanobacteria, 143
Piezometric level, groundwater level, 13–14, 64
Pimephales promelas, 220
Pinaceae, 28, 35
Piscivores, 216, 220
Pithart, D., 138
Plagues, 231, 246
Planktivores, 215–221
Plankton, 137–169, 216, 220
Planktonic community, 138
Planktonic food web, 138
Plankton taxonomic structure, 170
Planktothrix agardhii, 152
Plant dynamics, 68–69, 75
Pleistocene, 24–27, 35–37
Poaceae, 28–29, 33, 35–38
Podiceps cristatus, 203
Podiceps nigricollis, 203
Poiani, K.A., 66
Point pollution, 89
Policy makers, 14, 79
Polistena, 244
Pollen analyses, 28
Pollinic palynomorphs, 24
Polunin, N.V.C., 177
Polygonum, 33, 37
Pons, A., 24
Population
  changes, 90
  density, 14, 95–96, 100–101, 103, 265
  dynamics, 138, 264, 266
Porometer, 47
Portero, J.M., 9
Potamogeton, 33, 35, 37–38
Potamogeton lucens, 180–181
Potthoff, A.J., 216, 219, 220, 224
Pou, A., 62
Prairie Pothole, 220
Predators, 169, 170, 200, 223, 264
Preston, E.M., 98
Pribán, K., 66, 68
Price, J.S., 69
Primary producers, 70–71, 116, 155, 166–167, 177–179, 215, 217, 218, 222
Príncipe Don Juan Manuel, 200
Priority habitat for waterfowl, 231
Procambarus clarkii, 183, 223–224, 231, 249, 261, 264–265
Production, 87, 110, 131, 167, 177–178, 189–191, 218, 222, 223, 250, 261, 265
Propagule banks, 183, 266
Proteobacteria, 141–142
Puente Navarro dam, 6–7, 17, 111, 235, 245
Puente Navarro outflows, 60
Pu, M., 192
Pumpkinseed. *See Lepomis gibbosus*

Q

Quaternary period, 24, 25, 35, 36
Quercus, 28, 29, 31, 33, 35–37

R

Rader, R., 264
Radiation, 47, 66, 168
Radiometric date, 31
Rain and groundwater fed wetlands, 89, 91, 92
Rainfall, precipitation, 6, 9, 13, 45, 47–51, 53, 55–65, 76, 78, 79, 89, 131, 216, 270
Rain-fed agriculture, 16
Rain fed wetlands, 91, 98
Ramirez, J.I., 9
Ramos, J., 33
Ranunculaceae, 33, 35, 37
Ranunculus peltatus, 181, 182
Ranunculus trichophyllus, 181
Ratio of emergent macrophyte cover to open water area, 66
Rawlik, P.S., 68
Reber, A., 267
Recolonization, 183
Red-crested pochards. *See* Netta rufina
Reddy, C.N., 130, 131
Reed. *See Phragmites australis*
Reed-cover, 68, 74, 75, 131, 188, 191, 206, 210
Reeder, B.C., 128
Reed, J.R., 221
Re-establishment of submerged vegetation, 221
REGATA. *See* Gradual Restoration of Las Tablas

Rehabilitation, 215–226, 232, 259, 262, 273, 274
Reille, M., 24, 27
Relaciones topográficas, 8, 200
Relative humidity, 47
Renault-Miskovsky, J., 27
Rendell, H.M., 28, 29
Rendón, M.A., 203
Resilience, 206, 262, 267, 271, 274, 275
Resurrection ecology, 266, 267
Reynolds, C.S., 138
Rhône River Delta, 128
Riansares catchment, 88, 91, 95–87
Ribeiro, M.D., 176, 192
Richards, K.S., 131
Richardson, C.J., 4
Richness, 3, 70, 81, 110, 138, 149, 150, 155, 169, 181, 182, 184, 198, 200, 202, 203, 205, 206, 265
Ricker, W.E., 198
Riebsame, W.E., 86
Riera, S., 33, 35
Rijaca, 247
Rind, D., 62
River discharge, 4, 17, 53, 54, 64–66, 127, 190
River flow, 47, 60, 89, 91
River network, 89
Roach. *See Rutilus rutilus*
Robarts, R.D., 141, 144, 169
Rodewald-Rudescu, L., 178
Rodrigo, M.A., 137–170, 265
Rodríguez, C.F., 59, 190, 223
Rojo, C., 137–170
Roman colonization, 243–244
Roman occupation, 33
Rosell, J., 94
Rosenberg, M.S., 179
Rosenberry, D.O., 50, 51, 53, 61
Rossell, J., 13, 14
Rotifers, 140, 155, 160, 161, 164, 169, 220
Rott, E., 14
Rowsell, R.D., 61
Ruddiman, W.F., 23
Ruggiero, A., 216
Ruidera catchment, 87, 88, 91–97
Ruiz de la Hermosa, C., 201, 205, 206
Ruiz Zapata, M.B., 23–38
Runoff, 6, 55, 64, 104, 110, 123, 129, 232
Ruppia maritima, 181, 182
Rutilus rutilus, 220

S

Sacocornetea fruticosi, 11
Sáez-Royuela, R., 111, 113, 176
Sahuquillo-Herráiz, A., 12
Saline soils, 33, 184
Salinity, 33, 38, 66, 70, 98, 109–114, 175, 232, 250
Salinization, 4, 85, 109, 111–113, 176, 232
Salix, 35, 36
Saltmarsh, A., 191
Sammalkorpi, I., 216
Sánchez-Carrillo, S., 3–18, 45–79, 85–104, 109–131, 175, 178, 179, 222, 247, 257–275
Sánchez-Goñi, M.F., 24
Sánchez-Ortega, A., 264
Sánchez Soler, M.J., 201
Sánchez-Vila, X., 233
Sander vitreus, 220
Sanjiang Plain, 98
Santamaría, L., 64
Sanz-Donaire, J.J., 9, 47
Sarria, A., 8
Scheffer, M., 138, 217
Schrage, L.J., 216, 219
Schulte, L., 24
Schweinfurth, U., 271
Scrubland, 89
Seabloom, E.W., 179, 192
Seasonal assemblages, 138
Seasonal pattern, seasonal fluctuation, 50, 51, 53, 128, 144
Secondary producers, 138, 140
Sedimentation, 31, 110, 111, 123, 128–131, 191, 216, 222, 224, 226
Sediment traps, 111
Semi-arid, 3, 4, 9, 13, 46, 50, 51, 61, 62, 66, 69, 70, 104, 110, 137
Semi-natural shrub community, 87
Sen, A., 271
Service sector, 15, 90, 95, 99–102
Settier, J., 200, 246
SGDGOH, 11
Shallow lakes, 61, 128, 215–218, 221, 222, 224, 226
Shannon, C.E., 140
Shannon-Wiener diversity index, 140–141, 150, 156, 157
Shapiro, J.V., 216, 221, 222
Sherr, E.B., 169
Shovelers. *See Anas clypeata*
Siltation, 4, 109, 110, 128–131, 176, 232, 233
Sklar, F.H., 188, 190
Socioeconomic challenge, 271–275
Socioeconomic indicators, 90, 92–95
Soil hydraulic properties, 54, 59
Soil types, 9

Sokal, R.R., 140
Sommaruga, R., 141, 144, 169
Søndergaard, M., 216
Sophocleous, M., 55
Souch, C., 68
Sousa, A., 232
Spanish Armada, 244
Spanish Civil war, 246
Spanish droughts, 62
Spanish Geological Survey (IGME), 6, 10–14, 47, 64 89, 113
Spanish wetlands, 4, 229– 233
Sparks, R.E., 71
Spatial correlograms, 178
Spatial heterogeneity, 124–128, 130, 137, 138, 140, 144, 155, 160, 170, 191–192
Species diversity, 71
Sphingobacteria, 141
Spiers, A.G., 4
16S rRNA gene, 164
Standardized precipitation index (SPI), 45, 47, 62–65
Steppes, 11, 23, 36, 37
Sterna albifrons, 203
Stockner, J., 144, 166, 169
Stockwell, C.A., 267
Sub-Atlantic Cold Period, 36, 37
Subsurface flow wetlands, 270
Subsurface seepage, groundwater seepage, 53, 59
Sulphate, 113
Sulphide compounds, 113
Summer desiccation, summer dry-out, 75, 78
Sunflower, 78, 89, 181
Sustainable human use, 259
Swanson, S.K., 53

T

Tablas de Arenas de San Juan, 91
Tablas de Daimiel, 4–8, 10, 11, 17, 23–38, 46, 48–52, 54–56, 58, 63, 64, 72, 87, 110, 115, 116, 118, 123, 128, 130, 138, 139, 141, 144, 158, 166–169, 176, 179, 181–183, 188–191, 197–208, 210, 211, 215–216, 218–226, 230–236, 241–253, 257–275
Tablas del Záncara, 91
Tablas de Villarrubia, 91
Tablas de Villarta de San Juan, 91
Tablazo, 7, 8
Tachybaptus ruficollis, 203
Tamarix, 33, 187
Tamarix canariensis, 182, 184, 187
Tamarix gallica, 182, 187
Tang, Z., 104
Tarjuelo, J.M., 15
TASG eolian sequence, 25, 29, 30
Tátrai, I., 219
TD paleorecord, 27
Teals. *See Anas crecca*
Temporary ponds, 3, 12, 87, 91
Ter Braak, C.J.F., 90
Teresa Aldea, 250
Terriza, C.A., 241
Testa, J.M., 115
Testudinella patina, 155, 159
Thermophylus. *See Quercus*
Thesiger, W., 247
Thoms, M.C., 129
Thornthwaite potential evapotranspiration, 50
Tigris and Eufrates, 4, 247
Typha latifolia, 182
Thypha domingensis, 8, 181, 184
TN:TP ratio, N:P ratio, 140, 168
Tockner, K., 138
Tolypella hispanica, 181, 182
Top-down and bottom-up, 168
Top-down manipulations of fish communities, 216, 220
Total nitrogen, 111, 115–117, 119, 121, 123–126, 129, 140, 165, 168, 218, 219
Total organic carbon, 111, 115–118, 120, 124–127, 129, 140, 168, 169
Total phosphorus, 111, 116–119, 122–130, 140, 165, 166, 168, 179, 183, 199, 218, 219
Train, S., 168, 169
Transpiration, 47, 66–70, 72, 73, 175, 270
Trenberth, K.E., 62
Trigo, R.M., 62
Trophic cascades, cascading effects, 200, 216, 217, 220, 221, 226
Trophic relationships, 168, 170
Tropidophoxinellus alburnoides, 199
Tubuliflorae, 33
Tudesque, L., 116
Turbid state, 217
Turner, A.M., 216, 219
Turner, K., 104
Turner, M.G., 192
Tuschl, P., 67
Tuteja, N.K., 104
Typha, 8, 33, 35, 37, 38, 177, 181, 182, 184, 185, 188, 190, 248, 249
Tzedakis, P.C., 24

U

UNESCO, 8–12
United Nations Environment Programme, 50
Unsustainable economic development, 46
Upper Guadiana basin, 8–12, 14–16, 64, 78, 79, 86–88, 91–99, 103, 104
Upper Guadiana Special Plan, 17, 234
Urban, N.H., 188
Urban water consumption, 16
Urban water supply, 16
Usmanova, R.M., 61
Uuemaa, E., 104

V

Valdeolmillos, A., 24, 25, 28, 29, 31
Valdés, B., 27
Valero-Garcés, B.L., 24
Valladares, F., 230
Van del Valk, A.G., 61, 68, 71
Van der Kamp, G., 62
Vanellus vanellus, 203
Van Nes, E.H., 138
Vargas, M.J., 220
Vargo, S.M., 131
Vegas, J., 24, 29
Vegetation community structure, 222
Vegetation patches, 184
Vegetation services, 222
Velayos, M., 90
Verdugo, M., 47
Vertebrate ecology, 265–266
Vicente, E., 90
Vicente-Serrano, S.M., 62
Viladomiu, L., 13, 14, 94
Villanueva, C., 201
Villarrubia de los Ojos, 6, 115, 116, 119
Villarrubia de Santiago, 29
Vineyards, 15, 89
Virus ecology, 265
Visser, J.M., 191
Vollenweider, R.A., 179
Vörös, L., 166
VV, A.A., 230
Vymazal, J., 190

W

Wade, P.M., 255
Wagner, C., 138
Walleye. *See Sander vitreus*
Wang, G.P., 116, 124
Wang, L., 104
Wang, Y., 86
Ward, J.V., 138
Wardrop, D.H., 129
Warm conditions, 29, 31, 37
Warm Medieval Period, 35, 36, 38
Warm Roman Episode, 33, 36, 37
Wastewater, 4, 104, 115, 116, 119, 124, 128, 129, 139, 233, 234, 236, 250
Water
  column aeration, 216
  consumption by irrigation, 63–64
  cycle components, 59, 60, 71
  deficit, 17, 50–52
  inputs, 17, 50, 51, 53, 56–58, 60, 61, 71, 76, 113, 119, 128, 139, 179, 182, 198, 231, 250, 270
  level fluctuation, 61
  policies, 79, 232
  pollution, 16, 109, 110, 113–115, 176, 178, 179, 182–184, 191, 198, 200, 206, 232
  quality, 4, 17, 74, 86, 90, 100–104, 109–131, 176, 178, 179, 183, 189, 191, 192, 197, 199, 205, 206, 216, 218–220, 222–226, 232–234, 236, 249–251, 261, 264, 266
  scarcity, 47, 62, 75, 184, 234
  shortage, 46, 109, 206, 273
  source, 6, 8, 13, 17, 46, 54, 56, 60, 87, 89, 91, 179, 217, 232, 236, 261, 270
  storage, 6, 46, 52, 55–57, 59–61, 63
  supply, 13, 16, 46, 55, 111, 234, 263
  surplus, 4, 78
  temperature, 66, 167
Water budget, water balance, 46, 55, 59, 66, 71
Water diversions, water transfer, 234, 270
Waterfowl, 8, 66, 74, 123, 176, 200, 202, 203, 205, 231, 232, 244, 246, 253, 266, 268
Water Framework Directive EU, 104, 232, 265
Water management, hydrological management, 46, 79, 259
Water-mill, 6, 33, 38, 231, 242, 244–246, 250, 253, 261
Water-powered oil mills, 246
Weaver, W., 140
Weider, L.J., 266
Weisse, T., 167
Weller, M.W., 66
Weltzin, J.F., 62
Wenthworth Group, 275
Wetland
  buffer, 94

conservation, 17, 46, 100, 101, 104, 229–236, 258, 262, 267, 268
Wetland (*cont.*)
disappearance, 46, 66, 95, 100, 101, 103, 104, 216
health, 18, 232
lost, 4, 10, 100
management, 46, 75, 104, 230– 233
substrate composition, 53
Wetland Centre of La Mancha, 245, 248, 253
Wetland eutrophication abatement, 216, 218
Wetzel, R.G., 110, 131
Wheat, 16, 89
Whillans, T.H., 216, 221
White-headed duck. *See Oxyura leucocephala*
Wiemeyer, S.N., 114
Wigley, T.M.L., 78
Wilcox, D.A., 221
Wilhite, D.A., 62
Winter, T.C., 46, 50, 51, 53, 55, 59, 61, 66
Woo, M-K., 55, 61
Workforce, 90, 95
Wu, J.,

X
Xantophyceae, 144, 146
Xianghai Natural Reserve, 116

Y
Yang, X., 169
Younger Dryas, 29, 32

Z
Zacatena, 244, 247
Zalidis, G.C., 232
Záncara river, 9, 16, 91
Zedler, J.B., 70, 71, 216, 262
Zhang, J., 98, 104
Zignematales, 147
Zimmer, K.D., 219
Zooplankton, 138, 155–163, 167–170, 215, 217, 218, 220, 221
Zooplankton assemblages, 155, 160
Zwart, G., 141

MIX
Papier aus verantwortungsvollen Quellen
Paper from responsible sources
FSC® C105338

If you have any concerns about our products,
you can contact us on
ProductSafety@springernature.com

In case Publisher is established outside the EU,
the EU authorized representative is:
Springer Nature Customer Service Center GmbH
Europaplatz 3, 69115 Heidelberg, Germany

Printed by Libri Plureos GmbH
in Hamburg, Germany